Der Mensch im Kosmos: Lebenswelten und Kosmologien.

Man within the Cosmos: Lifeworlds and Cosmologies.

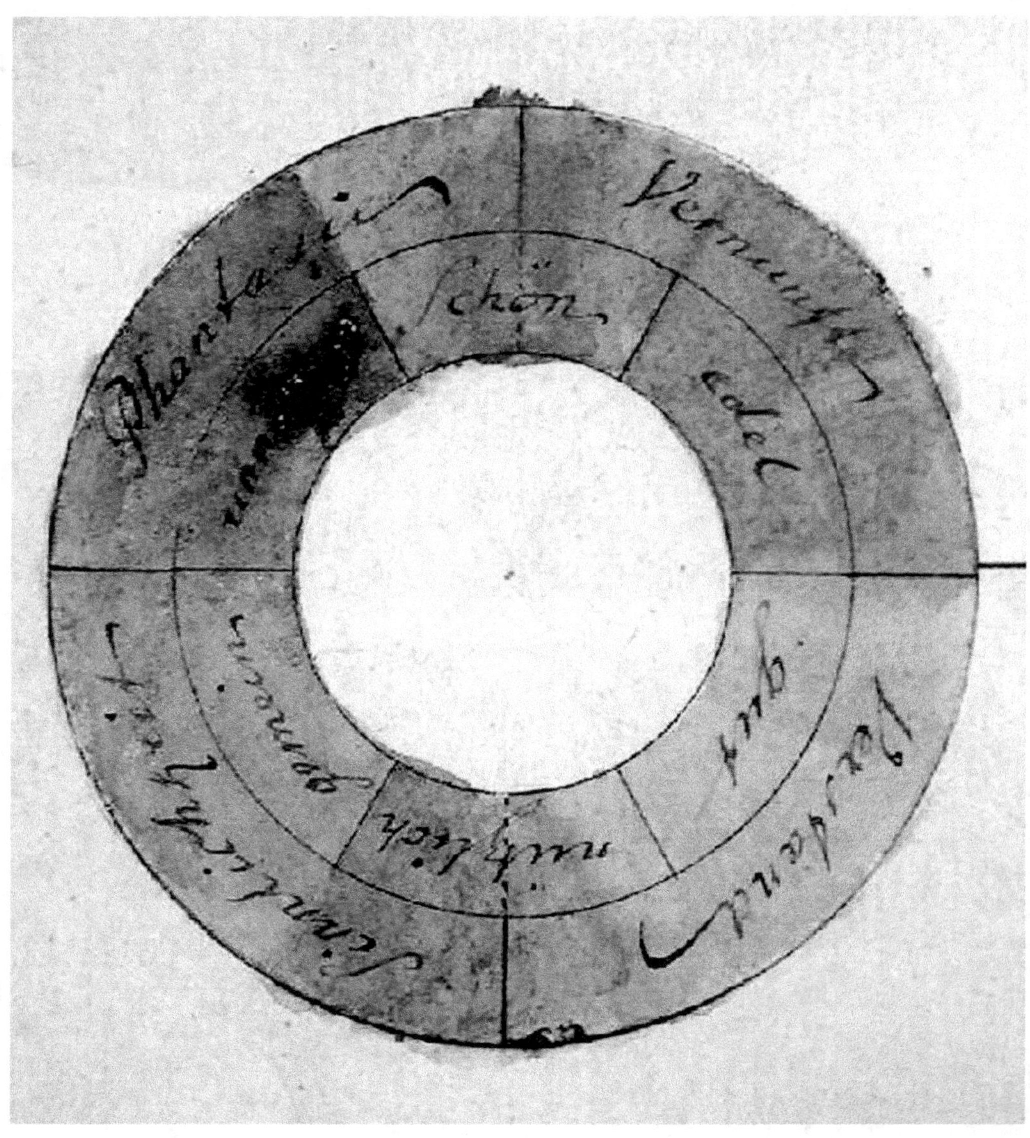

Abbildung 0.1:
Farbenkreis, aquarellierte Federzeichnung von Goethe, 1809

Original: Freies Deutsches Hochstift – Frankfurter Goethe-Museum. Das Schema illustriert das Kapitel „Allegorischer, symbolischer, mystischer Gebrauch der Farbe“ in Goethe’s Farbenlehre. Umschrift: (innerer Ring) [rot] „schön“ [orange] „edel“ [gelb] „gut“ [grün] „nützlich“ [blau] „gemein“ [violett] „unnöthig“ (äusserer Ring) [rot-orange] „Vernunft“ [gelb-grün] „Verstand“ [grünblau] „Sinnlichkeit“ [violett-rot] „Phantasie“. (CC).

Nuncius Hamburgensis
Beiträge zur Geschichte der Naturwissenschaften
Band 53

Wolfschmidt, Gudrun (Hg.)

Der Mensch im Kosmos

Lebenswelten und Kosmologien

Man within the Cosmos

Lifeworlds and Cosmologies

Tagung der *Gesellschaft für Archäoastronomie* in Weimar 2023

Ahrensburg bei Hamburg: tredition 2024

Nuncius Hamburgensis
Beiträge zur Geschichte der Naturwissenschaften

Hg. von Gudrun Wolfschmidt, Universität Hamburg,
AG Geschichte der Naturwissenschaft und Technik
(ISSN 1610-6164).

Der Titel „Nuncius Hamburgensis" wurde inspiriert von „Sidereus Nuncius" und von „Wandsbeker Bote".

Wolfschmidt, Gudrun (Hg.): Der Mensch im Kosmos – Lebenswelten und Kosmologien – Man within the Cosmos: Lifeworlds and Cosmologies. Proceedings der Tagung der Gesellschaft für Archäoastronomie in Weimar 2023. Hamburg: tredition (Nuncius Hamburgensis – Beiträge zur Geschichte der Naturwissenschaften; Band 53) 2024.

Abbildung – Cover vorne: Mensch im Kosmos (Collage: Michael A. Rappenglück), vgl. Abb. 17.6, S. 334

Frontispiz: Goethes Farbenkreis (1809), (CC)

Abbildung – Cover hinten: Sonnenaufgang im Römischen Haus in Weimar, 23.6.2023 (Foto: Katrin Cura).

AG Geschichte der Naturwissenschaft und Technik, Hamburger Sternwarte,
Universität Hamburg, Bundesstraße 55 – Geomatikum, 20146 Hamburg, Germany
`https://www.fhsev.de/Wolfschmidt/GNT/home-wf.htm`

Dieser Band wurde gefördert von der Gesellschaft für Archäoastronomie.

Verlag: tredition GmbH, An der Strusbek 10, 22926 Ahrensburg, Germany
ISBN: 978-3-384-10351-2 (Hardcover), 978-3-384-10350-5 (Softcover),
978-3-384-10352-9 (e-Book), © 2024 Gudrun Wolfschmidt.

Inhaltsverzeichnis

Vorwort: Der Mensch im Kosmos: Lebenswelten und Kosmologien
Wolfschmidt, Gudrun (Hamburg) 12

EINFÜHRUNG INS THEMA 16

1 Between Worlds and Realities: Cosmic Openings, Passages, and Pathways in the Ancient Cultures' Worldview and Cosmopraxis
Michael A. Rappenglück (Gilching) 18

1.1 The lifeworld as chronotopos: places and paths 22
1.2 Where and when: orientation, navigation, and chronology 22
1.3 The first-person perspective: the body, the conscious being, the place, the centre, and the horizon of perception 24
1.4 Ontologies and perspectives 24
1.5 Openings, passages and paths in natural and artificial enclosures and landscapes 25
1.6 Anthropological framework and cosmological code: Openings, paths, passages, stations, and spheres of existence 25
1.6.1 Places and paths of transition and transformation 26
1.6.2 Openings at the edge of sky, earth, and underworld 29
1.6.3 Openings and paths in the underworld: subterranean natural and artificial spaces (caverns, crypts) 31
1.6.4 Orbits around the World Mountain 35
1.6.5 The path of Sun, Moon, and planets: the ecliptic and the zodiac 36
1.6.6 Another path in the sky: the zodiacal light 39
1.6.7 The Golden Gate of the ecliptic: the Pleiades and Hyades open star clusters 41
1.6.8 The Milky Way: nocturnal path to the beyond 42
1.6.9 The rainbow: diurnal the path to the beyond 48
1.6.10 Travelling is dangerous: clashing rocks, magic formulas, guardians, soul guides, and spirit helpers 50
1.6.11 The world axis and the celestial pole: path and gateway to transcendental spheres 54
1.6.12 The Gate to Heaven at the Top of a Person's Head 59

1.7 Concluding remarks 61
1.8 Literatur 61

2 Ancient Astronomies – Ancient Worlds
Clive Ruggles (University of Leicester, UK) 76
2.1 Literatur 78

3 Von der Höhle zum Himmel: Kulturastronomie & Kulturelle Kosmologie
Michael A. Rappenglück (Gilching) 80

KULTURGESCHICHTE DER STEINZEIT UND BRONZEZEIT 84

4 Stonehenge and the Major Standstill Moon
Clive Ruggles (University of Leicester, UK) 84
4.1 Literatur 86

5 Die Stehenden Steine von Callanish: Ein 3D-Datensatz für *Stellarium*
Georg Zotti (Wien, Österreich), Victor Reijs, Emma Rennie & Alistair Carty 88
5.1 Einleitung 90
5.2 Der Mondwendezyklus 92
5.3 Laserscan 94
5.4 Ein 3D-Modell für Stellarium 95
5.4.1 Callanish für alle! 98
5.5 Ausblick: Photographische Justierung 99
5.6 Literatur 101

6 Höhenheiligtum am Pfitscher Sattel, Bedeutung und Entwicklung
Roland Gröber (Leverkusen) 104
6.1 Einleitung 106
6.2 Der Weg zum Höhenheiligtum 106
6.3 Im Außenbereich der Kultstätte 108
6.4 Die Kultstätte, Umfassungsmauer und Innenbereich 113
6.5 Der „große Stein“ und die Kammern 116
6.6 Schriften auf einigen Schalensteinen 120
6.7 Zusammenfassung 125
6.8 Literatur 126

7 Engraved Stones from burial mounds in Franconia point to observations of the Sun, of Orion and of the Pleiades
Siegfried Hess (Berlin, Röttenbach) & Thomas Storch (Hemhofen) 128
7.1 Introduction and Preliminary Remarks 130
7.1.1 Typical Patterns of the Engraved Stones 130
7.2 Observations of the Sun 131
7.2.1 Sunrise over a Hill 131
7.2.2 Remarks on sites and landscape 134
7.2.3 Solar Observatory 135
7.2.4 Further hints for solar observations 135
7.3 Star Maps 137
7.3.1 The Pleiades or Seven Sisters 137
7.3.2 Orion 139
7.4 Concluding Remarks 142
7.4.1 Relations to other regions 142
7.4.2 Attempts to date the engravings 142
7.5 Literatur 144

8 Im zyklischen Rhythmus des Kosmos liegt das Schicksal der Menschen – Die bildhafte Zeit im „Heydnischen" Grab von Göhlitzsch-Halle/Leuna
Klaus Albrecht (Kassel) 146
8.1 Fundgeschichte und bisherige Interpretationen der Innenverzierungen 149
8.2 Beschreibung und neue Interpretationen der Innenverzierungen 153
8.2.1 Darstellungen „Köcher, Axt, Pfeil und Bogen" auf Stein (c) . 156
8.2.2 Die vier Jahreszeiten auf Stein (c) 157
8.2.3 Monatszyklen auf Stein (b) 162
8.2.4 Jahreskalender auf Steinen (d) 165
8.2.5 Wandstein (e) im Osten, Wandstein (f) im Westen 168
8.2.6 Durchlaufende Friese mit Dreiecken auf den Steinen 171
8.3 Vergleichbare Beispiele für Ornamentik mit astronomischem Bezug . 174
8.4 Religiöse Bezüge 181
8.5 Literatur 187

ANTIKE KULTUREN 190

9 Man, Myth, Cosmos –The Sun and Solar Eclipses in Ancient Mesopotamia (Second and First Millennia BC)
Anna Paule (Linz, Österreich) 190
9.1 Introduction 192
9.1.1 Methodology 193

9.2 The Imagery of the Sun. An Introduction to Solar Imagery from Ancient Mesopotamia . . . 194
9.2.1 The Imagery of the Sun as a Divine Person. Evidence from Ancient Mesopotamia . . . 197
9.2.2 The Imagery of the Sun as a Divine Person. Family Ties and Astral Character . . . 198
9.3 The Imagery of the Sun in Cosmology. Important Sources of Research 199
9.3.1 The Imagery of the Sun in Cosmology. Astronomical Interpretations . . . 200
9.4 The Imagery of the Sun in Divination Literature . . . 206
9.5 Conclusions . . . 208
9.6 Literatur . . . 209

10 Rätsel der alten Astronomie
Jörg Bäcker (Gummersbach) 214
10.1 Literatur . . . 216

11 Der kretisch-minoische *Diskos von Phaistos*, ein herausragendes Artefakt der Archäoastronomie auf Basis einer dreiteiligen Kosmologie „Geist-Leben-Sache“
Hermann Wenzel (München) 218
11.1 Einführung . . . 220
11.2 Analyse und Entzifferung des *Diskos von Phaistos* . . . 221
11.3 *Diskos von Phaistos* – Arithmetik und Astronomie . . . 226
11.3.1 Das *Große Jahr der siderischen Planetenperioden* . . . 234
11.3.2 Drei Geschlechter – G-L-S . . . 239
11.3.3 Finsternis-Perioden . . . 240
11.3.4 Der Dorn und die Endpunkte . . . 241
11.4 Frühgeschichtliche und Antike Quellen mit Bezügen zum *Diskos von Phaistos* . . . 241
11.5 Fazit . . . 242
11.6 Literatur . . . 242

12 Iconography, Science, and Lightning Figures
Albrecht Ploum (Nijmegen, Netherlands) 244
12.1 Introduction: Iconography in Science . . . 246
12.2 Australian Lightning Figures and other figures . . . 247
12.3 Atmospheric phenomena: Red Sprites, Elves, Blue Jets and Gigantic Jets . . . 248
12.4 Pictorial views of Transient Luminous Events . . . 249

12.5 Climatological or geophysical opportunity for potential observation of the Red Sprite phenomena . . . 250
12.6 The ability of rock-painters to observe the Red Sprite phenomena . . . 251
12.7 Morphological resemblance between Lightning Figures and sprite phenomena . . . 255
12.8 Mythological relation with Sprites . . . 257
12.9 Conclusion . . . 259
12.10 Follow up research of Albrecht Ploum . . . 260
12.11 Literature . . . 261

KULTUR- UND ARCHITEKTURGESCHICHTE VOM MITTELALTER BIS ZUM KLASSIZISMUS 264

13 Die Trelleborgen in Dänemark: Eine astronomisch-geometrische Siedlungsplanung der Wikinger
Andreas Fuls (TU Berlin) 264
13.1 Die Wikingerzeit in Dänemark im 10. Jahrhundert . . . 266
13.2 Die Ringburgen der Wikinger . . . 266
13.2.1 Trelleborg . . . 268
13.2.2 Fyrkat . . . 269
13.2.3 Aggersborg . . . 272
13.2.4 Nonnebakken . . . 272
13.2.5 Borgring . . . 272
13.2.6 Funktion der Ringburgen . . . 272
13.3 Geographische Lage der Ringburgen . . . 273
13.3.1 Der Breitengrad von Trelleborg . . . 274
13.3.2 Grosskreis 1: Trelleborg, Fyrkat und Aggersborg . . . 275
13.3.3 Grosskreis 2: Haithabu, Jelling und Aggersborg . . . 275
13.4 Astronomie von Trelleborg . . . 275
13.5 Astronomie von Fyrkat . . . 279
13.6 Konstruktion von Trelleborg . . . 282
13.6.1 Das verwendete Längenmaß in Trelleborg . . . 282
13.6.2 Astronomische Konstruktion des Grundrisses . . . 283
13.7 Teppich von Bayeux . . . 283
13.8 Zusammenfassung . . . 284
13.9 Literatur . . . 285

14 Die Nordturmkapelle im Erfurter Dom
Burkard Steinrücken (Recklinghausen) 288
14.1 Literatur . . . 292

15 Kosmische Lebenswelten auf Himmels- und Erdgloben in Weimar und Gotha
Harald Gropp (Heidelberg) 294
15.1 Literatur 296

16 Ist das Römische Haus im Park an der Ilm in Weimar astronomisch verankert? Teil II: Vertiefung der archäoastronomischen Befunde
Ettore Ghibellino (Weimar) 298
16.1 Vorgeschichte 300
16.2 Tagung in Weimar 2023 301
16.3 Erstmalige Beobachtung des Verlaufs der Sonnenstrahlen in der Cella im Erdgeschoss zur Sommersonnenwende 2023 bei Sonnenaufgang . 303
16.3.1 Erwartungen 303
16.3.2 Sonnenstrahlenbeobachtung im Blauen Salon 2023 307
16.4 Ausblick 313
16.5 Fragen an den Verfasser anlässlich der Tagung in Weimar 316
16.6 Forschungsantrag Römisches Haus: Wissenschaftlicher Nachweis archäoastronomischer Verankerungen 317
16.7 Literatur 322

MENSCH IM KOSMOS – VERGANGENHEIT UND ZUKUNFT 324

17 Homo sapiens and the Universe – Where do we come from? Where are we going to?
Jaak Aru (Tartu, Estland) 324
17.1 Introduction 326
17.2 Where do we come from? 326
17.2.1 Miller-Urey-Experiment (1953) 326
17.2.2 Panspermia – Life is seeded by comets or meteors 327
17.2.3 Life in the universe – Life on another planets 329
17.3 Where are we going? 329
17.3.1 How NASA plans to build on other planets – Biological teleportation 329
17.3.2 Mohava Desert Experiment 330
17.3.3 Aim of such Interventions 330
17.4 Summary 331
17.5 Literatur 332

Programm der Tagung *Der Mensch im Kosmos* 334
18.1 Tagung Gesellschaft für Archäoastronomie in Weimar, 21.–25. Juni 2023 335

Autoren 341

Nuncius Hamburgensis 351

Personenindex 361

Abbildung 0.2:
Wappen von Thüringen

(Foto: Gudrun Wolfschmidt in Weimar)

Vorwort: Der Mensch im Kosmos: Lebenswelten und Kosmologien

Wolfschmidt, Gudrun (Hamburg)

Das Buch „*Der Mensch im Kosmos – Lebenswelten und Kosmologien*“ (Man within the Cosmos – Lifeworlds and Cosmologies) präsentiert die Vorträge der Tagung der *Gesellschaft für Archäoastronomie* in Weimar 2023 in 17 Kapiteln.[1] Die Tagung vom 22. bis 25. Juni 2023 wurde organisiert in Kooperation mit Anna Amalia und Goethe Freundeskreis e.V., Weimar, VHS / Stiftung Klassik.

Johann Wolfgang von Goethe (1749–1832) war sehr an Naturwissenschaften interessiert; man denke an die Chemie, Vulkanismus oder an das Phänomen der Farben, das er auch physiologisch und psychologisch untersucht; die physikalische Optik Isaac Newtons (1642–1726 jul./1643–1727 greg.) lehnte er allerdings ab. Er erforschte den Mikrokosmos mit dem Mikroskop (vgl. Abb. 0.3), zum Beispiel seine Gesteins- und Mineralien-Sammlung, sein Herbarium oder sogar Pantoffeltierchen. Im Makrokosmos beobachtete er auch astronomische Erscheinungen – Lichtphänomene wie Zodiakallichter, leuchtende Nachtwolken, Meteore. Meteoriten waren damals ein „hot topic“. Allerdings nahm Goethe an, dass sie sich in der Erdatmosphäre bilden im Gegensatz zu Ernst Florens Friedrich Chladni (1756–1827), der 1794 in seinem wichtigen Werk die innovative, heute noch gültige Theorie veröffentlichte, dass Meteoriten Körper sind, die aus dem Weltall durch die Atmosphäre auf die Erde kommen.

Der ausführliche, einführende Beitrag *Zwischen Welten und Wirklichkeiten: Öffnungen, Passagen und Wege in der Kosmologie und Kosmopraxis* von Michael Rappenglück spannt einen weiten kulturellen und thematischen Bogen für die Verortung des Menschen in Raum und Zeit für archaische und antike Kulturen. Als Abendvortrag bot er noch das Thema an: *Von der Höhle zum Himmel: Kulturastronomie & Kulturelle Kosmologie.*

Ein Highlight war der Vortrag von Clive Ruggles (University of Leicester, UK) zu seinem aktuellen Forschungsgebiet *Stonehenge and the Major Standstill Moon.* Als Public Invited Lecture / Öffentlicher Gastvortrag gab er einen Überblick zur Kulturgeschichte der Astronomie *Ancient Astronomies – Ancient Worlds.*

Die nächsten Beiträge widmeten sich der *Kulturgeschichte der Astronomie der Steinzeit und Bronzezeit.*

1 Hier das Booklet of Abstracts, hg. von Michael Rappenglück: `http://www.archaeoastronomie.org/content/abstractbooks-der-jahrestagungen/`.

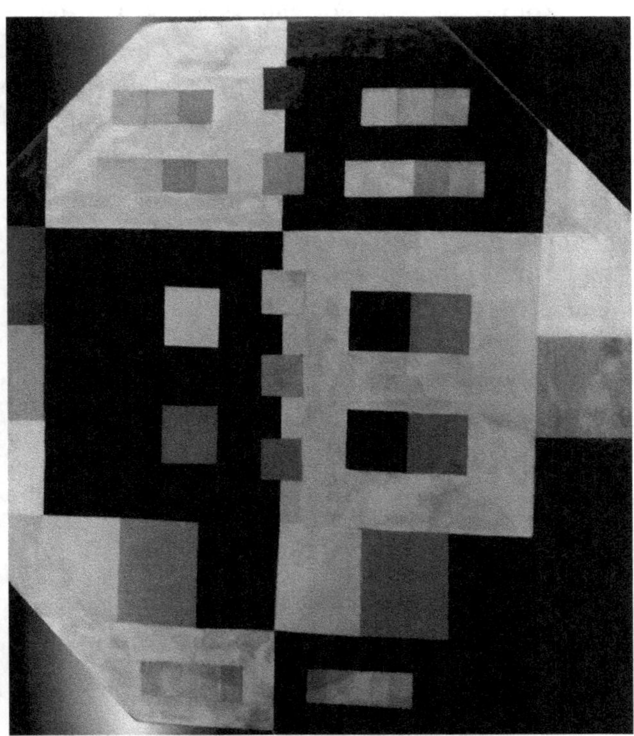

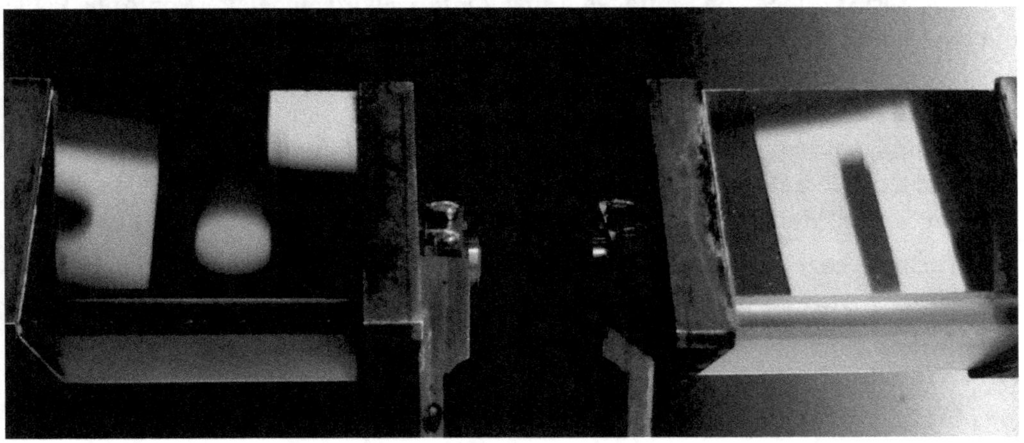

Abbildung 0.3:
Goethes Reisemikroskop, George Adams, 2. Hälfte 18. Jh.
Farbkarte und Farbuntersuchungen Goethes, Goethe-Klassik-Stiftung Weimar

(Fotos: Gudrun Wolfschmidt)

Georg Zotti stellte sein Forschungsprojekt vor: *Die Stehenden Steine von Callanish: Ein 3D-Datensatz für Stellarium.* Die Frage ist, ob der neolithische Steinkreis Callanish, Insel Lewis vor Schottland, mit systematischer Mondbeobachtung in der Vorgeschichte assoziiert ist. Stellarium erlaubt die realistische Simulation der Mondwenden und ihrer Wirkung zwischen den Steinen.

Roland Gröber präsentiert zusammenfassend seine 20jährige Forschung zum *Höhenheiligtum am Pfitscher Sattel – Bedeutung und Entwicklung*, ein herausragendes Beispiel zur Kultur der frühen Bronzezeit und für Südtirol. Er diskutiert die Fragestellung: Handelt es sich bei der Anlage um ein astronomisches Observatorium und/oder eine Kultstätte / Höhenheiligtum?

Siegfried Hess & Thomas Storch untersuchen in ihrem interessanten Beitrag *Engraved Stones from burial mounds in Franconia point to observations of the Sun, of Orion and of the Pleiades*, ob gravierte Steine aus fränkischen Grabhügeln auf Beobachtungen der Sonne, des Orion und der Plejaden hinweisen. Als mögliche Datierung wird 2000–3000 v. Chr. angegeben.

Klaus Albrecht zeigt in seinem Beitrag *Im zyklischen Rhythmus des Kosmos liegt das Schicksal der Menschen – Die bildhafte Zeit im „Heydnischen" Grab von Göhlitzsch-Halle/Leuna*, dass hier eventuell Mondzyklen in Zusammenhang mit zyklischen Vorgängen im Sonnenjahr gebracht werden.

Die nächste Gruppe von Beiträgen *Antike Kulturen* wird eingeleitet von Anna Paule mit dem Thema *Man, Myth, Cosmos – The Sun and Solar Eclipses in Ancient Mesopotamia (Second and First Millennia BC).* Sie untersucht Quellen zur Wahrnehmung der Sonne und der Sonnenfinsternis im alten Mesopotamien, um Hinweise auf mögliche Beobachtungen von Sonnenfinsternissen zu bekommen. Auch Jörg Bäcker thematisiert die mesopotamische Astronomie: *Rätsel der alten Astronomie*. Er beschäftigt sich mit Sonne, Mond und Venus im Gilgamesch-Epos.

Hermann Wenzel untersucht den *Kretisch-minoischen Diskos von Phaistos, ein herausragendes Artefakt der Archäoastronomie auf Basis einer dreiteiligen Kosmologie „Geist-Leben-Sache"*. Als Hypothese schlägt er eine mögliche mathematisch-astronomische Deutung vor.

Albrecht Ploum bietet ein interessantes, bisher kaum beachtetes Thema *Iconography, Science, and Lightning Figures*, Blitzfiguren und andere atmosphärische Phänomene in Gemälden der australischen Aborigines.

Es folgen Beiträge zur *Kultur- und Architekturgeschichte vom Mittelalter bis zum Klassizismus*. Andreas Fuls beginnt mit fünf Ringburgen der Wikingerzeit des 10. Jahrhunderts: *Die Trelleborgen in Dänemark: Eine astronomisch-geometrische Siedlungsplanung der Wikinger*. Er geht der spannenden Frage nach, ob ihre ungefähre

Abbildung 0.4:
Sonnenaufgang in Goethes Römischen Haus in Weimar am 23.6.2023 (vgl. Abb. 16.2)

(Fotos: Katrin Cura)

Ausrichtung nach Norden zufällig ist oder ob sie mit der Beobachtung von Auf- und Untergängen der Sonne und des Mondes in Zusammenhang steht. Für Trelleborg (und Fyrkat) konnte verifiziert werden, dass vom Mittelpunkt der Anlage aus gesehen der Konstruktionsplan eine Symmetrie aufweist, die mit Auf- und Untergangsrichtungen der Sonne zur Tag- und Nachtgleiche sowie zur Sommer- und Wintersonnenwende in Verbindung steht.

Burkard Steinrücken untersucht geometrisch das ursprüngliche Lichtereignis, den Einfall des Sonnenlichtes durch die runde Öffnung und die Hinterleuchtung des Kopfes einer Marienfigur in der *Nordturmkapelle im Erfurter Dom*. Durch den Anbau des gotischen Chores wurde diese Öffnung bereits um 1350 verdeckt. Von Interesse ist zudem die Ausrichtung des romanischem Doms auf den Sonnenaufgang in den Tagen um den 15. August – Mariä Himmelfahrt.

Harald Gropp behandelt das interessante Thema *Kosmische Lebenswelten auf Himmels- und Erdgloben in Weimar und Gotha*. Im Zentrum stehen die Globen von Johannes Schöner (1477–1547), Professor für Mathematik am Nürnberger Egidiengymnasium, dessen Erdgloben zu den ersten gehören, auf denen Amerika verzeichnet wurde.

Es folgt der eindrucksvolle Beitrag von Ettore Ghibellino zu seinem Forschungsprojekt in Weimar *Ist das Römische Haus im Park an der Ilm in Weimar astronomisch verankert? Teil II: Vertiefung der archäoastronomischen Befunde*. Glücklicherweise war die Beobachtung des Sonnenaufgangs im Römischen Haus von Goethe möglich – ein beeindruckendes Schauspiel. Es handelt sich um das erste klassizistische Gebäude Weimars im Stil eines dorischen Tempels.

Der letzte Beitrag steht unter dem Thema *Mensch im Kosmos – Vergangenheit und Zukunft*. Jaak Aru beschäftigt sich mit dem Thema *Homo sapiens and the Universe – Where do we come from? Where are we going to?* Entstand das irdische Leben auf der Erde oder durch kosmische Objekte wie Meteoriten und einen weiteren evolutionären Prozess.

Die Exkursionen boten die Möglichkeit, Weimar und seine Kunst und Kultur näher kennenzulernen. Begonnen wurde mit einem Stadtrundgang mit einer kompetenten Führung. Auch die Besichtigung des Museums für Ur- und Frühgeschichte war eingeplant. Viele haben vor und nach der Tagung weitere der 30 Museen in Weimar besucht, das Goethe-Nationalmuseum mit Wohnhaus am Frauenplan, Goethes Gartenhaus, das Schiller Museum und Wohnhaus sowie das Stadtmuseum. Natürlich durfte das Bauhaus-Museum Weimar, das Haus Am Horn, das Musterhaus von Georg Muche für die Bauhausausstellung (1923), das Haus Hohe Pappeln für Henry van de Velde und das Museum Neues Weimar nicht fehlen. Es war alles in allem eine gelungene Tagung, die diese Proceedings widerspiegeln.

Abbildung 0.5:
Lyonel Feininger (1871–1956): Kathedrale des Sozialismus,
Holzschnitt zum ersten Bauhaus-Manifest (April 1919)

(Foto: Gudrun Wolfschmidt in Weimar)

Abbildung 1.1:
Der Bogen der Milchstraße über La Silla, Paranal, Chile
The Arc of the Milky Way over La Silla, Paranal, Chile

Courtesy: P. Horálek/ESOESO, ESOcast85,
`https://www.eso.org/public/images/ann16037a/`

Between Worlds and Realities: Cosmic Openings, Passages, and Pathways in the Ancient Cultures' Worldview and Cosmopraxis

Michael A. Rappenglück (Gilching)

Abstract

For archaic and ancient cultures, the lifeworld appeared spatiotemporally organised. It formed a chronotopos, a term that was first applied to biology and then to literary studies in connection with general relativity, but which excellently reflects the view of archaic and ancient cultures. The cosmological model 'located' the human being in the (life) world. It also provided a 'housing' where he could move and develop according to his given abilities. With the temporal component, it also incorporated transformation, development, and synchronisation processes. For the archaic and ancient cultures, it was apparent to identify in the cosmic housing, analogous to rooms in human dwellings, different areas of spatiotemporal efficacy that interact and influence each other in the system of heaven, earth, and the underworld (Rappenglück 2013, 2014). Regarding the wandering stars, especially the sun and the moon, various peoples developed the image of asterisms and the year points on the ecliptic as certain dwelling stations and gates in the sky (Huxley 2000). Similarly, the places of the rising and setting of the sun and moon and the celestial bodies on the horizon were understood as openings (a kind of holes, doors, gates, or windows) often localised in world mountains supporting the sky vault, in the world cave or the world house (Rappenglück 2013). Analogously, the people thought of areas in the underworld that the sun, the moon, the other wandering stars, and fixed stars and asterism pass on their way from setting to rising. These stations could be dangerous but also peaceful. Some cultures had the idea of a network of paths through the underworld, for example, the concept of tunnel caves or a liana netting. On the earthly plane, people closely followed the periodic movement of the sun, moon, planets, stars, and asterisms between the extreme places of their rising and setting on the horizon. People regarded these locations as openings, doors, gates, windows (Chevalier and Gheerbrant 1996). Across cultures and eras, people used openings in the natural formation of the landscape and artificial structures, such as windows, doors, roof openings, skylights, smoke outlets, etc., to observe and aim at sun, moon, fixed stars, and asterisms, or other wandering stars, important for timekeeping, orientation, and symbolism (Albani 1994; Rappenglück 2013). They paid particular attention to the natural or artificially produced alternation of light and shadows during the daily or annual cycle of the sun. Sometimes, for example in the apocryphal Enoch (Jacobus 2015), people associated these ideas with the gnomon's measuring instrument. The design of the openings, entrances, and passages of the buildings and the settlements often followed cosmological and cosmopractical concepts of the respective culture (Jiang 2014; Rappenglück 2005). One looked at the building as a cosmos model or regarded the cosmos as analogous to a building (Rappenglück 2013, 2014). For cultures, the passage of the stars and the Milky Way through the openings told stories of their world and other worlds.According to the views of many ancient cultures, the cosmos consisting of different spheres of reality, which they perceived as "other" worlds, enfolds the human living world (Rappenglück 2014). They imagine these areas as unique spaces that are not accessible to ordinary perception but exist in a sense parallel to the everyday world. However, in their view, these realms are not entirely closed off from each other. Occasionally under certain circumstances (place, time, perceiving entity), the enclosures may become partially translucent and even open to access. That also includes the motif of the clashing rocks (Rappenglück 2005). Then openings offer passages between the worlds at particular places and times, possible for certain entities: plants, animals, qualified humans, spirit beings, demigods, and gods (Reiche 1993; Schafer 1986; Coomaraswamy und Coomaraswamy 1997). The only paths run over two passing points described as passages, channels, holes, openings, ravines, doors, or gates. The passing points must sit on unique locations of the earthly landscape. People regarded the intersection of the ecliptic with the equator as on passage gate. They identified the other gate with the crossing of the Milky Way with the ecliptic (or the zodiac). There was the view that the Milky Way mediates a path between worlds (Rappenglück 2017). The intersection of the zodiac (ecliptic) and the Milky Way (galactic equator) remains stable in time, unaffected by precession. According to ancient tradition, the paths and passages between heaven and earth are particularly effective at the time of the solstices and equinoxes, when these annual points come to lie in the Milky Way in an alternation lasting thousands of years. Finally, some people recognised another special opening in the celestial pole (north/south), especially when a pole star was at this location. That was the most significant transition place from the earthly to a transcendent sphere. In certain epochs, the celestial pole (possibly marked by a pole star) laying in the Milky Way opened an extraordinary path to other worlds. The talk provides a systematic overview about the astronomical foundations of the ideas of archaic and ancient cultures about the openings and paths as components of their cosmology and cosmopraxis. The study follows an integral methodology (Rappenglück 2022).

Zusammenfassung: Zwischen Welten und Wirklichkeiten: Öffnungen, Passagen und Wege in der Kosmologie und Kosmopraxis

Für archaische und antike Kulturen erschien die Lebenswelt raum-zeitlich organisiert. Sie bildete einen *Chronotopos*. Das kosmologische Modell verortet den Menschen in der Welt. Es bot ihm eine Behausung, in der er sich entsprechend seinen gegebenen Fähigkeiten bewegen und entwickeln konnte. Die zeitlichen Komponente umfasste Transformations-, Entwicklungs- und Synchronisationsprozesse. Für die archaischen und antiken Kulturen war es naheliegend, in der kosmischen Behausung, analog zu Räumen in menschlichen Behausungen, verschiedene Bereiche raumzeitlicher Wirksamkeit zu identifizieren, die im System von Himmel, Erde und Unterwelt interagieren und sich gegenseitig beeinflussen. In Bezug auf die Wandelsterne, insbesondere Sonne und Mond, entwickelten verschiedene Völker das Bild der Sterngruppen und Jahrespunkte auf der Ekliptik als Aufenthaltsorte und Tore am Himmel. In ähnlicher Weise wurden die Orte des Auf- und Untergangs von Sonne und Mond und die Himmelskörper am Horizont als Öffnungen verstanden, die oft in den das Himmelsgewölbe stützenden Weltenbergen, in der Weltenhöhle oder dem Weltenhaus lokalisiert waren. Auf der irdischen Ebene verfolgte man aufmerksam die periodische Bewegung von Sonne, Mond, Planeten, Sternen und Sterngruppen zwischen den extremen Orten ihres Auf- und Untergangs am Horizont und betrachtete diese Orte als Öffnungen, Türen, Tore, Fenster. Man nutzte Öffnungen in der natürlichen Gestaltung der Landschaft und in künstlichen Strukturen wie Fenster, Türen, Dachöffnungen, Oberlichter, Rauchabzüge usw., um Sonne, Mond, Fixsterne und Asterismen oder andere Wandelsterne, die für Zeitmessung, Orientierung und Symbolik wichtig sind, zu beobachten und anzuvisieren. Die Gestaltung der Öffnungen, Eingänge und Durchgänge der Gebäude und der Siedlungen folgte oft kosmologischen und kosmopraktischen Vorstellungen der jeweiligen Kultur. Man betrachtete das Gebäude als Kosmosmodell. Für die Kulturen erzählte der Durchgang der Sterne und der Milchstraße durch die Öffnungen Geschichten über ihre Welt und andere Welt(en). Dazu gehört auch das Motiv der klappenden Struktur. Öffnungen bieten an bestimmten Orten und Zeiten für Pflanzen, Tiere, qualifizierte Menschen, Geistwesen, Halbgötter und Götter Durchgänge zwischen den Welten. Die Pfade verlaufen über Übergangspunkte (Schnittpunkte Horizont – Himmelsäquator – Ekliptik – Milchstraße – Zodiakallicht), die als Durchgänge, Kanäle, Löcher, Öffnungen, Schluchten, Türen oder Tore beschrieben werden und die sich an einzigartigen Stellen der irdischen Landschaft befinden. Für viele Kulturen weltweit war die Milchstraße der auffälligste und bedeutendste Weg zwischen den Welten. Ebenso wichtig war die besondere Öffnung am Himmelspol, vor allem wenn sich ein Polarstern an diesem Ort befand, die den Weg in eines jenseitige, transzendente Sphäre symbolisierte. Die Studie gibt einen systematischen Überblick über die astronomischen Grundlagen der Vorstellungen archaischer und antiker Kulturen über die Öffnungen und Wege als Bestandteile ihrer Kosmologie und Kosmopraxis.

1.1 The lifeworld as chronotopos: places and paths

In the understanding of most ancient and some of today's cultures that are still halfway close to nature, place and time (4D) belong together and form qualitative events of the respective lifeworld and cultural cosmology (Rappenglück 2024). Space is not a "container", and time is not simply causally structured. The 4D events structure a network of interactions, a web of life. The temporal component also includes transformation, development, and synchronization processes. The lifeworld appears as a chronotopos (Rappenglück 2024).

1.2 Where and when: orientation, navigation, and chronology

Orientation, navigation and time determination are closely linked abilities that allow living beings, especially highly mobile animals and humans, to find their way in the living world using different but mutually supportive methods (orientation), i. e. to determine and remember their location, to move from one place to another and to determine the best route to a destination, taking into account possible drifts (navigation), as well as to record changes, transformation and the relationship between events (chronology) (Rappenglück 2019). Migrating creatures, especially if they must cover long distances, need coherent information about their home and destination and the bearing, i. e. a 'map' (Rappenglück 2019). Equally important are certain external sensory stimuli that indicate useful destinations for the needs of the living being towards which it moves: they form the 'compass'. 'Maps' is a memorization aid that links information from various sensory stimuli from several locations and the routes between them (Rappenglück 2019). They can be designed as a two-dimensional image or a three-dimensional model (Rappenglück 2019). However, it is also possible to 'tell' an extremely functional location and route description (Rappenglück 2019). Proven compass signals in the animal kingdom, in insects, fish, birds and mammals, include the Earth's magnetic field, the Sun, the Moon, certain stars and asterisms, the Milky Way, the circumpolar asterisms, the polarization of celestial light (position of the Sun and position of the celestial pole), unique landmarks and smell (Rappenglück 2019). The different compasses are hierarchically graded and linked to each other (Rappenglück 2019). There are backup, calibration, and standby systems (Rappenglück 2019). A special two-dimensional positioning system and internal clocks determine geographical longitudes (Rappenglück 2019).

Some of the orientation and navigation abilities in some animal species also exist more or less clearly in humans or are very likely to exist in humans but have not yet been sufficiently confirmed (Rappenglück 2019). Scientific studies prove a magnetoreception sense, the possibility of perceiving the patterns of the polarized skylight (Haidinger's brush), the inclusion of the Sun's azimuth for walking in a straight line and the existence of various internal clocks as time and pulse generators (Rappenglück 2019).

In addition, people can use language and writing to expand natural abilities by linking stories with orientation and navigation to 'directions' (story-tracks) and storing them externally on different materials in the form of 'maps' (story-maps) (Rappenglück 2019). These expansions certainly date back to the Middle Palaeolithic, i. e. to *Homo neanderthalensis sapiens* 300,000 years ago and early *Homo sapiens* 200,000 years ago (Rappenglück 2019).

Behavioural, neurophysiological, and linguistic research has shown that humans use a spatial, mostly visual and, in addition to animals, a verbal memory strategy to orient and navigate themselves (Rappenglück 2019). A holistic and static and, in addition, a sequential and dynamic approach ("dual-coding theory") is used to recognise locations, directions and movements (Rappenglück 2019). The spatial reference systems consist of a coding from object to object, called allocentric, and a coding from object to subject, called egocentric (Rappenglück 2019). The first appears in the form of cards. The second appears in path integrations (*dead reckoning*). In cognitive neuroscience, both correspond to semantic and episodic memory (Rappenglück 2019). While the former abstracts knowledge about the world, the latter records individual subjectively experienced events (Rappenglück 2019).

Through locomotion and orientation, humans iteratively transform the sensory perceptions of environmental signs and their spatiotemporal relationships that relate to their point of view into a continuous system of inner and outer representations that manifest themselves as the respective "home world" and other "worlds" (Rappenglück 2019). People on the move, who are nomadic, semi-nomadic or peripatetic, are more accustomed to constantly changing the point of reference and adapting to new "worlds" compared to largely sedentary people, the rural and urban dwellers (Rappenglück 2019). They are trained to understand time as sequential areas of spatial perceptions, as a series of viewpoints connected by specific paths (Rappenglück 2019). The paths contain transformation rules from one perspective to another that ensure the spatiotemporal coherence of experience (Rappenglück 2019). It is discovered and constructed by 'walking through something' or 'going through something'. This could be denoted as "hodological space" (Bollnow 2011, S. 153–164) which can be understood as a preference for egocentric over allocentric coding. This refers to the space created by a network of paths (Rappenglück 2019).

Since at least the Middle Palaeolithic, people have found chronological aids in the rhythms of living beings (phenophases of fungi, plants, animals, and humans), the tides, the climate, and the stars. They sought to correlate these with each other and relate them to individual developments (biological, psychological), collective (sociological, economic) processes, and spiritual experiences (Ginzel 1906; Ruggles 2015). They expressed their chronotopic knowledge in symbols, myths, and rituals and later in proto-scientific and scientific terms.

1.3 The first-person perspective: the body, the conscious being, the place, the centre, and the horizon of perception

The human body, the conscious being, and the worldly horizon of perception collectively shape each other (Rappenglück 2024). The first-person perspective defines the centre of space-time and, thus, the centre of a subject's world. The limits of the senses' abilities and practical, rational, and emotional understanding determine the respective horizon of perception, the ontology. Since the bodily centre with the ego perspective (Rappenglück 2024) is a topocentric system of the world (cosmos), events (places with times) are initially more essential than abstractions such as spaces and times (Rappenglück 2024). The human body embodies the centre, its location, and its own time and, at the same time, the starting point of all experience and knowledge (Rappenglück 2024) with the embodiments (things), phenomena and forces of the world. In corporeality and moving bodies, individuals and cultures experience dwellings and landscapes (Rappenglück 2024).

1.4 Ontologies and perspectives

From a cultural anthropological perspective, the different life worlds correspond to various ontologies, each representing the view of human communities/societies on their respective ecosystems. Living beings (plants, animals) and the manifestations of inanimate matter could have the status of persons (Rappenglück 2024). Communication between these and humans would be possible according to the old view. The first person's perspective can be changed: One can experience one's own world differently or other worlds in the place of other beings. Changes in perception and consciousness appear as journeys into other realities. Associated with this is the idea of many perspectives of different entities in the world (Rappenglück 2024). In ancient cultures that led to the concept of numerous worlds with entities and forms of existence within them, which have their own characteristics but interact with each other more or less intensively.

1.5 Openings, passages and paths in natural and artificial enclosures and landscapes

The cosmological model locates people in the (living) world. It also provides an "enclosure" in which they can move and develop according to their abilities: Localization, materiality, orientation, and navigation (Rappenglück 2022, 2024). Housing, locally concentrated, e. g. caves, tents, houses, cult buildings, graves or in landscapes spatially, stabilizes individual and collective existence in a fixed reference system (cultural cosmology), which is built around the subject (individual, collective) and forms the centre and horizon of space, time, forms, perception, evaluations, order and meaning: a sphere of culture (Rappenglück 2024). A world enclosure is formed in which we can linger and embed ourselves and find home and meaning.

Natural (e. g. caves, terrain) and artificial enclosures (e. g., tents, houses, temples, protective walls) have rock faces or walls as an essential component (Rappenglück 2013, 2014b). They separate, enclose, and protect space for a certain period. Openings in them, e. g., crevices, passages, doors, portals, windows, wells, shafts, roof openings, smoke vents, etc., are practically necessary (Rappenglück 2013, 2014b). They offer places of entry and exit, passage, and transition from one area to another (Lehner 2006). They allow space to be accessed horizontally and vertically, with special construction elements such as ropes, ladders, stairs, etc. (Lehner 2006). They regulate lighting and the exchange of material, water, air, and heat. They are also an element of necessary protection systems. When selecting openings in natural enclosures and creating openings in artificial enclosures, attention was also paid to intentional orientations, which were determined by practical considerations, e. g., light incidence, wind direction, rain tendency, local temperature, focus of protection and ritual-spiritual requirements. This practice also required a certain basic knowledge of the rhythms of life, the shape of the landscape, the prevailing weather phenomena and the times set by celestial phenomena. In addition, openings, passages and paths in natural and artificial enclosures and landscapes also provided a cultural-cosmological code that made anthropological and spiritual experiences understandable in symbolic, mythical, and ritualistic terms (Lehner 2006; Rappenglück 2014b; Ruggles 2015).

1.6 Anthropological framework and cosmological code: Openings, paths, passages, stations, and spheres of existence

The view of openings, paths, passages, stations, and other spheres of existence is based on certain human experiences that have been present and handed down in cultures for millennia. These include the various types of states of consciousness (daydream, trance, meditation, dream, psychosis, ecstasy, near-death experiences), the experience of dealing with the psychosomatic and social phases of human development (Rappenglück 2024), individuation, transformation, metamorphosis and the perception of the life worlds and different sensory abilities of other living beings. The vectorial character of existence ("arrow of time"), reflected in hodological space, is also significant. Finally, near-death experiences (Rappenglück 2024) and, from the perspective of ancient cultures, encounters with entities from paranormal realms, humanoid or non-human supernatural beings, e. g. souls, spirits, angels, demons, or divine beings (Rappenglück 2024), the idea of other realities and the possibilities of movement between them.

According to the ideas of many ancient cultures, the cosmos, consisting of various spheres of reality that they perceived as "other" worlds, encompasses the human living world (Rappenglück 2024). They imagine these areas as unique spacetimes that are not accessible to ordinary perception but exist in some way parallel to the everyday world. However, from their point of view, these spheres are not completely sealed off from each other. Under certain circumstances (place, time, perceiving entity, character prerequisites of the human being) the enclosures can become partially transparent and access points recognisable. Passages back and forth between the worlds are then possible for plants, animals, humans, spirit beings, demigods and gods who are exceptionally qualified through sensory perception, valuable action, or spirituality (Rappenglück 2024).

This leads to the idea of heavenly paths with stations, a hierarchy and a network of other worlds, possible passages between the worlds through openings and transitional areas, semi-permeable membrane(s) as borders of other realities, a transcendent beyond and realm(s) of souls and divine beings. Birth, development, death and continued existence are thematized in the symbolism, myths and rituals that refer to the opening and closing of the transitions, the passage, the way stations and other, perhaps also final, abodes.

1.6.1 Places and paths of transition and transformation

Voyages can take place either "horizontally" (2D) across land, sea, and sky and through the underworld or "vertically" (3D) between the greatest depth and height, from caves or sea depth to mountain peaks and the realm beyond the last sphere of stars. The travel sections can be combined. The cosmic "travel routes" (fig. 1.2) include the paths of the Sun (ecliptic) and of the other wandering stars (Moon, five classically known planets) through certain zodiacal asterisms (stations, houses), especially the path of the Moon through peculiar asterisms (Moon stations or houses), the zodiacal light (under very rarely given conditions of its observability), the Milky Way with its special positions to the observer's point of view (as a ring in the horizon, in bridge position transverse and vertical), the rainbow, and finally the voyage along the zenithal or the polar axis (Weidner 1931; Gundel, W. 1936; Gandz 1943; Weinstock 1949; Yampolsky 1950; Van der Waerden 1952–1953; Erlenmeyer, M.-L. & H. Erlenmeyer 1958, S. 55–57, 59-60; Jobes, G. 1962; Jobes, G. & J. Jobes 1964; Hübner 1982, 2002a, 2002b; Dean-Otting 1984; Sehgal 1989; Strätz 1996; Marinatos 2001; Hammond 2007; Schweizer 2010; Kelley and Milone 2011; Ruggles 2015; Rappenglück 2005, 2015, 2017, 2019, 2020c, 2024; Latura 2018; Prendergast 2021). On the one hand, cultures have similar and standard views because they are determined by the general course of nature, and on the other hand, there are different variants based on the respective ontological perspectives and local natural conditions.

Certain concrete and abstract transitions exist in the cosmological imagination of cultures. Some refer to time-independent exact locations, others to time-dependent larger areas. Among the first are the horizon, the cardinal points, the zenith, and the nadir. The latter include the concrete rising and setting areas of the celestial objects (Sun, Moon, planets, Milky Way), the location of the celestial poles determined by the precession of the equinoxes (abstract) and, if present in an epoch, the concrete location of a pole star among the stars, the intersection of the ecliptic and equator (abstract), the intersection of the Milky Way and zodiac (abstract) or zodiacal light (concrete), the abstractly conceived lunar node or the movable radiant of meteor streams (concrete). The zodiac signs and the lunar stations form slightly extended transitions fixed with asterisms (concrete). The permanent appearances of the Milky Way and the zodiacal light belong to the concrete extended paths in the sky. The course of the Sun (equinoxes, solstices, analemma), Moon (different months, lunar nodes), and planets (sidereal and synodic orbits, apparent retrograde motion, pentagram of Venus, elongation, opposition, conjunction, etc.) characterize the paths in the Zodiac. Rainbows or other meteorological phenomena (e. g. crosses and columns of light, halos, etc.), which, according to the ancient view, mainly belong to the upper world, appear only temporarily perceptible and unstable.

The paths, which can be narrow paths or wide roads, run through transitional places that are described as openings, holes, ravines, chasms, doors, gates, portals, windows, houses, palaces, tombs, oculi, false doors or gates, spirit holes, shafts, wells, springs, sluices, maelstroms, whirlpools, smoke vents, reflective surfaces (natural, artificial), crystals, rings, perforated disks or through the orifices of individual human beings (vagina, mouth, anus, fontanel, etc.) as well as mythical cosmic beings (Lehmann 1945, S. 5–6, 11, 13, 16, 19, 27; Krupp 1997, S. 267; Tate 1999, S. 176, 179, 184; La Riva Gonzalez 2000, S. 179; Hammond 2007; Nordberg 2009; Norelius 2017; Rappenglück 2023, 2024). The transit points are located at unique points in the earthly, subterranean, or heavenly spheres. The design of the openings, entrances and passageways of the buildings and settlements often followed cosmological and cosmopracti-

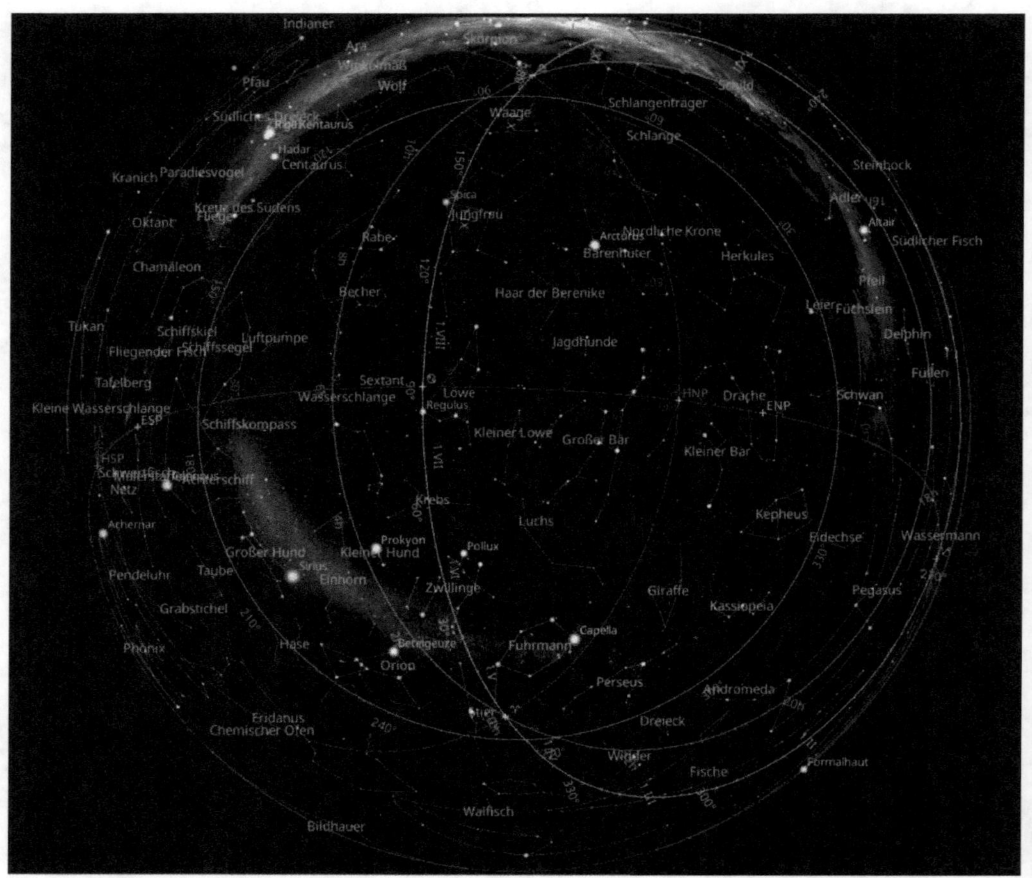

Figure 1.2:
A 360° view of the sky 2,800 BC in a Lambertian azimuthal projection with essential "celestial paths" (celestial equator, ecliptic, Milky Way) and points (the northern celestial pole at Thuban (α Dra, the ecliptic pole in today's constellation Dragon).

(Created with Stellarium 24.1. Courtesy: Michael A. Rappenglück)

cal ideas of the respective culture (Snodgrass 1985; Krupp 1997, S. 20–21, 47, 56, 82, 108, 137–138, 200, 269, 300, 309-310; Lethaby 2004; Lehner 2006; Jiang 2014; Mauvieux, Reinberg, and Touitou 2014; Rappenglück 2013, 2014b, 2023; Ruggles 2015). The building was regarded as a model of the cosmos and vice versa.

1.6.2 Openings at the edge of sky, earth, and underworld

Some pre- and protohistoric cultures believed that the twilight arcs in the morning and evening indicate a separation of the sky from the earth and, thus, as it were, an "opening" of the cosmos leading in the beyond. This view is closely related to the motif of the folding structure, of which there is a good example in an Algonquin (North America) myth (Rappenglück, Barbara 2017, S. 185). The Milky Way's striking position (fig. 1.3) as a ring lying on the horizon (depending on location and epoch) may also have led to the idea of an opening and closing aperture in the horizon (Rappenglück 2017, S. 196–197). In particular, people regarded the (extreme) locations of the periodic rising and setting of the Sun, Moon and some stars and asterisms on the horizon as special openings, a kind of holes, doors, gates or windows, which were often located in the world mountains or a ring of mountains supporting the celestial vault, in the world cave or the world house (Chevalier & Gheerbrant 1996, S. 191, 422–423, 529–531, 806, 925, 1112; Huxley 2000; Rappenglück 2007, 2013, 2014b; Mauvieux, Reinberg & Touitou 2014; Rappenglück 2018b, 2020a, 2020b; Therik 2023).

In the *Odyssey* (Od. 10.82–86, 11.14–19, 24.11–14), Homer (second half of the 8th century/first half of the 7th century BC) refers to "the distant gate", by which he means the area of sunrise in the east. The "gate of the Sun" is in the west and marks the sunset (Nakassis 2004, S. 224–25). Hesiod (before 700 BC) recognizes a single gate of "night and day", which is located on the horizon ("edge of the earth") at the entrance to the underworld, where the sky bearer *Atlas* (the whole world axis) also stands (Nakassis 2004, S. 227).

In connection with the openings in the cosmic building, individual mountains (at the centre or a special cardinal position), a set of two mountains (forming a gate next to each other or facing each other) or a ring of mountains along the complete horizon play an essential role in the myths of cultures worldwide (Rappenglück 2020c). The Sun rises and sets not only through an opening in a single world mountain supporting the celestial vault but, according to some traditions (ancient Mesopotamian, Egyptian, Greek and Indian cultures), also from a "gate" formed by two world mountains (Nakassis 2004, S. 227–29).

Mountains of the rising and setting of the Sun (of the Moon and planets) also exist in Sumer (Jensen 1890, S. 38, 188, 236–237, 241; Heimpel 1986, S. 140–46). The world mountain Măsu (Jensen 1890, S. 316–317) stands at the boundary between the upper light and lower dark spheres of the world. There is the place of the Sun's entry and exit into and out of the interior of the cosmos (Foster, Frayne & Beckman 2001; Lauber 2008, S. 68–69).

According to Sumerians, Babylonians and Akkadians at dawn, the Sun god Šamaš works his way out of the underworld into the mountains (on the edge of the earth) with a "saw". There are the "doors of heaven". At dusk, he descends through analogous

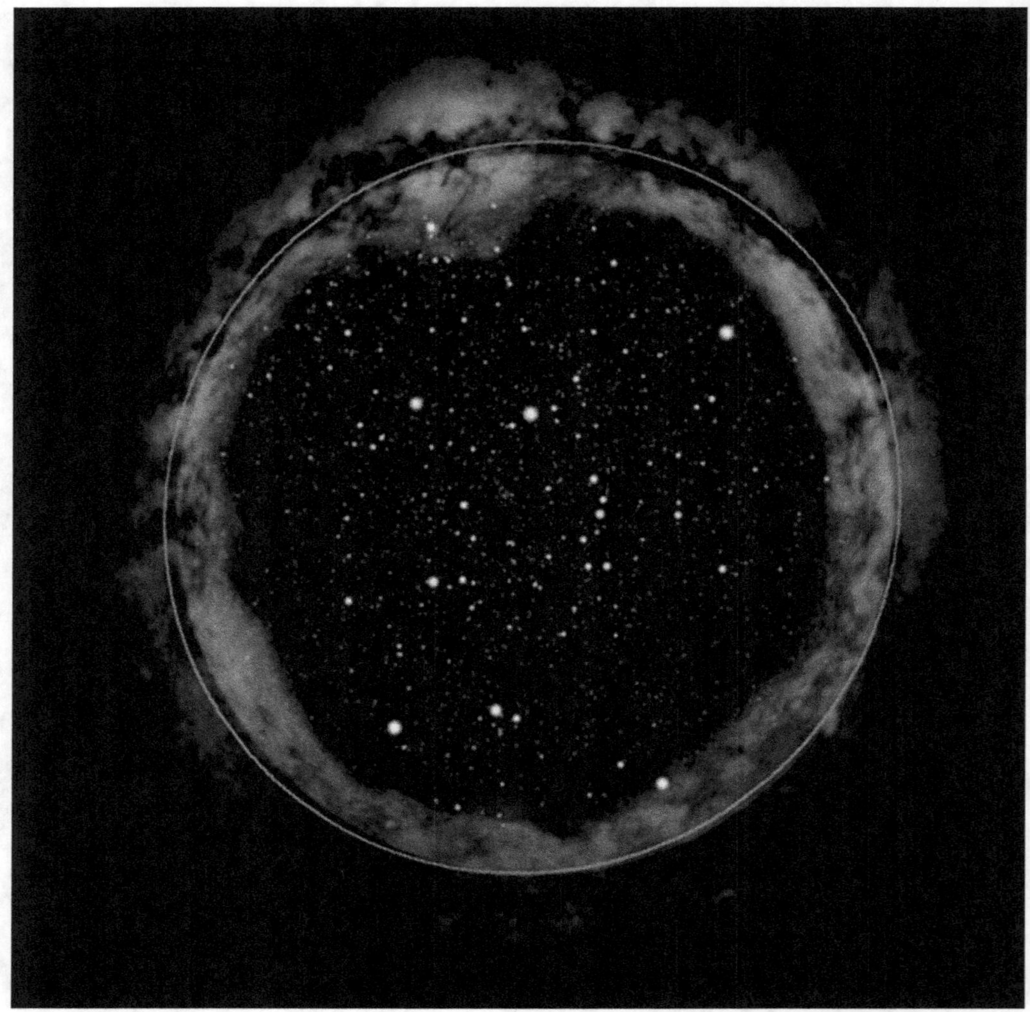

Figure 1.3:
An example of the ring position of the Milky Way in the horizon (latitude: −50°, 2,500 BC)

(Created with Stellarium 24.1. Courtesy: Michael A. Rappenglück)

doors back into the subterranean sphere, which is referred to as his (private) chamber kummu (Heimpel 1986).

In the imagination of the Yanomami (Ironasitéri), South America (Becher 1984, S. 459–60, 1974, S. 2–3, 209), there are two celestial spheres, the lower of which is supported by two inselbergs that protrude from the earth's disk. The openings of the underworld to the earth and the sky lies in this eastern and western world mountain. The Moon, the highest entity in the Yanomami cosmos and at the same time the abode of the deceased, and the Sun move to the left (phase course of the Moon) through a "hole" dug by an armadillo in the time in the inselbergs into the underworld and out of it again.

In Hopi mythology, the Sun enters a kiva in the northwest and exits through a door (trapdoor) to the northeast (Leeming 2010, S. 132–33).

These ideas all initially move within the system of the horizon. For the people the question then very quickly arises as to where the Sun, the Moon and the other celestial bodies are located after they set and before they rise, whether there is a path between the two places and what this way is like. There are two possible explanations: The stars pass under the visible world (concept of underworld paths) or they move around it and are obscured, for example (in the northern hemisphere), by mountains in the north. Some cultures also believe that the Sun moves in the opposite direction in the beyond as it does in this world. They expressed these ideas in symbols and buildings (Nordberg 2009, S. 51–52). This idea is presumably due to the view that processes are reversed in the afterlife.

1.6.3 Openings and paths in the underworld: subterranean natural and artificial spaces (caverns, crypts)

For archaic people caves open paths into other realities of the cosmos (Rappenglück 2007, 2014b). There different spaces and times are linked in a multidimensional structure. The cave is the place where the centre of the world, as well as the entrances, passages and exits to the lower, middle, and upper worlds, can be found. This is where the difficult and perilous passage between the different space-time worlds, the "other worlds", is achieved via mostly narrow passages. Caves resemble "gates" of transition and change. They offer an opening through which it is possible to pass from one cosmic region to another, from heaven to earth or from earth to the underworld, and vice versa.

According to the Greek philosopher Parmenides (c. 520/515-460/455 BC) (fr. 1, 11), there were two gates separating the paths of night and day (equinoxes?), and thus, at the same time, those of appearance and being (Russo 2002). These gates were also known as "horns" (Russo 2002). These transitions separate the dream world (night) and everyday life (day), the human and divine worlds, this world, and the hereafter.

In Japanese mythology, the Sun goddess Amaterasu Omikami withdrew into the celestial grotto, whereupon the world sank into darkness. Two other goddesses attract her by pointing to the presence of a superior goddess outside the world grotto. As proof of their claim, however, they hold a mirror up to Amaterasu Omikami in which she sees herself. Amaterasu Omikami then emerges through a door in the cosmic cave and the world is given light again (Akima 1993, S. 155). This Japanese version is part of a widespread myth complex that seems to go back thousands of years (Witzel 2015).

The Homeric epic of the *Odyssey* (8th /7th century BC) can be understood as a solar journey that leads in a nocturnal section through the gates of the underworld, which are imbued with a special cosmic symbolism (Hammond 2007, S. III, 51–57; III, 60, IV, 11–20). The journey as a kind of cosmic voyage, more as an earthly seafaring, seems to be also connected with the Egyptian conception (Marinatos 2001).

During its passage through the underworld (fig. 1.4), the Sun has to pass through several paths and stations with passages ("gates"), as demonstrated, for example, by the Sun god Re in twelve-night hours with his barque in the ancient Egyptian underworld book Amduat, documented in the New Kingdom (1550–1070 BC [18th h to 20th Dynasties]) (Hornung 1967, 1998; Lieven 2007; Richter 2008; Schweizer 2010). The origins of the Amduat probably lie in the Old Kingdom, in the 6th Dynasty (2347–2216 BC). Similar underworld topographies can also be found in the Book of Gates (The Book of the Underworld Ones Who Assist Osiris), 1300 BC, end of the 18th Dynasty), in the Book of Caves, 13th century BC, 19th Dynasty, in the Two Way Book, Middle Kingdom, 11th dynasty (2137–1994 BC) and in the Tomb Book (Speech of the 12 Tombs, Pap. Berlin P. 3006), 21st dynasty (1075–652 BC) (Werning 2019; Hornung 2002; Hermsen 1991; Hornung 1979/1984). The length of the underworldly path travelled by the Sun god Re in the Book of Amduat is astonishing, and its origin is unexplained at 39,000 km, corresponding to the circumference of the earth.

In Tenga Bithnua (The Ever-new Tongue), a manuscript probably written in the 9th century, Egyptian ideas reappear together with those from certain apocalyptic and gnostic texts, for example, the "Apocalypse of Enoch" or the gnostic text "The Vision of St. Paul" (Carey 1994, S. 14–16). Every night, from setting to rising, the Sun passes through twelve stations of the underworld.

For the Sumerians, Babylonians, and Akkadians, the Sun god had to travel through the underworld to emerge from it in the morning after entering it at dusk (Heimpel 1986). In the inner place of heaven is the "White House" of the Sun god, his residence, his resting place and the "bosom of heaven" (Jeremias 1913, S. 36–37). The texts describe the gates (babu) and portals (abullu) of heaven, mechanically detailed (Heimpel 1986, S. 132–139). The scriptures also state that the Moon (Babylonian-Assyrian: Sin), Venus (Ištar) and Saturn (Ninurta) pass through them (Heimpel 1986, S. 134–139). It is particularly interesting that the stars, apart from the circumpolar ones, also disappear from the visible into the invisible sphere through "gates" assigned to them on the horizon or reappear from the latter into the former (Heimpel 1986, S. 139–40). This means that the azimuths of the and those of selected fixed stars were observed in the (natural) horizon circle.

Figure 1.4:
Map of the netherworld from the coffin of Gua,
from Deir el-Bersha, Egypt, 12th Dynasty, 1985–1795 BC

(London, British Museum (EA30839), Public Domain)

In the Sumerian myths, Inanna (Venus; in the Akkadian version: Ištar) descends into the underworld and must pass through seven gates (transitions of the seven heavenly spheres) until she reaches the centre of the seventh-gated underworld palace of her sister Ereškigal, the mistress of the underworld, at the greatest depths (Jeremias 1913, S. 65–66; Lütge 2008, S. 254–258). In the ancient Oriental cultures, Venus, the moving star, opens the gates of heaven at heliacal rising (Erlenmeyer, M.-L. & Erlenmeyer, H. 1958, S. 57). For the Maya (Mesoamerica), Venus precedes the Sun as its guardian during the night (Dütting & Schramm 1984, S. 8).

In ancient Egypt, the Sun's journey through the underworld (uterus) could also represent the phases of embryonic development (caverns/stations) up to birth (sunrise), at the same time linked to a mythical cosmogony (Renggli 2002). Other mythical cultural traditions, e. g. the Maya, Aztecs (Mesoamerica), the Aborigines (Australia) or the ancient Greeks speak of the fact that the celestial bodies as well as individual humans and entire tribes are created in the subterranean spaces, which act as a cosmic "birth cave", too (Rappenglück 1999, S. 284–85, 2007, S. 66, 68, 73–75, 2014b, 2018a, 2023). Animals, human beings, or chimaeras generated in the earth's womb cave emerged from the darkness through an opening in the rock into daylight on the earth's surface. The event of the primordial beginning is repeated every day and every year when the Sun rises from the rocky cave of the "underworld" or sinks into it. Similar ideas can already be assumed for the symbolism of the cave in the Upper Palaeolithic, 50,000 to 12,000 years ago (Rappenglück 2007, 2014b). Here, it becomes clear that the individual life journey is a path of individuation and the soul journey (Rappenglück 2024) is in the background of the astronomical-cosmological-cosmogonic symbolism and myths: The course of the stars during the day illustrates earthly life, at night, existence in the afterlife. In some traditions, a new cycle is also associated with reincarnation.

In any case, the areas in the underworld that the Sun, the Moon, the planets, the fixed stars, and constellations pass through on their way from setting to rising could be energy-sapping, tense, chaotic and dangerous but also invigorating, relaxing, harmonious, and peaceful places (Rappenglück 2024).

In Taoism (China), the idea of the "grotto heavens" has existed since at least the 4th century AD, in whose depiction of ancient cosmology and cosmogony merge with psychosomatic and spiritual concepts (Rappenglück 2018a, 2023). Journeys through a cave paradise are described as an underworld reflecting the sky and forming a real and essential unity (yin/yang). The cave skies contain Suns, Moons and stars, the palaces of the divine immortals, and the registers of life and death. They form an infinite labyrinthine system of galleries that converge at one point and are designed like doors. At the centre is a golden city in a paradisiacal landscape that follows a time frame different from that of the earthly and cosmic. In the cosmos, in the earthly landscape and the human body, there is, corresponding to each other, a network of "cave galleries" (tunnels, channels) through which life energy (*qi*) flows (Rappenglück 2018a, 2023). The transition is only possible through a narrow passage (womb and tomb). The journeys are transits on a psychosomatic and spiritual level, a transformation (from birth to rebirth to immortality) and a transition to transcendence. This network of "grotto heavens" is also one of communication. The Buddhist worldview knows a network formed from the effective currents of the cosmic breath (*prana*) (Rappenglück 2018a). This "pneumatic" network of effects permeates the universe and, thus, also the human body. It resembles arteries and veins in a cosmic living being. The "breath lines" of the network set the shortest possible time and length, measured by *Prāṇa* (breath) and *Vāyu* (wind). The nodes in the pneumatic net of action are places (Marma) in the body of *Puruṣa* (macrocosmic) and correspondingly in the human being (microcosmic). In each of the nodes and in node complexes, the material cosmos manifests. The Manichaeans also described a network of artery-like channels connecting the celestial spheres and the zodiac with living beings (Rappenglück 2018a, 2023). The Amazonian peoples have very similar ideas of a cosmic network of paths that is present everywhere and can only be perceived by the sensitive (Rappenglück 2018a; Cipolletti 2019; Rappenglück 2023). In the mythical lore of these cultures, the world tree with its branches and root network is a prototypical example of this concept. Moreover, the cosmos has a dynamic "sponge-like" structure interspersed with "winds". Finally, a very similar conception exists in Manichaeism (Gulácsi & Beduhn 2011).

1.6.4 Orbits around the World Mountain

According to ancient beliefs, the planets, in ancient view including the Sun and Moon, the stars and the Milky Way orbit the world mountain (Rappenglück 2020a, 2020c). The Pole Star and/or circumpolar asterisms, e. g. the Big Dipper, are above the mountain peak. Sometimes, the Sun is seen as a sunbird that nests on the summit of the world. Or the wheellike Sun moves daily or annually over the summit or around it. The relationships of the Sun, Moon, and asterisms to certain mountain slopes are considered. The openings (gates) in the mountain through which the Sun, Moon, and stars rise, set, or pass and the areas illuminated by the Sun and Moon were important.

According to an Iranian tradition, Mount *Tërag* rises to the highest height in the centre of the cosmos (Rappenglück 2020a, 2020c). In addition, the Harbour (Elburs, Alburs, Albors Mountains) forms a ring-shaped mountain range that borders the disc-shaped world. It embodies the entire cosmos. For a few hours during the day, the Sun stands behind the gigantic cosmic mountain and is shaded by it. This creates the night. The World Mountain has 360 (= 12 months with 30 days) openings in its flanks, 180 (= half a year with 360 days) on the east and 180 on the west. These look like gates. The Sun and the Moon, the fixed and wandering stars, regularly appear in and disappear from them. The areas illuminated by the Sun and Moon ("continent") were also important (MacKenzie 1964, p. 517–521). Over five of these, the celestial objects light up a second time, resulting in 365 gates (= whole days in the solar year). A related system, but without reference to the World Mountain, is recorded in the Book of Lights or Astronomical Book (1 Enoch, 72–82), dated to at least the 3rd century BC (Rappenglück 2020a, 2020c). It is noteworthy that the gate system is not only related to a horizontal system but also to the projection of an ecliptic system (zodiac) onto the horizon. One could also take an ecliptical view instead of a horizontal one. The ancient constellation of the dragon (Dra) winds around the assigned northern ecliptic pole with a first turn. Researchers have speculated that this "hole" could have been perceived by ancient cultures (Santillana & Dechend 1993, S. 131, 378). Clear evidence of knowledge of the northern ecliptic pole has not yet been provided. According to Tibetan tradition, the palace on the summit of the world mountain is also inhabited by 360 deities or mountain deities are surrounded by 360 attendants (Rappenglück 2020a, 2020c). Similar views, but without numerical details of the gates for the passage, have been handed down (Rappenglück 2020a, 2020c) from Anaximenes of Miletus (ca. 610 – ca. 546 BC) to Kosmas Indikopleustes (484–577 AD). According to Chinese tradition, Mount *K'un-lun* is shaped like an inverted cone with a gate on each of its four sides (Rappenglück 2020a, 2020c).

The ring-shaped construction of the megalithic structure of Stonehenge (Avon near Amesbury, southern England, 51°10′44″ N, 1°49′35″ W), 3,100 BC to 1,600 BC, especially with the construction phase of the "gates" of vertical sarsen stones, covered and connected with horizontal lintel stones, Stonehenge 3 II (2,600–2,400 BC), makes one think of the concrete realization of the observation technique and mythical meaning described much later in the Book of Lights or the Astronomical Book (1 Enoch, 72–82). Similar concepts could also underlie some ring ditch sites, e. g. Goseck (near Goseck, Burgenlandkreis, Saxony-Anhalt, 51°11′54″ N, 11°51′52.5″ E, 4,900–4,700 BC), Pömmelte (near Zackmünde, Barby, Salzlandkreis, Saxony-Anhalt, 51°59′49.4″ N, 11°47′58.9″ E, 2,335–2,050 BC) or Pranhartsberg 2 (Sitzendorf an der Schmida, Hollabrunn district, Lower Austria, 48°35′25.19″ N, 15°59′11.44″ E, 4,700–4,500 BC). The temple of the Assyrian king Sennacherib (705–681 BC) in Assur also had "gates" (Jeremias 1913, S. 126; Huxley 2000), including the sunrise at the vernal equinox ("Gate of the Firmament"), stars in the northern part of the sky ("Gate Path of Enlil") including the circumpolar stars, the Chariot Star, stars in the southern part of the sky ("Gates of Igigi"), whereby the rising and southernmost positions were taken into account.

1.6.5 The path of Sun, Moon, and planets: the ecliptic and the zodiac

Regarding the wandering stars, especially the Sun and Moon, various cultures developed the image of asterisms, stations, and the points of the year on the ecliptic as specific locations and gates in the sky (Gundel, H. G. 1893–1980b; Brown 1899–1900, S. 63, 67, 97, 147, 221, 232, 267, 286, 314, 360, 361; Erlenmeyer, M.-L. & H. Erlenmeyer 1958, S. 55–57, 59–60; Gundel, H. G. 1893–1980b; Strätz 1996; Huxley 2000; Kelley & Milone 2011; Ruggles 2015; Rappenglück 2020c).

The *Enūma eliš* (V, 9), probably Dynasty IV (2nd Isin), 1153–1022 BC, already describes how Marduk opens gates on both sides of the ribs of the cosmic chaos being *Tiamat* (Heinrich, Mitto & Jiménez 2021). Since Tiamat creates 11 monsters, one could think of a form of zodiac.

On the one hand, these classifications were based on practical considerations of time calculation, especially of purposes for the calendar. On the other hand, the zodiac and similar divisions, e. g., systems of houses and decans, symbolically illustrate the individual journey through life as a path of individuation and the journey of the soul in the beyond. Thus, the voyage of life and the way of existence in the beyond is made comprehensible from the perspective of cultures of the time (Müller 2009, S. 220–221). The 12 gates of the heavenly Jerusalem (Acts 21:21), designed as jewels, correspond to the stations of the Zodiac (Boll 1914, S. 39; Lütge 2008, S. 536). In Egypt, the system of 36 deans had existed since at least around 2,100 BC during the 9th /10th dynasty (Gundel, W. 1936; Gundel, H. G. 1893–1980a; Lieven 2007). The decans were assigned to the stages of life in their course: birth, life ("ascent", especially "work" and "descent"), death, and a new birth (Lieven 2007). Among the Maya and Mesoamerican cultures in general is the idea of the "flower world", an otherworldly paradise along the Sun's path at the ecliptic (Taube 2004, S. 69–70). Since at least Hellenistic times, one recognized the great divine cosmic being with its limbs and organs appearing in the zodiac and the decans attributed to the human body parts and organs (Zodiac Man, Homo Signorum) related to melothesis (Gundel, W. 1936, S. 238; Gundel, H. G. 1893–1980b, 1893–1980a).

An unusually detailed explanation of the gates in the zodiac and their classification in an ancient cosmology comes from Manichaeism (Henning 1948; Gulácsi & Beduhn 2011, S. 75, 76, 78, 79, 91, 92, 98). 40 angels hold 10 celestial spheres, and in each of the 4 gates in the cardinal directions, there are 12 gates (synodic months)[1] in the solar year, six of which are arranged to the left and six to the right of the world axis, and 14 gates (half sidereal month; 14-day period). The 12 gates (one year) were divided into 6 thresholds (360 days). Half a threshold comprised 30 bazaars (synodic lunar month of 30 days). Each bazaar consisted of 12 rows, and each row consisted of 2 sides (24 hours). Each side consisted of 180 stalls (one stall: 4 minutes). In each stable were nature spirits and demons of the four directions (one direction: 1 minute), divided into male and female (0.5 minutes), locked up. All the nature spirits and demons in the zodiac were woven back and forth like roots, veins, and limbs (Henning 1948, S. 313,

1 In the brackets I interpret the passage astronomically; angles or times can be used alternately.

315–316). This (coordinate grid) was guided through a "hole" at the northern celestial pole and attached to an upper permanently moving wheel (rotation), which is mounted below the seventh celestial sphere.

In the Iranian tradition and Mithraism, the spheres of the seven wandering stars (geocentric) were regarded as gates through which a soul must pass to detach itself completely from its material form until it reaches its true essence in the eighth sphere. This gradual path also corresponded to the degrees of initiation into the mysteries (Cumont and Gehrich 1903, S. 105–106; Burkert 1963, S. 110–115; Lütge 2008, S. 613).

Some board games transform the existential journey from the mythical-symbolic version into rituals (Rappenglück 2024). The ancient Egyptian Mehen game, for example, which dates to around 3,000 BC, thematizes the journey of the Sun god Re through the underworld along the 'road of the (snake god) Mehen' (Rappenglück 2024). Mehen is the protector of Re and the deceased pharaoh and every soul (Ba) on their difficult and perilous journey through the stations of the underworld. In Mehen, there were gates on the serpentine path to the Sun god in the centre, with whom the soul wanted to unite (Kendall 2007, S. 41–42). Some led the bad to their doom, but the good, who knew the right passwords to pass through, reached their destination. The Egyptian Senet, which can be traced back to around 3,050 BC, also uses the journey of the deceased soul (Ba) through the realms of the underworld in a playful way to prepare the living and the dead for their new existence in union with the Sun god Re. In later times (New Kingdom), the game is related to the death and new life of the god Osiris (associated with the asterism of Orion) and, in addition to the Sun, also to the Moon, the asterism of Orion and the decans.

Some hopscotch games in the "Heaven and Hell" game are also about journeys between heaven and the underworld associated with the seasonal course of the Sun and archaic simple cosmological models (Rappenglück 2024). The walk and the dance (with ball game [Sun]) through a labyrinth at Easter in a Christian cathedral (Ball Game Order of the Chapter of Auxerre, 18.4.1396 (Rappenglück 2024) also symbolizes the cosmic journey of the Sun and the path taken by Christ to redemption from the mortal world, which everyone should follow.

The Indo-Tibetan karma game *Moksha Patamu* (also Mosksha Patamu: "path to salvation", Gyan chauperljnan chauper: "game of wisdom"; modern Western variant: "snakes and ladders"), based on philosophical-cosmological aspects of Hinduism, Buddhism, and Jainism, in which the player moves through different spheres of the cosmos and must deal with good and destructive processes, is about liberation from karma (Rappenglück 2024). There is also a variant of the game in Sufism (Rappenglück 2024). This focuses on the Sufi's path from life in the world to union with God.

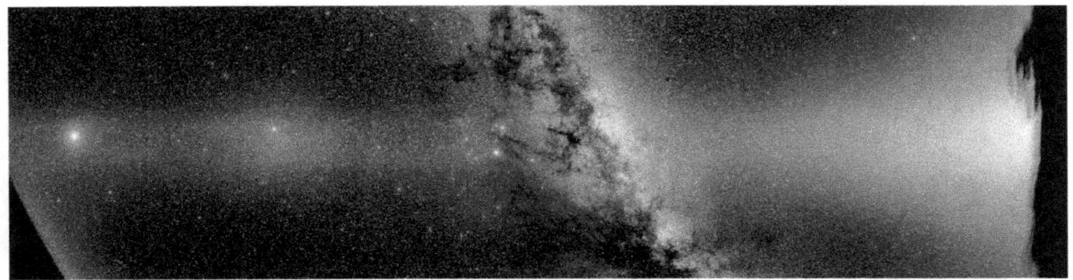

Figure 1.5:
Photo of False Dawn, *Gegenschein*, and the other of the zodiacal light band, which the Milky Way visually crosses.
It was taken at ESO's La Silla Observatory, Chile (2400 m ASL).

(Courtesy: ESO/Petr Horálek, CC BY 4.0)

1.6.6 Another path in the sky: the zodiacal light

In the discussion (Redhouse 1878; Gandz 1938, S. 17–46, 1943; Latura 2014, 2018) is yet another "path" in the sky: the zodiacal light (fig. 1.5). This very delicate light, caused by reflection, scattering, and refraction of sunlight, by interplanetary dust particles in the plane of the ecliptic, is extremely difficult to observe with the naked eye as a pale cone of light around the equinoxes in temperate northern or southern latitudes. In the tropics, however, it can be seen at the equinoxes all year round with excellent seeing if only a narrow crescent Moon (old light/new light) is present at night or, even better, if it is a new Moon at night. Alexander von Humboldt already reported this (Humboldt 1845, S. 142–144). According to him, under excellent conditions, it can reach at least the brightness of the Milky Way (Humboldt 1845, S. 144). It may well have been possible for observers in ancient civilizations, in ancient Egypt, among the Hebrews and Arabs, to observe the zodiacal light. In ancient Egypt, the god Horus-Sapdi, a figure of the sky god Horus, controlled the zodiacal light. For example, this light phenomenon is called "False Dawn" in Islamic culture (Redhouse 1878; Gandz 1943, 1938, S. 22–29). On the other hand, aurora events or meteorological phenomena (e. g. columns of light, noctilucent clouds, atmospheric dust) and occasionally bright comet tails could also be confused with the zodiacal light. However, the observation of zodiacal light does not immediately lead to the conceptual idea of a "celestial path" analogous to the permanently observable Milky Way, as is claimed in some studies (Latura 2014, 2018). If the role of the zodiacal light in ancient cultures were comparable to that of the Milky Way – the idea of a path between this world and the spheres beyond – the zodiacal light would have to be visible in symbols, myths, and rituals, and even then, in a special terminology in the traditions. So far, there is hardly any evidence for this, perhaps except in the Islamic cultural world (Redhouse 1878; Gandz 1938, 1943). In addition, the zodiacal light appears cone-shaped to the naked eye in contrast to the meandering path of the Milky Way. Photographs show the entire band of zodiacal light around the ecliptic. The philosopher Plato (347–327 BC) writes extensively about the intersection of the ecliptic and the Milky Way, which he compares to the large Greek letter *Chi* (Chi). The zodiacal light would give the abstract structure of the ecliptic a certain temporal-concrete appearance. However, Plato does not mention it.

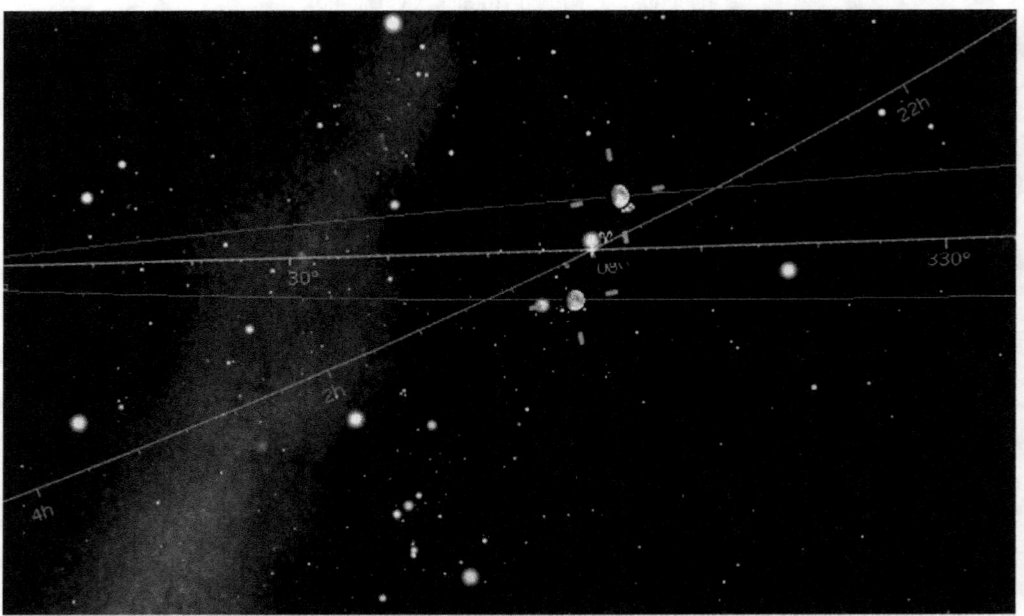

Figure 1.6:
The "Golden Gate of the Ecliptic", at the equinox ca. 2,600 BC.
The Moon's orbit area is shown by the Pleiades (+5.145°) above,
and the Hyades (−5.145°) below the ecliptic in 18.61 years.

(Created with Stellarium 24.1. Courtesy: Michael A. Rappenglück)

1.6.7 The Golden Gate of the ecliptic: the Pleiades and Hyades open star clusters

The Pleiades (M 45) and the Hyades (Mel 25) open star clusters lie on both sides of the ecliptic and the zodiac. The classical variable stars – Sun, Moon, and the planets Mercury to Saturn – move between the two open star clusters. The positions of the Pleiades and Hyades in the orbit of the Moon and the ecliptic are special in several respects (Rappenglück 2008, S. 14–20). Over 18.61 years, the Moon's orbit moves from the Pleiades (i: $+5.145°$), above the ecliptic, to the Hyades (i: $-5.145°$), below the ecliptic and back again (fig. 1.6). In the first case, the Moon moves through the Pleiades, and in the second, the Hyades. This also means that some of the stars of the respective open star cluster are covered. In the case of the Pleiades, this is particularly impressive, as the apparent size of M 45 (1.83°) corresponds to more than three apparent full Moon diameters (0.49° – 0.56°). The full Moon is, therefore, significantly smaller than the open star cluster, even if it doesn't look like it at first glance. It is equally impressive when the Moon (0.49° – 0.56°) covers the bright star Aldebaran (α Tau; 0.75–0.95 mag), which is located at an ecliptic latitude (2024) of –5.46558°. Several millennia in the past, the duration of the occultation was even longer due to Aldebaran's proper motion. Around 2,500 BC, Aldebaran was at an ecliptic latitude of 5.29337°: this meant longer occultations by the Moon. The extreme positions of the Moon in the Pleiades and Hyades correspond to the Major and Minor Solstices on the horizon. Attentive observers would have been able to calculate the duration of the Saros period (for predicting solar and lunar eclipses; 18.029 years) from 223 synodic months ($\approx$242 draconitic and $\approx$239 anomalistic months) after an eclipse.

Another special feature is that the brightest star in the sky after the Sun and the full Moon, Venus, also passes through the open star cluster of the Pleiades at long intervals, just like the moon. It then looks as if the Pleiades are being extended by an eighth extremely bright star, which is impressive. Finally, a special annual point (equinoxes, solstices) can also occur between the two open star clusters because of precession. When several such events coincided, this was undoubtedly why ancient cultures saw this area in the sky as having a particular function. It is evident, for example, that the two open star clusters were seen as "catching nets", as the moon was "caught" in them, or as "sieves" (Venus in the middle of the Pleiades) (Santillana & Dechend 1994, S. 151, fn. 49, 153, 161–163; Berezkin 2005a, S. 88; Rappenglück 2008, S. 29; Berezkin 2010). Other people recognized an "opening" here (Berezkin 2010) through which the classical wandering stars had to pass. It was also known as the "Golden Gate of the Ecliptic". Virgil (70–19 BC) states, *"When the shimmering bull rings in the year with its golden horns"* (Georgica 1,217). This establishes a relationship between the Sun and the constellation Taurus and, therefore, to the two open star clusters of the Pleiades and Hyades. The sentence in Virgil is also very interesting for another reason: the bull with the golden horns heralds the year. This began with the equinox in spring. At the time of Virgil, however, the point of the vernal equinox was in the constellation Pisces (Psc). This is, therefore, a visual image from an older time: around 3,000 BC, the vernal equinox passed the star Aldebaran (approx. 5.8° distance) in the constellation Taurus (Tau) and thus entered the astral "gate". Around 2,600 BC, the vernal equinox was exactly on an abstract connecting line between Aldebaran and the Pleiades. At around 2,300 BC, the vernal equinox reached its smallest distance from the Pleiades (approx. 3.9°). This fits in very well with the four cardinal signs (Taurus, Leo, Scorpio/Eagle [Paranatellonta to Scorpio), Man [Aquarius]) with the corresponding points of the year on the ecliptic, which were of great importance in the cultures of the Middle East, especially in ancient Mesopotamia.

In the traditions of the ancient cultures of Mesopotamia, India, China, and elsewhere, there was the idea of asterisms on the ecliptic as gates through which the changing stars must pass, especially the Sun and Moon. The gate of the Sun was referred to as golden, that of the Moon as silver, and the metals gold and silver were assigned to them (Xavier Pena 2020). There is also the idea of the silver gate of the moon and the golden gate of the Sun with reference to the metals that are assigned to the orbits, i. e. the respective spheres ("gates") of the seven wandering stars (Gundel, Wilhem & Hand Georg Gundel, 1893–1980b, 2163, G, 1). Sometimes, the solstices and the equinoxes are associated with the corresponding gates.

There is also an architectural reference to the idea of the Golden Gate. In the Temple of Solomon in Jerusalem, there was the "Golden Gate" on the east side (Morgenstern 1929, 1948, S. 376, 378, 417, 422–425, 429, 459-461, 464–466, 472–473, 491–492, 1964) which was linked to the rising of the Sun at the equinoxes, and an associated ritual of closing and opening. The tradition dates to the 12th century but is probably much older (Morgenstern 1929, S. 15–20; Lethaby 2004, S. 142–163). Archaeoastronomical findings (Reidinger 2002, 2004, 2005) prove that the temple axis is aligned roughly east-west, and that the rays of the rising Sun illuminated the Holy of Holies in the west from the entrance in the east on the spring and autumn equinoxes in the 10th century BC.

1.6.8 The Milky Way: nocturnal path to the beyond

In myths worldwide, the Milky Way is regarded as the path between the earthly and the other spheres of the cosmos ("worlds"), which, for example, the souls and spirits of animals, humans (especially rulers, heroes, and shamans), still in life (ecstasy, dream) or after death (fig. 1.7), and even the gods take (Curtis 1907, S. 34, 134; Boll 1914, S. 32–34, 39–43, 72; Jackson, A. V. W. 1925, S. 246; Hassrick 1964, S. 297; Powers 1975, S. 53, 93; Murie 1981, S. 12; Sommarström 1987, S. 222–223; Krupp 1995; Esmailpour 2007, S. 173–174; López & Giménez Benítez 2008; Rappenglück

Figure 1.7:
A longhouse from Palau. It shows the ancestors' path along the Milky Way through various cosmic levels.

(Museum am Rothenbaum. Kulturen und Künste der Welt (MARKK), *Ethnological Museum*, Hamburg. Courtesy: Michael A. Rappenglück)

2017). They sought to come into contact and communicate with the point of origin of creation, to remain there or to bring the elixir of life to Earth from there. Depending on its location (world mountain, world tree, liana, path, bridge, river etc) the Milky Way connects and separates heaven and earth, the "land of the dead" with the "land of the living" and is thus closely linked to the stations of existence in life, but also before and after.

The way along the Milky Way has a moral component. The Yazidis have a concrete bridge, the Silat Bridge (Pira Silat), in the Lalish Valley (Iraq), which architecturally realises the transition from a profane to a sacred realm (temple) (Rappenglück 2024). The landscape around the bridge is associated with a cave, the Holy Mount Arafat, a sacred stone, two sacred springs and an important pilgrimage there. This idea is linked to the Chinvat bridge (Avestan: Cinvatô Peretûm, "bridge of judgment/beam-shaped bridge") of Zoroastrianism. The bridge offers a wide entrance into the positive sphere ("heaven") of the afterlife for those deceased who did good, while it narrows down to a narrow edge for those who did evil. The latter then falls from the bridge into the negative sphere ("hell"). In medieval tales of journeys to the afterlife, the crossroads is repeatedly thematized as an overall assessment of good and bad deeds in life, often associated with the symbol of the bridge, the ladder, and other symbols of the path as a transition from this world to the afterlife (Dinzelbacher 1973; Dinzelbacher 1986, 2019; Rappenglück 2024). Behind this are also near-death experiences Dinzelbacher 2019; Rappenglück 2024).

The exclusive paths run over two passing points described as clefts, gorges, passages, channels, openings, holes, gates, or doors. A unique situation occurs when, because of precession (approx. 25,920 years) over the millennia, the celestial pole(s), perhaps even close to a bright star, and/or the colure points are in the Milky Way (fig. 1.8). People have always regarded these times as particularly effective for the passage between different spheres of reality ("worlds"). If, depending on the time of year and night, the ends of the Milky Way correspond to certain fixed points on Earth, primarily the cardinal points, then living beings, especially humans, spirits, and divine beings, whether good or evil, are permitted to move from earth to heaven and back (Jeremias 1913, S. 58, 60). Thus, the Milky Way leads to the highest or deepest cosmic regions, passing through the Zenith and Nadir or, in certain epochs of the precession period, contains the celestial poles (sometimes with polestars located there; fig. 1.8).

The intersection of the zodiac (ecliptic, possibly with zodiacal light) and the Milky Way (galactic equator) is characterized by the fact that the precession of the equinoxes does not influence it and therefore remains stable over time. The present-day constellations of Taurus (Tau)/ Gemini (Gem)/ Cancer (Cnc) and opposite Scorpius (Sco)/ Sagittarius (Sgr)/ Capricornus (Cap) are located nearby.

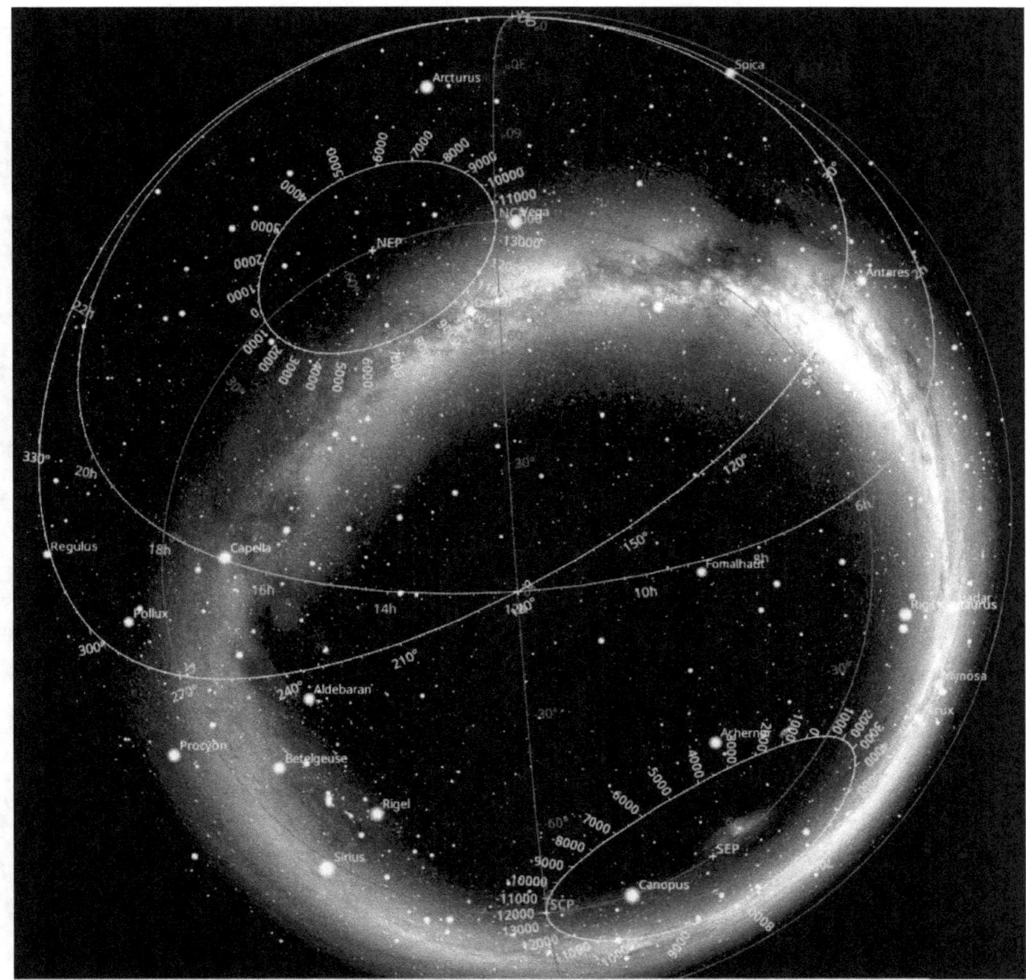

Figure 1.8:
Both poles were in the Milky Way around 12,025 BC:
the northern celestial pole α Lyr (0.02 mag; Vega) and the
southern celestial pole ν Pup (3.18 mag) / α Car (−0.62 mag; Canopus).

(Created with Stellarium 24.1. Courtesy: Michael A. Rappenglück)

Some cultures, e. g., the Greco-Roman and the Indian in Vedic tradition, identified these intersections with two cosmic gates (Reiche 1993; Rappenglück 2017, S. 193). In the *Odyssey* (13, 102–112), Homer talks about two gates of a grotto of the nymphs, which are located on the meridian line (Riedweg et al. 2019, S. 13). People descend into the world through the gate towards midnight (in the north). The immortals, gods, and souls pass back and forth through the gate at noon (south). It is closed to mortals. The Greek Platonist Numenios of Apameia (mid-2nd century) speaks (frg. 31/32 and 35) of the Sun gates located in the constellations of Cancer and Capricorn, the Milky Way as the path of souls and the people of dreams (Alt 1998). He refers to older ideas of the Pythagoreans. The Syrian Neoplatonist Porphyrios of Tyre (around 233 – between 301 and 305) writes in his work Περὶ τοῦ ἐν Ὀδυσσείᾳ τῶν Νυμφῶν ἄντρου (*Perì tou en Odysseía tōn Nymphō̄n ántrou*, Greek. *The cave of the nymphs*) thereof (Alt 1998; Riedweg et al. 2019; Topp 2019, S. 124–126, 129, 136–138; Krupp 1995, S. 427) that the two openings in the rock grotto open very special paths: The northern one allows the souls of humans to descend into the earthly world, while the southern one allows the gods to ascend into the heavenly world. Similarly, the Sun moves in a day and a year. It reaches its highest position daily at noon in the south, its lowest at midnight in the north and annually at the solstice in summer and winter. The two exits and entrances to the cave are assigned to these daily and annual minimum and maximum positions of the Sun. This is probably why the opening is called "heaven": it is the passageway to it. The souls/gods descend through the constellation Cancer (Cnc) in the north ("the north gate") and ascend through the constellation Capricorn (Cap) in the south ("the south gate"). The northern and southern areas in the cosmic cave are highly significant: there are the sacred "gates" of the heavens, where the cloud-swirling winds arise, which flow through these gaps as north and south winds. These winds carry the souls: the north wind takes them out of the rocky vault, and the south wind carries them into it. The forces of time, shown in the image of the Horai (deities of the hours) (Iliad V 751), indicate whether the cosmic passages open or close, which again brings up the motif of the "folding structure" (Roscher 1884–1965, Sp. 2712–2713). In addition to these two entrances in the rocky vault of the cosmos cave, which lie on the meridian and indicate the direction north-south, there are two more: they are located on the first vertical along the direction east-west: in the sky, they correspond to the points of the equinoxes. According to other views, there are three gates: in Scorpius (Sco)/Cancer (Cnc)/Leo (Leo) and Aquarius (Aqr)/Pisces (Psc) (Reiche 1993). Later, the Greek Neoplatonist Proclus (412–485) also deals with this topic (Alt 1998, S. 480). The Samoyedic shamans tell of the two opposing openings of the world cave in the north and the south (Eliade 1989, S. 41). The Tabwa (Africa) believe that the souls of the dead follow the Milky Way until they reach a particular place, a cave, southwards at the end of their seemingly boundless path (Roberts 1981, S. 32–33). It is under the control of a subterranean spirit being. Souls can ascend from this grotto back into the earthly world.

In the Aztec imagination, the Milky Way is a double-headed serpent from whose two mouths the gods Tezcatlipoca and Quetzalcoatl emerge (Brundage 1981, S. 146–147, 243). Quetzalcoatl (Venus) appears (birth) from a gaping opening in the abdomen (branching of the Milky Way between the constellations Sagittarius and Scorpio) of a goddess (Milky Way). Venus passes through the Milky Way, interpreted as crossing the underworld (death) (Milbrath 2013, Position 4931–4937). In ancient Egypt, the sky goddess Nut (Milky Way) has two openings with her mouth and vagina, which the Sun god Ra passes through on his way (Lütge 2008, S. 583–84). The Tenga Bithnua (The Ever-new Tongue), a manuscript probably written in the 9th century, tells that the Milky Way, a mighty river, the World Stream (Greek: okeanós οκεανός), runs in a ring around the disc-shaped earth. In it, the gate to the underworld is located at the deepest point in the cosmos in the North (Carey 1994, S. 16–17).

The Way of St. James from Santiago de Compostela ("Sanctus Jakobus in Campo stellae" ["in the field of stars"] to Cape Finis Terrae (end of the world) is an example of the earthly transmission of the spiritual cosmic path of the Milky Way (Rappenglück 2024). It is about the path of souls from this world to paradise in the afterlife, which was localized in the West at the end of the world (Finis Terrae). The journey's end for the pilgrims is not Santiago de Compostela, but Cape Finisterre (Cabo de Finisterre in Spanish, Cabo Fisterra in Galician), around 90 km away. Pilgrims burn their clothes there to mark their transformation from this world to the next. A dog, which stands for the constellation of the Great Dog (CMa) with the brightest star, Sirius α CMa, accompanies the apostle. There are also pilgrimages in India (Sanskrit: yātrā, departure, journey, procession), for example, the Vishveshara Antargriha Yātrā, which follows the path of the Milky Way with its terrestrial image, the river Ganges (gaṅgā: Goddess of the River Ganges and Milky Way), the topographical representation of the psychosomatic body (chakra teaching) and a path of meditation and self-knowledge with the goal of salvation from this world (Rappenglück 2024).

The "bifurcation" of the Milky Way between today's constellations of Sagittarius/Scorpio or Ara (Ara)/Lupus (Lup), which is caused by dark clouds (dust) as we know today, was particularly conspicuous and therefore highly significant for the cultures: There was the entrance into the Underworld. Luiseño, Pawnee and Cherokee (North America) myths tell that a star at the end of the Milky Way, probably Antares (α Sco, 1.09 mag), the ghost star, where it branches into two paths, receives the souls of the deceased (Rappenglück 1999, 142, 329). The Maya (Mesoamerica) located around today's constellation Sagittarius, the portal to the Otherworld Xibalbá (Maya: Place of fear), and they recognized a skeletal serpent called "white-boned snake, which is approximately the today constellation Scorpion (Sco), too (Milbrath 1999, S. 101, 444–445). (Milbrath 1999, S. 558). This entity testifies to the state of a transformation. The Mayans also think that a reptilian monster existed here on the way to the underworld (Thurston 2010/2011, 15, 151, 153, 156, 157). The Eastern North American natives also know the asterism of the Serpent (area of Scorpius) at this point in the Milky Way (Lankford 2007, ch. 10).

Abbildung 1.9:
Rainbow and Milky Way: a common path by day and night

(Courtesy: Collage by Michael Rappenglück, based on Wikipedia: Double rainbow in full size in the Wrangell-St. Elias National Park, Alaska (Eric Rolph, CC BY-SA 2), and Milky Way above Lake Sticada, Croatia (Šime Barešić, CC0))

1.6.9 The rainbow: diurnal the path to the beyond

Some cultures regarded the rainbow as a path from the earth to the sky and back again, visible during the day as a "bridge" or "snake" (Hooykaas 1956, S. 293–294, 298–299, 308–311; Lowenstein 1961; Eliade 1967, S. 211, 1989, S. 132–135, 138, 173, 353, 354, 490; Blust 2000; Chevalier and Gheerbrant 1996, S. 122, 691, 783–786; Leeming 2010, S. 220; Penprase 2011; Schuetz-Miller 2012; Witzel 2015, S. 22; Dinzelbacher 2019, S. 125; Blust 2023) while the Milky Way shows its nocturnal counterpart, or vice versa (fig. 1.9). In the view of cultures, the rainbow and the Milky Way (conceived as a river) belong to a single cosmic being, which has a polar, sometimes hermaphroditic, twin structure, e. g. among the Aborigines, Australia (creator, rain snake) (Worms & Petri 1998, S. 158; Maddock 1978, Reprint 2011, S. 15, 17–18), in the tradition of Benin people (Africa) (Chevalier and Gheerbrant 1996, S. 848, 909) or Norse myths (Bifröst/Bilröst between Midgard (earth) and Asgard (world of the gods) (Ohlmarks 1941; Simek 1993, S. 36; Lindow 2002, S. 81). The idea that the rainbow is a snake biting its tail also fits in with this. The rainbow and the Milky Way share the same Ouroboros motif (Leeming 2010, S. 110; Rappenglück 2017). The two ends of a rainbow gave the impression of a two-headed cosmic being, e. g. in the tradition Kwakiutl (North America) (Schuetz-Miller 2012, S. 371). In place of the heads, as with the Korku (India), two termite mounds can take the form of world mountains, in which a serpent (world serpent) lives, enabling the ascent to heaven (Biersack 2011, S. 219, 221, 131–133; Vahia, Halkare & Dahedar 2016, S. 221, 231). With the Chumash (North America) the rainbow in the middle world bridges the highest mountain of Santa Cruz Island with the mainland mountains (Penprase 2011, S. 129 und Abb. 4.27a). For the Hopi, the Sun moves over the rainbow between two kivas (Bäcker 1989, S. 45). In Indian and Khmer traditions, the Makara monster, made of parts of a dolphin and crocodile or fish and elephant (or elephant, turtle, fish, and crocodile), roughly assigned in the sky to today's constellation Capricorn (Cap), spits fertile water from its mouth, producing the rainbow (Snodgrass 1985, S. 292–296, 304, 351; Chevalier and Gheerbrant 1996, S. 423, 628, 785). Occasionally there is such a monster at both ends of the rainbow.

Among the Toraja (Sulawesi Island, Indonesia), one end of the rainbow is connected to the house of origin and spans the cosmic barana tree (banyan), the world axis (Waterson 2009, S. 151).

For many peoples worldwide (Hooykaas 1956, S. 292, 294–295, 297–299, 307–321; van Dooren 1987, S. 264; Eliade 1989, S. 78, 117–118; Stein 1990, S. 189–190; Chevalier and Gheerbrant 1996, S. 784; Blust 2000, S. 532–533; Penprase 2011, S. 127 und Abb. 4.26) the rainbow forms the boundary and, simultaneously, a path and a way of communication and action between the lower, middle and upper cosmic realms or two parallel worlds, this one and a transcendent reality (the beyond). In Indonesian myths (Hooykaas 1956, S. 292, 294–295, 297–299, 307–321), the rainbow, a coloured water snake (naga) pours life-giving and healing water from a waterspout (opening) at the highest point of the sky. A small passage (channel, tunnel) leads to the upper world, the transcendence, indicated by a ruby. The rainbow acts as a gate to the beyond, which can be reached using a soul boat. The Tamang (Nepal) shaman's initiation hut (gufa) at the top has an opening corresponding to the fontanel. It is furnished with red and white cloths symbolizing the rainbow, along which the free soul moves to transcendence and vice versa (Peters 1989, S. 36, 125). Something similar can also be found among the Buryats (Eliade 1989, S. 118).

In many cultures, the seven-coloured rainbow resembles the seven planetary paths (realms, spheres) and looks like a special kind of ladder from earth to sky and vice versa (Snodgrass 1985, S. 284–92; Eliade 1989, S. 121, 134; Stein 1990, S. 193; Chevalier and Gheerbrant 1996, S. 214, 581, 584, 784, 786; Schuetz-Miller 2012, S. 359). For the Dogon (Africa), the chameleon (colours) is the rainbow bridge between heaven and earth (Chevalier & Gheerbrant 1996, S. 181).

1.6.10 Travelling is dangerous: clashing rocks, magic formulas, guardians, soul guides, and spirit helpers

There is only a brief period of transparency, openness and accessibility before the paranormal spheres close off again and become inaccessible. An apt symbolic-mythical expression for this phenomenon can be found, for example, in the motif of folding structures, e. g. in opening and closing crevices, in rocks, animal mouths, volcanoes, doors (fig. 1.10), rotating crosses, and others (Santillana & Dechend 1993, S. 289–292; Rappenglück, Barbara 2005, 2017; Lankford 2007, S. 219, 227–238; Rappenglück 2008, S. 30; Dinzelbacher 2019, S. 112). The adaptation of the form (change of form, metamorphosis, shapeshifting, change of mask and/or clothing) is necessary (Rappenglück 2024). For a successful transition between other spheres of being, helping entities with a positive attitude is needed as soul guides (psychopomp, spirit helpers, angels, etc.) and companions: Animals, the deceased, spirits, and gods (Rappenglück 2024). Finally, the places of passage (earthly, underworldly, heavenly) are guarded by anthropomorphic animals or hybrid beings (chimaeras) (Rappenglück 2024). Depending on the cultural tradition, guardian figures stand at or near the rising and setting points of the stars (Sun, Moon, planets, stars), especially at the points of the solstices and equinoxes on the horizon, at the cardinal points and the celestial sectors, the highest or lowest point of the cosmos (zenith – nadir), the celestial poles, the ecliptic pole, the intersections of the ecliptic as well as the celestial equator and additionally the Milky Way, especially if the most important points of the year and/or the celestial poles lie in it, the four annual points on the ecliptic, the night hours, the stations of the zodiac and the course of the moon, the sectors of the decans, the lunar nodes and the orbits of the planets (Cumont & Gehrich 1903, S. 105–106; Wallis Budge 1904, S. 360–361; van Buren 1947, S. 324–325; Winston 1966, S. 204; Doeringer 1982, S. 315–321; Santillana & Dechend 1993, S. 221; Lehner 2006; Hammond 2007, S. V, 19–24; Lütge 2008, S. 841). A few selected examples illustrate the ideas below.

The Mayan Witz monster embodies the opening in the transcendentally conceived "flower mountain", which as a world mountain embodies the connection between the subterranean water sphere and the heavenly spheres in which the deceased reside (Carrasco & Hull 2002, S. 29; Klein, Guzman, Mandell, Stanfield & Volpe 2002, S. 390; Taube 2004; Ishihara 2008; Guernsey 2010, S. 87–88; Schwerin 2011).

In Babylonian tradition, the guardians of the celestial ocean are located at certain gates (Jeremias 1913, S. 29, 57–58, 133). It remains unclear which places in the sky

Abbildung 1.10:
Two door panels of a burial chamber in Sichuan, Chian, Eastern Han Dynasty (24–220 AD), show the tiger and the dragon, constellations in the west and east

(Museum Rietberg, Zurich, Switzerland. Courtesy: Michael A. Rappenglück)

are common (zodiacal constellations, intersection of the zodiac and celestial equator or others). The city of the dead in the underworld also has seven guardians of the gates (Jeremias 1913, S. 67). In Zoroastrianism, however, Sirius (Tishtrya; α CMa, -1.46 mag) is one of the good guardians of the heavens (Jeremias 1913, S. 202).

In Roman culture, perhaps going back to Etruscan ideas, the god Janus was the guardian of the passage of the year. However, there are variants with double and quadruple faces, showing that this god relates to the sky, the course of the Sun, the meridian, the cardinal points, the world axis, and alternating paths (Erlenmeyer, M.-L. & H. Erlenmeyer 1958, S. 58; Taylor 2000). The leontocephalic god of the Mithras Mysteries (fig. 1.11) with a key, which is a gnomon (Merkelbach 1994, S. 144–145), a symbol of the world axis (Cooper, Middell, G. & M. Middell 1988, S. 8, 166–167), signifying knowledge, initiation, liberation, opens and closes the gates of the zodiac (zodiacal stations) and of the spheres of wandering stars to the souls for embodiment and disembodiment (Jackson, H. M. 1985, S. 19–20). This happens especially at the summer solstice (ianua coeli; Capricorn; divine realm) and winter solstice (ianua inferni; Cancer; human realm). However, with his key, he opens the eighth sphere of the fixed stars and gives access to the realm behind/above, the hyperouranios topos (Ulansey 2003). Mithras, the leptocephalic God Ianus, or later Peter, all have the powers of opening and closing, binding, and loosening.

In the myths of Eurasia and the Americas, the most important guardian figures on the Milky Way celestial path include the Canidae mythical variants (Berezkin 2005b). The motif already existed in Indo-European culture in the form of the "hell or sky dog", which guard the entrance to path from this world to the afterlife at the bridge, which is probably the Milky Way. It turns out that there were two (four-eyed) dogs (one dark and one light) at either end of the river (Lincoln 1979; Dirven 2009, S. 66–67). According to American myths, a dog asterism, sometimes recognized as Sirius (α CMa, -1.46 mag) and Antares (α Sco, 1.09 mag), is located at each end of a tree trunk (zodiac?) that bridges a river (the Milky Way). A path of dogs (wandering stars in the zodiac?), marked with stars (zodiacal asterisms), adjoins the way to the Otherworld (Milky Way). The Quechua-Huanca put a dog between the feet of a deceased person in the grave (Berezkin 2005b). The Tibetans also know a sky dog as a guardian (Hummel 1961).

Two scorpion people (female and male) guard the access to the world mountain Măsu (Jensen 1890, S. 316–317) at the boundary between the upper light and lower dark spheres of the world. They guard the entry and exit of the Sun into and out of the interior of the cosmos (Foster, Frayne, and Beckman 2001; Lauber 2008, S. 68–69). The Serpent Mount, Adams County, Ohio, USA, ca. 320 BC (or 1070 AD), an earthwork of the Adena culture, figures the one asterism in the constellation Scorpio, which guards the entrance to the underworld as a great serpent and soul guide (Romain 2019, S. 62, 65, 68, 72–73, 77, Romain 2021).

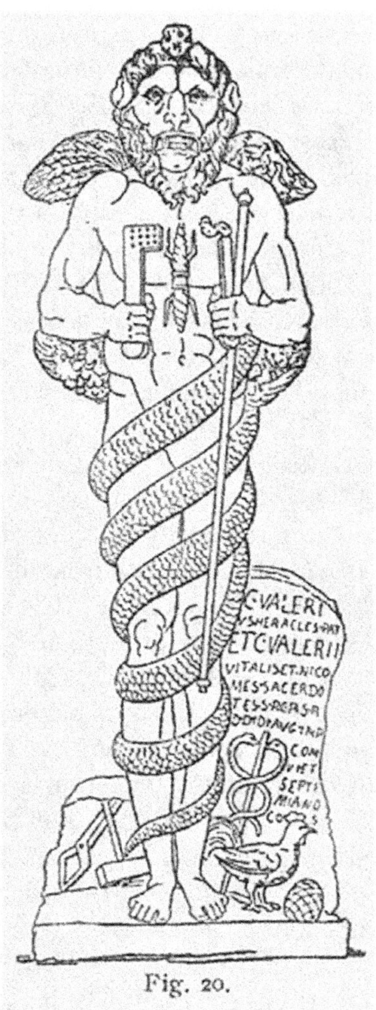

Fig. 20.

Abbildung 1.11:

The Mithraic, Leontocephalic God (resembling Aion) is the cosmic gatekeeper and ruler of fate. He has four tiny wings, each showing a particular image (shoulders: a dove and a swan, and ears; hips: grapes and two palms), symbolising the seasons. A thunderbolt can be seen on his chest. He holds a key with twelve holes (the months) in his right hand, interpreted as a gnomon, and in his left, a sceptre. With these, he opens the gates through which the adept passes on his ascent through the seven degrees, the gate of the Sun's rising and setting, or the yearly gate of Janus. A serpent winds six times around the body and rests its head on the head of the god; it marks the spiral path of the Sun from one solstice to the next. At his feet, there are further images: a pair of tongs, a hammer, a caduceus, a cock, and a pinecone.

(Drawing of a white marble statue, originally coloured red, initially in the Fagan Mithraeum, Ostia Antica, now Vatican Library. Cumont & Gehrich: The Mysteries of Mithra, 1903, figure 20, p. 105)

In Manichaeism, there are various guardian figures on pillars at the left and right edge of the celestial sphere, at the entrance/exit to the outer, middle, and inner rampart spanning the world (Gulácsi & Beduhn 2011, S. 65, 75, 79, 81, 84, 91). The guardian on the inner wall regulates access to the air sphere. Angelic beings stand at the four gates of the cardinal directions. There are also 32 gates around the highest plateau on the World Mountain. These represent 32 cities built for the descendants of the fallen "Guardians". The number 32 corresponds to a further subdivision of the compass rose (Uhden 1935, S. 14–15, 17; Tolmacheva 1980, S. 183–86).

In Pawnee (North America) lore, the North Star acts as the gatekeeper for the dead at the entrance to the Milky Way. The South Star, being the god of the whirlwinds at the end of the path, carries the souls into the afterlife, heralds the sphere of the spirit world (Chamberlain 1982, S. 113).

In Zoroastrianism, the star Wanand (Vega), the General of the West (MacKenzie 1964, S. 513; Boyce 1975, S. 376) on the world mountain Alburz guards the (horizontal) gates through which the Sun moves daily and protects the (vertical) path across the moon and star realm to the Sun into paradise from the entry of evil beings (Henning 1942, S. 231; Lütge 2008, S. 276, 536, 656, 703, 841, 849). Haftoreng (Ursa Major, UMa), the general of the North, guards the Gates of Hell from an outbreak of the countless demons imprisoned there (Henning 1942, S. 232, 233).

For the Chumash (Saint-Onge SR., Johnson & Talaugon 2009, S. 33, 35, 36, 43, 52), there are guardian stars (from today's constellation Ursa Major; UMa) near the North Star (celestial pole). In the Mithras liturgy (Greek magical papyrus PGM, IV.475–834.2; probably early 4th ca. AD), there are seven female and seven male guardian stars at the celestial pole and for the world axis, perhaps the little and the big dipper (Dieterich 1910, S. 31, 34, 70–80; Panaino 1995/1996, S. 194).

Finally, there is a giant serpent or dragon asterism near the northern celestial pole, well-known as the dragon, which guards the apple of the Hesperides (Staal 1988). It has already been discussed on various occasions whether the ecliptic pole was known to cultures several millennia ago, e. g. in Mesopotamia, China, or elsewhere (Bouché-Leclerq 1899, S. 40, 122; Santillana & Dechend 1993, S. 131, 378; Maeder 2011). So far, there is no conclusive evidence for this. However, the star Thuban (α Dra, 3,67 mag), which, e. g. in ancient China (Pankenier 2015, S. 27, 2004), was observed as a pole star around 2800 BC, certainly played an important role in the assigned guardian function.

1.6.11 The world axis and the celestial pole: path and gateway to transcendental spheres

According to ancient beliefs, the ascent and descent take place using various vehicles: World tree (including lotus, vine, liana), world mountain, cosmic river, light co-

Abbildung 1.12:
A Mapuche shaman's ladder for ascending to the heavens (Mapuche, Chile)
(Musée du Quai Branly in Paris. Courtesy: Michael A. Rappenglück)

lumn, light ray, Milky Way, zodiacal light, rainbow, ladder (fig. 1.12), staircase, tower, step pyramid, rope, certain animals (birds, spider, fish etc.) (Eisler 1910, S. 299–301; Gandz 1938, 1943; Zerries 1952, S. 220–221; Sehgal 1989; Brenk 1993, S. 153; Liebenberg 2016, S. 178–182; Hammond 2007, S. III, 63–64, V, Appendix 1,2–1,6; Rappenglück 1999, 2005, 2007, 2009, 2014a, 2014b, 2017, 2020b; Dinzelbacher 2019, S. 110–128).

The passage goes through vertically successive openings, described as holes, gates, doors, chimneys, wells, etc., from the greatest depth to the greatest height, the limits of a limited cosmos (Rappenglück 1999, S. 245–252; Hammond 2007, S. III, 63–64; Leeming 2010, S. 35). These openings are "narrow" and must be "tunnelled through". The Milky Way, identified with the World Tree and with the World Mountain, breaks through the spheres of the cosmos and thus creates a path for souls and spirits of animals, humans, still in life (ecstasy, dream) or after death, and gods between this side and the other side (including the netherworld). Shaman drums (Rappenglück 1999), for example, show the "openings" to the three or more "worlds" visually and concretely (metal rings). Some natural places (e. g. caves, mountains) were selected, and some dwellings and cult buildings were designed so that vertically arranged openings, usually focusing on the ultimate top and bottom, made the idea tangible (Snodgrass 1985; Rappenglück 1999; Lethaby 2004; Lehner 2006; Rappenglück 2014b, 2020c, 2013; Ruggles 2015).

The World Mountain provides access to one or more other realities outside the ordinary world (Rappenglück 2020a, 2020c). Up there is the realm of bliss and eternal existence. These other worlds can be reached either through special openings at the base, the flanks of the world mountain or within via a cave system (Rappenglück 2020a, 2020c), at its summit (in the form of a supernatural realm, palace, garden, etc.) or from there further upwards through the zenith or the celestial pole to a transcendent sphere above the highest heaven (Rappenglück 2020a, 2020c). In the World Cave within the World Mountain, the same treasures of the world beyond, the Otherworld, await them as on the summit of the World Mountain. The path to the innermost part of the world mountain and the path to its summit, to the outermost part of the cosmos, lead from a material to a spiritual space, one into the other (like a Klein bottle ([originally Klein's surface]; Klein 1882). Both end in a transcendental sphere in which the sphere of immortality opens.

Rituals of (helical), often clockwise (following the Sun's daily course) circumambulation of the world mountain or its architectural models, especially for pilgrims, set out the right way ascending and descending the world mountain (alternatively its artificial replicas) or entering the World Cave within (Rappenglück 2020a, 2020c). The ceremonies took place at certain sacred times and astronomically significant dates. Circumnavigating sacred mountains as a pilgrimage (fig. 1.13) conveys purification, meditation, and self-knowledge (Rappenglück 2020c, 2020a).

Abbildung 1.13:
The pilgrimage to heaven: stone staircase up Tai'shan (Mount Tai), Shandong Province, China

(Wikipedia: Charlie fong, Public Domain)

The North Star, usually today's (Polaris), is also the central opening of the sky for some cultures in the northern hemisphere (Rappenglück 1999, S. 246; Lankford 2007, S. 237). According to ancient Chinese traditions, at least since the Han Dynasty (206 BC–220 AD), the "Gate of Heaven" (t'ien-men), the central gate (Ch'ang-ho'a), is located above the one world mountain (k'ung sang) in the middle of the sky at the northern celestial pole. The Huainanzi (180–122 BC) identifies the "Gates of Heaven" with those of "the Palace of the Purple Palace of Formless Unity [Tao] (tzu-wei kung), the abode of the supreme god Shang-ti" (Henricks 1995, p. 73–74; Pankenier 2004). Often, it is associated with a star or asterism nearby. It is considered the highest of the nine gates, which mortals, particularly sensitive shamans, must pass on their soul voyage to reach the ultimate border of the cosmos and return to Earth. From there, the journey of the immortals in heaven begins (Major 1979, S. 69–70).

These views may go back to the end of the last ice age, as the Great One (Tao) is identified with the Pole Star in the Milky Way, a primordial, female, creative, formless being, the source (opening in the sky) of the celestial river (Milky Way) from which the cosmos originated (Allan 2003). The last bright pole star in the Milky Way was Vega (α Lyr) (Doeringer 1982, S. 315–321; Rappenglück 2017).

Myths of some peoples of the northern hemisphere support such an interpretation. The Shoshone (North America) tell that the souls ascend to the Milky Way and fly south along it until they encounter a lake in which a conical rock (World Mountain as axis) with an opening at its tip is centred (Curtis 1970, S. 82, 134, 186). Through this, they pass from this world to another parallel paradisiacal earth. Other people's myths say that at a time immemorial, the opening at the celestial pole, which is normally closed by the world axis as if by a plug, was removed by destroying the world mountain or the world column, cutting down the world tree, cutting a rope or doing something similar. This resulted in gigantic flooding of the earth because the water was able to flow out unhindered from the original sphere of primordial water existing behind the highest sky (Santillana & Dechend 1993, S. 196–205; Rappenglück 2017, S. 209–211, 2018b, S. 262–263, 2020b).

In China, the character for "Heaven's Gate" (Chang-ho) is the symbol of two Suns in a gate. This can be understood as follows: The Sun rises on the horizon through one portal and sets through another (Doeringer 1982, p. 321–322). These horizontal ones and the highest and lowest on the world axis form a single passage between the generative chaos (which forms a "hyperspace") and the cosmos of entities that emerge from it. This corresponds to similar ideas of the ancient Greek philosophers and writers, which have been characterized as unipolar-bipolar, combining a horizontal and a vertical view of the cosmic sphere (Heimpel 1986; Nakassis 2004). That way of presentation was already used in a kind of sky map in the cave of Lascaux (Montignac, Dordogne, France), around at least 16,500 years ago (Rappenglück 1999).

1.6.12 The Gate to Heaven at the Top of a Person's Head

In trance, ecstasy or at death, the soul leaves the body through an opening in the skull, the heavenly door, usually the fontanel, e. g. in Tamang (Nepal) shamanism. When clear visions are possible, the soul's journey through the nine cosmic realms starts with a nine-step ladder. This symbolizes the world axis analogous to the spinal column (Peters 1982, S. 26–27, 34). According to the Iban shamans (Borneo/Kalimantan), the soul travels to the summit of the world mountain Rabung, the afterlife, which is the home of the heavenly shamans (Sather 2008, S. 70). The summit corresponds to the top of the head with the anterior fontanel: an opening through which the soul leaves or enters the body. According to the Fang (Africa), the soul escapes to heaven through the fontanel (Mary 1988, S. 241, 243). According to certain yoga schools, the 7th chakra (Sahasrara) is at the head's crown (Mallinson & Singleton 2017). In it, there is the "opening of Brahma" (brahmarandhra), which is identified with the fontanel and is named sky-chakra (akasacakra) (Mallinson and Singleton 2017, S. 177–178, 201–202, 357). After awakening the Kundalini force at the lower end of the vertebral column, it passes through the six chakras along the spine and leaves the body in the seventh chakra at the fontanel, leading into transcendence (Mallinson & Singleton 2017, S. 178–179, 201–202). There is a comparable view in Tantrism (Flood 2022, S. 29–31). It becomes clear that a biological-psychic-parapsychic process of a journey through cosmic spheres is paralleled – body (vertebral column) and mountain (world axis) (Mabbett 1983) – until a state of the highest state of being is reached, which shines in a sun-like glow of light. The opening at the top of the skull, in a roof structure (opaion, chimney) and the dome of the cosmos (zenith, celestial pole) are symbolic-mythical expressions of the same idea of an opening into the transcendent sphere beyond (Stein 1990; Allan 1991; Coomaraswamy, Kentish & Coomaraswamy 1997, S. 192–241). In ancient China, during the Han dynasty (206 BC–210 AD), the stone sarcophagi resembled a "vessel" or "house" for the earthly body-soul (Confucian: po), mirroring the cosmos (Wu Hung 2012, S. 199, 202–205). The doors and gates depict the passage of the heavenly free soul (Confucian: hun) from this to a transcendent world. The perforated disk (Bi) (fig. 1.14), made from jade, which is the "stone of the sky", illustrates the heaven's gate tian-men, through which the deceased moves to the world mountain Kunlun, the place of the immortals (Dianzeng & Shuguan 1990; Lam 2019, S. 133–137). The symbolism of the Bi goes back (Childs-Johnson 2012) at least to the Liangzhu Culture (3300–2300 BC).

In the Christian Middle Ages, going back to antiquity (Rappenglück 2024), the opening of the path to transcendental reality is symbolically expressed by the alchemical fenestra aeternitatis ("window into eternity", equated with the philosopher's stone), the spiraculum aeternitatis ("air hole into eternity"), a term used by Gerhard Dorn, ca. 1530–1584 (Franz 1974, S. 261), or the mysterium fenestrae ("mystery of the win-

Abbildung 1.14:
Jade bi, tomb of King of Chu, Shizi Mountain (Xuzhou, province of Jiangsu, China), Western Han Dynasty, 2nd century BC, Nanjing Museum

(Courtesy: Mary Harrsch, CC BY-SA 4.0, Wikipedia)

dow") of the Kabbalah (Franz 1974, S. 260). Through this "hole", the timeless hereafter works its way into the spatiotemporally limited world. The Holy Spirit hole in the ceiling of some Christian churches, usually opened at Pentecost and otherwise closed, through which the Holy Spirit, represented in the symbol of the dove, descends from transcendence into the spatiotemporal world and onto the faithful, is a later symbolic and ritual expression of the millennia-old idea of the opening between the worlds that exists in many cultures. This hole in the sky/cosmos is probably related to near-death experiences. Hieronymus Bosch (ca. 1450–1516) created a visual expression of this in his painting *Ascent of the Blessed* (Gallerie dell'Accademia, Venice, Italy) as part of his work *Vision of the Afterlife* between ca. 1500–ca. 1504 (fig. 1.15).

1.7 Concluding remarks

The theme of cosmic openings, passages, and paths has three levels of meaning in ancient cultures worldwide. Firstly, it deals with practical problems and solutions for positioning in space (orientation and navigation) and time (time calculations). Secondly, the theme contains a geographical-astronomical-cosmological code in symbols, myths, and rituals for biological, psychological, and sociological, slow, and fast changes in the life development phases of the individual and collective. In a third perspective, however, this code also expresses experiences of the individual beyond ordinary sensory perceptions, which are closely linked to trance, ecstasy, near-death experiences, and mystical-spiritual states, regardless of how these are seen from today's perspective as biologically-medically founded, psychologically, thanatologically, or even parapsychologically evaluated. For the ancient cultures which thousands of years ago passed on individual and collective experiences ("knowledge"), initially exclusively and later predominantly, orally, symbols, myths, and rituals, as well as buildings and objects, were essential means of communication, not only in the respective present but also over long periods. Biological, geographical, and astronomical elements that appear continuously or periodically over time in the respective cultural cosmology thus offer an excellent opportunity to find anthropological subjective or intersubjective experiences expressed in the phenomena of the world and, from there, to understand better one's existence with its changes.

1.8 Literatur

AKIMA, TOSHIO: The Origins of the Grand Shrine of Ise and the Cult of the Sun Goddess Amaterasu Omikami. In: *Japan Review* **4** (1993), p. 141–198.

ALLAN, SARAH: *The Shape of the Turtle: Myth, Art, and Cosmos in Early China. SUNY series in Chinese philosophy and culture.* Albany, NY: State University of New York Press 1991.

ALLAN, SARAH: The Great One, Water, and the Laozi: New Light from Guodian. In: *T'oung Pao* **89** (2003), 4/5, p. 237–285.

ALT, KARIN: Homers Nymphengrotte in der Deutung des Porphyrios. In: *Hermes* **126** (1998), p. 466–487.

BÄCKER, JÖRG: Baatorsang, „Der keine Mühsal fürchtete". Eine Ewenkisch-Solonische Abenteuerreise. In: *Religious and Lay Symbolism in the Altaic World and Other Papers.* Proceedings of the 27th Meeting of the Permanent International Altaistic Conference, Walberberg, Federal Republic of Germany, June 12th to 17th, 1984, edited by KLAUS SAGASTER & HELMUT EIMER. Wiesbaden: Otto Harrassowitz 1989.

BECHER, HANS *Poré, Perimbó: Einwirkungen der lunaren Mythologie auf den Lebensstil von drei Yanonámi-Stämmen: Surára, Pakidái und Ironasitéri: Ergebnisse der 1970 durch-*

Abbildung 1.15:
Hieronymus Bosch (ca. 1450–1516): *Ascent of the Blessed*,
Part of Visions of the Hereafter, painted ca. 1500 – ca. 1504

(Wikipedia: Public Domain)

geführten Expedition nach Nordwestbrasilien-Völkerkunde-Abteilung des Niedersächsischen Landesmuseums Hannover. Hannover: Münstermann (Völkerkundliche Abhandlungen; 6) 1974.

BECHER, HANS: Die Symbolische Klassifikation von „links" und „rechts" bei drei Yanonámi-Stämmen in Nordwestbrasilien. In: *Indiana* **9** (1984), p. 457–71.

BEREZKIN, YURI: Cosmic Hunt: Variants of Siberian-North American Myth. In: *Folklore Electronic Journal of Folklore* **31** (2005a), p. 79–100.

BEREZKIN, YURI: 'The Black Dog at the River of Tears': Some Amerindian Representations of the Passage to the Land of the Dead and Their Eurasian Roots. In: *Forum for Anthropology and Culture* **2** (2005b), p. 130–170.

BEREZKIN, YURI: The Pleiades as Openings, the Milky Way as the Path of Birds, and the Girl on the Moon: Cultural Links Across Northern Eurasia. In: *Folklore Electronic Journal of Folklore* **44** (2010), p. 7–34.

BIERSACK, ALETTA: The Sun and the Shakers, Again: Enga, Ipili, and Somaip Perspectives on the Cult of Ain: Part One. In: *Oceania* **81** (2011), 2, p. 113–136.

BLUST, ROBERT: The Origins of Dragons. In: *Anthropos* **95** (2000), p. 519–536.

BLUST, ROBERT: *The Dragon and the Rainbow: Man's Oldest Story.* Leiden, Boston: Brill (Ancient languages and civilizations; Band 5) 2023.

BOLL, FRANZ: *Aus der Offenbarung Johannis: Hellenistische Studien zum Weltbild der Apokalypse.* Leipzig, Berlin: B. G. Teubner (Stoicheia – Studien zur Geschichte des antiken Weltbildes und der griechischen Wissenschaft; Heft 1) 1914.

BOELHAUVE, URSULA; KÜHNE-BERTRAM, GUDRUN; LESSING, HANS-ULRICH & FRITHJOF RODI (ed.): *Otto Friedrich Bollnow: Schriften. Mensch und Raum, Band 6.* Würzburg: Königshausen & Neumann 2011.

BOUCHÉ-LECLERCQ, AUGUSTE: *L'Astrologie grecque.* Paris: Ernest Leroux 1899.

BOYCE, MARY: *A History of Zoroastrianism.* Handbuch der Orientalistik 1. Abt., 8. Band: Religion, Erster Abschnitt, Lieferung 2, Heft 2A, Vol. 1. Leiden: Brill 1975.

BRENK, FREDERICK E.: A Gleaming Ray: Blessed Afterlife in the Mysteries. In: *Illinois Classical Studies* 18 (1993), S. 147–164.

BROWN, ROBERT: *Researches into the Origin of the Primitive Constellations of the Greeks, Phoenicians and Babylonians. 2 vols.* London: Williams & Norgate, 1899–1900.

BRUNDAGE, BURT CARTWRIGHT: T*he Phoenix of the Western World: Quetzalcoatl and the Sky Religion.* Norman: University of Oklahoma Press 1981.

BURKERT, WALTER: Iranisches bei Anaximandros. In: *Rheinisches Museum für Philologie*, Neue Folge, **106** (1963), 2, p. 97–134.

CAREY, JOHN: The Sun's Night Journey: A Pharaonic Image in Medieval Ireland. In: *Journal of the Warburg and Courtauld Institutes* **57** (1994), p. 14–34.

CARRASCO, MICHAEL DAVID & KERRY HULL: The Cosmogonic Symbolism of the Corbeled Vault in Maya Architecture. In: *Mexicon* **24** (2002), 2, p. 26–32.

CHAMBERLAIN, VON DEL: *When Stars Came down to Earth: Cosmology of the Skidi Pawnee Indians of North America.* Los Altos, California: Ballena Press (Ballena Press Anthropological Papers; 26) 1982.

CHEVALIER, J. & A. GHEERBRANT: *A Dictionary of Symbols.* London: Penguin (Penguin reference books) 1996.

CHILDS-JOHNSON, ELIZABETH: Speculations on the Religious Use and Significance of Jade Cong and Bi of the Liangzhu Culture. In: *Liangzhu Late Neolithic Jades.* Edited by THROCKMORTON FINE ART. New York: Eastwood Litho 2012.

CIPOLLETTI, MARÍA SUSANA: *Kosmospfade: Schamanismus und religiöse Auffassungen der Indianer Südamerikas.* Baden-Baden: Academia-Verlag (Studia Instituti Anthropos; 59) 2019.

COOMARASWAMY, ANANDA KENTISH & RAMA P. COOMARASWAMY: *The Door in the Sky: Coomaraswamy on Myth and Meaning.* Princeton, New Jersey: Princeton University Press 1997.

COOPER, JEAN C.; MIDDELL, GUDRUN & MATTHIAS MIDDELL: *Illustriertes Lexikon der traditionellen Symbole.* Wiesbaden: Drei-Lilien-Verlag 1988.

CUMONT, FRANZ & GEORG GEHRICH: *Die Mysterien des Mithra: Ein Beitrag zur Religionsgeschichte der Römischen Kaiserzeit.* Leipzig: B. G. Teubner 1903.

CURTIS, EDWARD S. ET AL.: *The North American Indian, being a series of volumes picturing and describing the Indians of the United States, and Alaska, Vol. 1.* Cambridge, Mass.: Cambridge University Press 1907.

CURTIS, EDWARD S.: *The North American Indian; being a series of volumes picturing and describing the Indians of the United States, the Dominion of Canada, and Alaska, Vol. 15.* Norwood, Mass.: Plimpton Press 1970.

DEAN-OTTING, MARY: *Heavenly Journeys: A Study of the Motif in Hellenistic Jewish Literature.* Frankfurt am Main: Peter Lang (Judentum und Umwelt; 8) 1984.

DIANZENG, ZHAO & YUAN SHUGUAN: Tianmen Kao'-Jianlun Sichuan Han Huaxiang Zhuan(Shi) De Zuhe Yu Zhuti: On the „Heavenly Gate“ as Well as the Composition and Theme of Sichuan Pictorial Bricks and Stones. In: *Sichuan Wenwu* **6** (1990), p. 3–11.

DIETERICH, ALBRECHT: *Eine Mithrasliturgie.* Leipzig und Berlin: B. G. Teubner (2. Auflage) 1910.

DINZELBACHER, PETER: *Die Jenseitsbrücke im Mittelalter.* Dissertationen, Universität Wien. Wien: Verband der Wissenschaftlichen Gesellschaften Österreichs 1973.

DINZELBACHER, PETER: The Way to the Other World in Medieval Literature and Art. In: *Folklore* **97** (1986), Issue 1, p. 70–87.

DINZELBACHER, PETER: Neue Quellen zur Jenseitsbrücke und Himmelsleiter. In: DINZELBACHER, PETER (ed.): *Vision und Magie: Religiöses Erleben im Mittelalter.* Leiden, Paderborn: Brill, Schöningh 2019, S. 99–128.

DIRVEN, LUCINDA: My Lord with His Dogs. Continuity and Change in the Cult of Nergal in Parthian Mesopotamia. In: *Edessa in hellenistisch-römischer Zeit: Religion, Kultur und Politik zwischen Ost und West.* Beiträge des internationalen Edessa-Symposiums in Halle an der Saale, 14.–17. Juli 2005. Hg. von LUTZ GREISIGER, CLAUDIA RAMMELT, JÜRGEN TUBACH & DANIEL HAAS. Würzburg, Beirut, Halle/Saale: Ergon Verlag Würzburg in Kommission; Orient-Institut Beirut; Universitäts- und Landesbibliothek Sachsen-Anhalt (Beiruter Texte und Studien; 116) 2009, S. 47–68, 371.

DOERINGER, FRANKLIN M.: The Gate in the Circle: A Paradigmatic Symbol in Early Chinese Cosmology. In: *Philosophy East and West* **32** (1982), 3, p. 309–324.

DÜTTING, DIETER & MATTHIAS SCHRAMM: Venus the Moon and the Gods of the Palenque Triad. In: *Zeitschrift für Ethnologie* **109** (1984), 1, p. 7–74.

EISLER, ROBERT: *Weltenmantel und Himmelszelt: Religionsgeschichtliche Untersuchungen zur Urgeschichte des antiken Weltbildes. 2 Bde.* München: C. H. Beck'sche Verlagsbuchhandlung Oskar Beck 1910.

ELIADE, MIRCEA: Australian Religions: An Introduction. Part II. In: *History of Religions* **6** (1967), 3, p. 208–235.

ELIADE, MIRCEA: *Shamanism: Archaic Techniques of Ecstasy.* London: Arkana Penguin Books 1989.

ERLENMEYER, MARIE-LOUIS & HANS ERLENMEYER: Über Griechische und Altorientalische Tierkampfgruppen. In: *Antike Kunst* **1** (1958), 2, p. 53–68.

ESMAILPOUR, ABOLQASEM: Manichaean Gods and Goddesses in a Classical Arabic Treatise: Kitāb Al-Radd-I ‚La'l-Zandiq Al-Laīn Ibn Al-Muqaffa'. In: *Central Asiatic Journal* **51** (2007), 2, p. 167–176.

FLOOD, GAVIN: The Purification of the Body in Tantric Ritual Representation. In: *Indo-Iranian Journal* **45** (2022), 1, p. 25–43.

FOSTER, BENJAMIN READ; FRAYNE, DOUGLAS & GARY BECKMAN: *The Epic of Gilgamesh: A New Translation, Analogues, Criticism.* New York: Norton (A Norton Critical Edition) 2001.

FRANZ, MARIE-LOUISE VON: *Number and Time: Reflections Leading Towards a Unification of Depth Psychology and Physics.* Evanston: Northwestern University Press 1974.

GANDZ, SOLOMON: Studies in Hebrew Mathematics and Astronomy. In: *Proceedings of the American Academy for Jewish Research* **9** (1938), p. 5–50.

GANDZ, SOLOMON: The Zodiacal Light in Semitic Mythology. In: *Proceedings of the American Academy for Jewish Research* **13** (1943), p. 1–39.

GUERNSEY, JULIA: A Consideration of the Quatrefoil Motif in Preclassic Mesoamerica. In: *Anthropology and Aesthetics* **57/58** (2010), p. 75–96.

GULÁCSI, ZSUZSANNA & JASON BEDUHN: Picturing Mani's Cosmology: An Analysis of Doctrinal Iconography on a Manichaean Hanging Scroll from 13th /14th Century Southern China. In: *Bulletin of the Asia Institute*, New Series **25** (2011), p. 55–105.

GUNDEL, HANS GEORG: Dekane. In: *Paulys Realencyclopädie der Classischen Altertumswissenschaft.* Neue Bearbeitung. Hg. von *Georg Wissowa, Wilhelm Kroll, Karl Mittelhaus, Konrad Ziegler & Hans Gärtner*. 83 Bde., XA (19). Stuttgart: Metzler 1893–1980a, S. 116–124.

GUNDEL, HANS GEORG: Zodiakos. Der Tierkreis in der Antike. In: *Paulys Realencyclopädie der Classischen Altertumswissenschaft.* Neue Bearbeitung. Hg. von GEORG WISSOWA, WILHELM KROLL, KARL MITTELHAUS, KONRAD ZIEGLER & HANS GÄRTNER. 83 Bde., XA (19). Stuttgart: Metzler 1893–1980b, S. 462–709.

GUNDEL, WILHELM: *Dekan und Dekansternbilder.* Glückstadt, Hamburg: J. J. Augustin (Studien der Bibliothek Warburg; XIX) 1936.

GUNDEL, WILHEM & HANS GEORG GUNDEL: Planeten bei Griechen und Römern. In: *Paulys Realencyclopädie der Classischen Altertumswissenschaft.* Neue Bearbeitung. Hg. von *Georg Wissowa, Wilhelm Kroll, Karl Mittelhaus, Konrad Ziegler & Hans Gärtner.* 83 Bde., XX, 2 (1950). Stuttgart: Metzler 1893–1980b, S. 2015–2185.

HAMMOND, ROSEMARY MARGARET: *Cosmological Motifs and Themes in the Odyssey of Homer, with some Antecedents in Minoan and Mycenaean Iconography.* PhD thesis, The Open University, Milton Keynes (2007), `https://oro.open.ac.uk/64051/1/13890121.pdf`, last access 15/12/2023.

HASSRICK, ROYAL B.: *The Sioux, Life and Customs of a Warrior Society.* Norman: University of Oklahoma Press 1964.

HEIMPEL, WOLFGANG: The Sun at Night and the Doors of Heaven in Babylonian Texts. In: *Journal of Cuneiform Studies* **38** (1986), 2, p. 127–151.

HEINRICH, ADRIAN C.; MITTO, TONIO D. N. & ENRIQUE JIMÉNEZ: Poem of Creation (Enūma Eliš) Chapter V He Made Positions for the Great Gods. In: *Electronic Babylonian Library* (2021), `https://doi.org/10.5282/ebl/1/1/2`.

HENNING, W. B.: An Astronomical Chapter of the Bundahishn. In: *The Journal of the Royal Asiatic Society of Great Britain and Ireland* **3** (1942), p. 229–248.

HENNING, W. B. A.: Sogdian Fragment of the Manichaean Cosmogony. In: *Bulletin of the School of Oriental and African Studies,* University of London **12** (1948), 2, p. 306–318.

HERMSEN, EDMUND: *Die zwei Wege des Jenseits: Das altägyptische Zweiwegebuch und seine Topographie.* Freiburg/Schweiz, Göttingen: Vandenhoeck & Ruprecht (Orbis biblicus et orientalis; 112) 1991.

HOOYKAAS, JACOBA: The Rainbow in Ancient Indonesian Religion. In: *Bijdragen tot de Taal-, Land- en Volkenkunde* **112** (1956), 3, p. 291–322.

HORNUNG, ERIK: *Das Amduat.* Wiesbaden: Harrassowitz (Ägyptologische Abhandlungen 13) 1967.

HORNUNG, ERIK: *Das Buch von den Pforten des Jenseits: Nach den Versionen des Neuen Reiches. 2 Bde.* Genf: Editions de Belles-Lettres (Ægyptiaca Helvetica [AH]; 7–8) 1979/1984.

HORNUNG, ERIK: *Die Nachtfahrt der Sonne: Eine altägyptische Beschreibung des Jenseits.* Zürich: Artemis & Winkler 1991, (Neuauflage) 1998.

HORNUNG, ERIK (Hg.): *Die Unterweltsbücher der Ägypter.* Düsseldorf: Patmos-Verlag (2. Auflage) 2002.

HÜBNER, WOLFGANG: *Die Eigenschaften der Tierkreiszeichen in der Antike: Ihre Darstellung und Verwendung unter besonderer Berücksichtigung des Manilius.* Wiesbaden: Steiner (Sudhoffs Archiv; Beihefte 22) 1982.

HÜBNER, WOLFGANG: Das Thema der Reise in der antiken Astrologie. In: *Pallas* **59** (2002a), p. 27–54.

HÜBNER, WOLFGANG: Tierkreis. In: *Der Neue Pauly: Enzyklopädie der Antike.* Stuttgart: Metzler 2002b, Sp. 553–563.

HUMBOLDT, ALEXANDER VON: *Kosmos. Entwurf einer Physischen Weltbeschreibung. Erster Band.* Stuttgart, Tübingen: J. G. Cotta 1845.

HUMMEL, SIEGBERT: Der Hund in der religiösen Vorstellungswelt des Tibeters II. In: *Paideuma – Mitteilungen zur Kulturkunde* **7** (1961), 7, p. 352–361.

HUXLEY, MARGARET: The Gates and Guardians in Sennacherib's Addition to the Temple of Assur. In: *Iraq* **62** (2000), S. 109–137.

ISHIHARA, REIKO: Rising Clouds, Blowing Winds: Late Classic Maya Rain Rituals in the Main Chasm, Aguateca, Guatemala. In: *World Archaeology* **40** (2008), 2, p. 169–189.

JACKSON, A. V. WILLIAMS: The Doctrine of Metempsychosis in Manichaeism. In: *Journal of the American Oriental Society* **45** (1925), p. 246–268.

JACKSON, HOWARD M.: The Meaning and Function of the Leontocephaline in Roman Mithraism. In: *Numen* **32** (1985), 1, p. 17–45.

JENSEN, P.: *Die Kosmologie der Babylonier: Studien und Materialien.* Straßburg: Karl J. Trübner 1890.

JEREMIAS, ALFRED: *Handbuch der Altorientalischen Geisteskultur.* Leipzig: J. C. Hinrisch'-sche Buchhandlung 1913.

JIANG, BING: *Chinese Gates of Late Imperial China in the Context of Cosmo-Religious Rituals. Book I.* Ph.D. thesis, School of Architecture, Faculty of Social Sciences, University of Sheffield 2014.

JOBES, GERTRUDE: *Dictionary of Mythology, Folklore, and Symbols. 3 vols.* New York: The Scarecrow Press, Inc. 1962.

JOBES, GERTRUDE & JAMES JOBES: *Outer Space: Myths, Name, Meanings, Calendars: From the Emergence of History to the Present Day.* New York, London: The Scarecrow Press, Inc. 1964.

JUNG, HERMANN & MICHAEL A. RAPPENGLÜCK (ed.): *Signaturen des Lebens: Bilder und Zeichen von Kosmos und Bios und Symbole des Alltags – Alltag der Symbole.* Frankfurt am Main: Peter Lang (Symbolon, N.F. 16) 2007.

JUNG, HERMANN (ed.): Symbolon: Gesellschaft für wissenschaftliche Symbolforschung e. V., Jahrbuch Band 20. Neue Folge. Frankfurt am Main: Peter Lang, 2017.

KELLEY, DAVID H. & EUGENE F. MILONE: *Exploring Ancient Skies: A Survey of Ancient and Cultural Astronomy.* New York, NY: Springer Science+Business Media LLC (2nd ed.) 2011.

KENDALL, TIMOTHY MEHEN: The Ancient Egyptian Game of the Serpent. In: *Ancient Board Games in Perspective.* Papers from the 1990 British Museum Colloquium, with Additional Contributions, edited by IRVING L. FINKEL. London: British Museum Press 2007, S. 33–45.

KLEIN, CECELIA F.; GUZMAN, EULOGIO; MANDELL, ELISA C.; STANFIELD, MAYA & JOSEPHINE VOLPE: The Role of Shamanism in Mesoamerican Art: A Reassessment. In: *Current Anthropology* **43** (2002), 3, p. 383–419.

KÕIVA, MARE; PUSTYLNIK, IZOLD & LIISA VESIK (eds.): *Cosmic Catastrophies. A collection of Articles: Proceedings of the European Society for Astronomy in Culture (SEAC) 2002.* Tartu, Estonia: Eesti Kirjandusmuuseumi rahvausundi ja meedia rühm 2005.

KRUPP, EDWIN CHARLES: Negotiating the Highwire of Heaven: The Milky Way and the Itinerary of the Soul. In: *Vistas in Astronomy* **39** (1995), p. 405–430.

KRUPP, EDWIN CHARLES: *Skywatchers, Shamans & Kings: Astronomy and the Archaeology of Power.* New York: Stewart, Tabori & Chang (Wiley popular science) 1997.

LA RIVA GONZALEZ, PALMIRA: Le Walthana Hampi ou la reconstruction du corps conception des la grossesse dans Les Andes du Sud du Pérou. In: *Journal de la Société des Américanistes* **86** (2000), p. 169–184.

LAM, HAU-LING EILEEN: Representation of Heaven and Beyond: The Bi Disc Imagery in the Han Burial Context. In: *Asian Studies; Vol. 7: Meaning and Transformation of Chinese Funerary Art during the Han and Wei Jin Nanbei Periods* (2019), No. 2, p. 115–151.

LANKFORD, GEORGE E.: *Reachable Stars: Patterns in the Ethnoastronomy of Eastern North America.* Tuscaloosa: University of Alabama Press (4th ed.) 2007.

LATURA, GEORGE: Plato's X & Hekate's Crossroads: Astronomical Links to the Mysteries of Eleusis. In: *Mediterranean Archaeology & Archaeometry* **14** (2014), 3, p. 37–44.

LATURA, GEORGE: The Zodiacal Light and Its Use in Cultic Practice. In: *Imagining Other Worlds: Explorations in Astronomy and Culture.* Edited by NICHOLAS CAMPION & CHRIS IMPEY. Ceredigion, Wales: Sophia Centre Press (Studies in cultural astronomy and astrology; Vol. 9) 2018, S. 193–202.

LAUBER, STEPHAN: Zur Ikonographie der Flügelsonne. In: *Zeitschrift des Deutschen Palästina-Vereins* **124** (2008), 2, p. 89–106.

LEEMING, DAVID ADAMS: *Creation Myths of the World: An Encyclopedia.* Santa Barbara, California: ABC-CLIO (2nd ed.) 2010.

LEHMANN, KARL: The Dome of Heaven. In: *The Art Bulletin* **27** (1945), 1, p. 1–27.

LEHNER, ERICH: *Ideen und Konzepte der Architektur in außereuropäischen Kulturen.* Wien: Neuer Wissenschaftlicher Verlag (NWV) im Verlag Österreich GmbH 2006.

LETHABY, WILLIAM RICHARD: *Architecture, Mysticism, and Myth.* Mineola, N.Y: Dover Publications 2004.

LIEBENBERG, DEON: The Rainbow-Snake and the Birds' Plumage: An Exploration of the Theme of the Continuous and the Discrete in Claude Lévi-Strauss's Mythologiques. In: *Journal of Folklore Research* **53** (2016), 3, p. 167–203.

LIEVEN, ALEXANDRA VON: *Grundriss des Laufes der Sterne: Das sogenannte Nutbuch.* Copenhagen: Museum Tusculanum Press (Carsten Niebuhr Institute of Ancient Eastern Studies – CNI publication: Carlsberg Papyri 8, Band 31) 2007.

LINCOLN, BRUCE: The Hellhound. In: *Journal of Indo-European Studies* **7** (1979), p. 273–86.

LINDOW, JOHN: *Norse Mythology: A Guide to the Gods, Heroes, Rituals, and Beliefs.* Oxford: Oxford University Press 2002.

LÓPEZ, ALEJANDRO MARTÍN & SIXTO R. GIMÉNEZ BENÍTEZ: The Milky Way and its Structuring Functions in the Worldview of the Mocoví of Gran Chaco. In: *Archaeologia Baltica* **10** (2008), p. 21–24.

LOWENSTEIN, JOHN: Rainbow and Serpent. In: *Anthropos* **56** (1961), 1./2., p. 31–40.

LÜTGE, MICHAEL: *Der Himmel als Heimat der Seele: Visionäre Himmelfahrtspraktiken und Konstrukte göttlicher Welten bei Schamanen, Magiern, Täufern und Sethianern. Iranische Spuren im Zostrianos von Nag Hammadi.* Habilitation, Georg August Universität Göttingen 2008. `http://homepage.ruhr-uni-bochum.de/Michael.Luetge/`.

MABBETT, I. W.: The Symbolism of Mount Meru. In: *History of Religions* **23** (1983), 1, p. 64–83.

MACKENZIE, D. N.: Zoroastrian Astrology in the „Bundahišn". In: *Bulletin of the School of Oriental and African Studies*, University of London **27** (1964), 3, p. 511–529.

MADDOCK, KENNETH: Introduction. In: *The Rainbow Serpent: A Chromatic Piece.* Edited by IRA R. BUCHLER & KENNETH MADDOCK. Berlin, The Hague: De Gruyter Mouton (World Anthropology – An Interdisciplinary Series, ed. by Sol Tax) 1978, Reprint 2011.

MAEDER, STEFAN: The Big Dipper, Sword, Snake and Turtle: Four Constellations as Indicators of the Ecliptic Pole in Ancient China? In: *Mapping the Oriental Sky.* Proceedings of the Seventh International Conference on Oriental Astronomy (ICOA-7), held on September 6–10, 2010 in the National Astronomical Observatory of Japan. Edited by TSUKO NAKAMURA, WAYNE ORCHISTON, MITSURU SÔMA & RICHARD STROM. Tokyo: National Astronomical Observatory of Japan 2011, p. 57–63.

MAJOR, JOHN S.: Notes on the Nomenclature of Winds and Directions in the Early Han. In: *T'oung Pao, Second Series* **65** (1979), 1/3, p. 66–80.

MALLINSON, JAMES & MARK SINGLETON: *Roots of Yoga.* London: Penguin Classics 2017.

MARINATOS, NANNO: The Cosmic Journey of Odysseus. In: *Numen* **48** (2001), 4, p. 381–416.

MARY, ANDRÉ: Le schème de la naissance à l'envers: scénario initiatique et logique de l'inversion (the 'Reverse Birth' Model: Initiation Process and the Logic of Inversion). In: *Cahiers d'Études Africaines* **28** (1988), 110, p. 233–63.

MAUVIEUX, BENOIT; REINBERG, ALAIN & YVAN TOUITOU: The Yurt: A Mobile Home of Nomadic Populations Dwelling in the Mongolian Steppe Is Still Used Both as a Sun Clock and a Calendar. In: *Chronobiology International* **31** (2014), 2, p. 151–56.

MERKELBACH, REINHOLD: *Mithras: Ein persisch-römischer Mysterienkult.* Weinheim: Beltz-Athenäum 1984, (2. Auflage) 1994.

MILBRATH, SUSAN: *Star Gods of the Maya: Astronomy in Art, Folklore, and Calendars.* Austin: University of Texas Press (Linda Schele Series in Maya and pre-Columbian Studies) 1999.

MILBRATH, SUSAN: *Heaven and Earth in Ancient Mexico: Astronomy and Seasonal Cycles in the Codex Borgia.* Austin: University of Texas Press (Linda Schele Series in Maya and pre-Columbian Studies) 2013.

MORGENSTERN, JULIAN: The Gates of Righteousness. In: *Hebrew Union College Annual* **6** (1929), p. 1–37.

MORGENSTERN, JULIAN: The Chanukkah Festival and the Calendar of Ancient Israel: VI the History of the Calendar in Israel During the Biblical Period (Continued). In: *Hebrew Union College Annual* **21** (1948), p. 365-496.

MORGENSTERN, JULIAN: The Cultic Setting of the „Enthronement Psalms". In: *Hebrew Union College Annual* **35** (1964), p. 1–42.

MÜLLER, MARKUS: *Beherrschte Zeit: Lebensorientierung und Zukunftsgestaltung durch Kalenderprognostik zwischen Antike und Neuzeit. Mit einer Edition des Oassauer Kalenders (UB/LMB 2° Ms. astron. 1).* Kassel: kassel university press GmbH (Schriften der Universitätsbibliothek Kassel – Landesbibliothek und Murhardsche Bibliothek der Stadt Kassel; Band 8) 2009.

MURIE, JAMES R.: *Ceremonies of the Pawnee Part I: The Skiri. Part II: The South Bands.* Ed. by DOUGLAS R. PARKS. Washington, D.C.: Smithsonian Institute Press 1981.

NAKASSIS, DIMITRI: Gemination at the Horizons: East and West in the Mythical Geography of Archaic Greek Epic. In: *Transactions of the American Philological Association* (1974–2014) **134** (2004), 2, p. 215–233.

NORDBERG, ANDREAS: The Grave as a Doorway to the Other World: Architectural Religious Symbolism in Iron Age Graves in Scandinavia. In: *Temenos* **45** (2009), 1, p. 35–63.

NORELIUS, PER-JOHAN: „Mahān Puruṣaḥ": The Macranthropic Soul in Brāhmaṇas and Upaniṣads. In: *Journal of Indian Philosophy* **45** (2017), 3, p. 403–472.

OHLMARKS, ÅKE JOEL: Stellt Die mythische Bifrost den Regenbogen oder die Milchstrasse dar? Eine textkritisch-religionshistorische Untersuchung zur mythographischen Arbeitsmethode Snorri Sturlusons. In: *Meddelande från Lunds astronomiska observatorium* **2** (1941), 110.

PANAINO, ANTONIO: Uranographia Iranica II: Avestan Hapta.Srū- and Mərəzu-: Ursa Minor and the North Pole? In: *Archiv für Orientforschung* **42/43** (1995/1996), p. 190–207.

PANKENIER, DAVID W.: A Brief History of Beiji 北极 (Northern Culmen), with an Excursus on the Origin of the Character Di 帝. In: *Journal of the American Oriental Society* **124** (2004), 2, p. 211.

PANKENIER, DAVID W.: Weaving Metaphors and Cosmo-Political Thought in Early China. In: *T'oung Pao* **101** (2015), 1/3, p. 1–34.

PENPRASE, BRYAN E.: *The Power of Stars.* New York, NY: Springer 2011.

PETERS, LARRY G. TRANCE: Initiation, and Psychotherapy in Tamang Shamanism. In: *American Ethnologist* **9** (1982), 1, p. 21–46.

PETERS, LARRY G.: Shamanism: Phenomenology of a Spiritual Discipline. In: *The Journal of Transpersonal Psychology* **21** (1989), 2, p. 115–37.

POWERS, WILLIAM K.: *Oglala Religion.* Lincoln, London: University of Nebraska Press 1975.

PRENDERGAST, FRANK: The North Sky and the Otherworld: Journeys of the Dead in the Neolithic Considered. In: *Advancing Cultural Astronomy.* Edited by EFROSYNI BOUTSIKAS, STEPHEN C. MCCLUSKEY & JOHN STEELE. Cham: Springer International Publishing (Historical & Cultural Astronomy) 2021, p. 141–165.

RAPPENGLÜCK, BARBARA: „Rushing through the clashing rocks" – Does this old motif have an astronomical meaning? In: KÕIVA, PUSTYLNIK & VESIK (eds.): *Cosmic Catastrophies.* Tartu 2005, S. 151–156.

RAPPENGLÜCK, BARBARA: Die Klappfelsen als Raum-, Zeit- und Bewusstseinsschwelle. In: JUNG, HERMANN (ed.): *Symbolon: Gesellschaft für wissenschaftliche Symbolforschung e. V.* Frankfurt am Main: Peter Lang (Symbolon, N.F. 20) 2017, S. 183–194.

RAPPENGLÜCK, MICHAEL A.: *Eine Himmelskarte aus der Eiszeit? Ein Beitrag zur Urgeschichte der Himmelskunde und zur paläoastronomischen Methodik, aufgezeigt am Beispiel der Szene in „Le Puits", Grotte de Lascaux (Com. Montignac, Dép. Dordogne, Rég. Aquitaine, France).* Dissertation, Universität München 1998. Frankfurt am Main: Lang 1999.

RAPPENGLÜCK, MICHAEL A.: The Pivot of the Cosmos: The Concept of the World Axis Across Cultures. In: KÕIVA, PUSTYLNIK & VESIK (eds.): *Cosmic Catastrophies.* Tartu 2005, S. 157–165.

RAPPENGLÜCK, MICHAEL A.: Copying the Cosmos: The Archaic Concepts of the Sacred Cave across Cultures. In: HERMANN JUNG & MICHAEL A. RAPPENGLÜCK (ed.): *Signaturen des Lebens: Bilder und Zeichen von Kosmos und Bios und Symbole des Alltags – Alltag der Symbole.* Frankfurt am Main: Peter Lang (Symbolon, N.F. 16) 2007, p. 63–84.

RAPPENGLÜCK, MICHAEL A.: The Pleiades and Hyades as Celestial Spatiotemporal Indicators in the Astronomy of Archaic and Indigenous Cultures. In: WOLFSCHMIDT, GUDRUN (ed.): *Prähistorische Astronomie und Ethnoastronomie.* Proceedings der Tagung am 24. September 2007 in Würzburg. Norderstedt: Books on Demand (Nuncius Hamburgensis; Band 8) 2008, S. 13–39.

RAPPENGLÜCK, MICHAEL A.: Heavenly Messengers: The Role of Birds in the Cosmographies and the Cosmovisions of Ancient Cultures. In: *Cosmology Across Cultures.* Proceedings of a Workshop Held at Parque de las Ciencias, Granada, Spain, 8–12 September 2008, edited by JOSÉ A. RUBIÑO-MARTÍN. San Francisco, California: Astronomical Society of the Pacific (Astronomical Society of the Pacific conference series; 409) 2009, S. 145–150.

RAPPENGLÜCK, MICHAEL A.: The Housing of the World: The Significance of Cosmographic Concepts for Habitation. In: *Nexus Network Journal* **15** (2013), 3, p. 387–422.

RAPPENGLÜCK, MICHAEL A.: The Cosmic Deep Blue: The Significance of the Celestial Water Sphere across Cultures. In: *Mediterranean Archaeology and Archaeometry* **14** (2014a), 3, p. 293–305.

RAPPENGLÜCK, MICHAEL A.: Weltgehäuse. Zur kosmographischen Symbolik von Höhle, Heiligtum und Haus. In: JUNG, HERMANN (ed.): *Symbolon: Gesellschaft für wissenschaftliche Symbolforschung e. V.* Frankfurt am Main: Peter Lang (Symbolon, N.F. 19) 2014b, p. 237—262.

RAPPENGLÜCK, MICHAEL A.: Voyages Guided by the Skies: Ancient Concepts of Exploring and Domesticating Time and Space Across Cultures. In: *SEAC 2011 Stars and Stones: Voyages in Archaeoastronomy and Cultural Astronomy.* Proceedings of the SEAC 2011 Conference. Ed. by FERNANDO PIMENTA, N. RIBEIRO, FABIO SILVA, NICHOLAS CAMPION, A. JOAQUINITO & L. TIRAPICOS. Oxford: Archaeopress (BAR international series; 2720) 2015, S. 16–24.

RAPPENGLÜCK, MICHAEL A.: Himmelsfluss, Ouroboros, Seelenweg und Orientierungshilfe. Die Milchstraße in Symbolik, Mythen und Ritualen. In: JUNG, HERMANN (ed.): *Symbolon: Gesellschaft für wissenschaftliche Symbolforschung e. V.* Frankfurt am Main: Peter Lang (Symbolon, N.F. 20) 2017, p. 185–213.

RAPPENGLÜCK, MICHAEL A.: The World as a Living Entity: Essentials of a Cosmic Metaphor. In: *Mediterranean Archaeology and Archaeometry* **18** (2018a), 4, p. 323–331.

RAPPENGLÜCK, MICHAEL A.: The World Tree: Categorisation and Reading of an Archaic Cosmographic Concept. In: *Journal of Skyscape Archaeology* **3** (2018b), 2, p. 253–267.

RAPPENGLÜCK, MICHAEL A.: Himmlische Reiseführer: Wie sich die alten Kulturen in Raum und Zeit orientierten. In: WOLFSCHMIDT, GUDRUN (ed.): *Orientierung, Navigation und Zeitbestimmung – Wie der Himmel den Lebensraum des Menschen prägt. Orientation, Navigation and Time Keeping – How the Sky Shapes People's Living Space.* Proceedings der Tagung der Gesellschaft für Archäoastronomie in Hamburg 2017. Hamburg: tredition (Nuncius Hamburgensis; Band 42) 2019, S. 221–267.

RAPPENGLÜCK, MICHAEL A.: A Catchy World Model: The Concept(s) of Cosmic Mountain(s) Used by Ancient Cultures. In: *Harmony and Symmetry: Celestial Regularities Shaping Human Culture.* Proceedings of the SEAC 2018 Conference in Graz. Edited by SONJA DRAXLER, MAX E. LIPPITSCH & GUDRUN WOLFSCHMIDT. Hamburg: tredition 2020, S. 136–145.

RAPPENGLÜCK, MICHAEL A.: Der Baum der Welt und der Welten: ein archaisches kosmographisches Modell. In: WOLFSCHMIDT, GUDRUN (ed.): *Maß und Mythos, Zahl und Zauber – Die Vermessung von Himmel und Erde. Measure and Myth, Number and Magic: Measuring Heaven and Earth.* Proceedings der Tagung der Gesellschaft für Archäoastronomie in Dortmund 2018. Hamburg: tredition (Nuncius Hamburgensis; Band 48) 2020, p. 54–88.

RAPPENGLÜCK, MICHAEL A.: Weltmodell und Lebensreise: Die Vorstellungen vom kosmischen Berg (der kosmischen Berge) in archaischen Kulturen (und heute). In: WOLFSCHMIDT, GUDRUN (ed.): *Himmelswelten und Kosmovisionen – Imaginationen, Modelle, Weltanschauungen. Sky Worlds and Cosmovisions – Imaginations, Models, Worldviews.* Proceedings der Tagung der Gesellschaft für Archäoastronomie in Gilching 2019. Hamburg: tredition (Nuncius Hamburgensis; Band 51) 2020, S. 296–329.

RAPPENGLÜCK, MICHAEL A.: Human Beings in Cosmic Lifeworlds: Anthropology, Ecospheres and Cultural Cosmologies. In: FABIO P. & LIZ HENTY (ed.): *Solarizing the Moon: Essays in Honour of Lionel Sims.* Oxford: Archeopress 2022, p. 157–181.

RAPPENGLÜCK, MICHAEL A.: Mensch, Lebenswelt und Welt-Organismus – Die Vorstellung vom kosmischen Lebewesen in verschiedenen Kulturen einst (und heute). In: WOLFSCHMIDT, GUDRUN (ed.): *Wandlungen in Raum und Zeit: Himmel – Heimat – Weltverständnis. Transformations in Space and Time: Heaven – Home – Understanding of the World.* Proceedings der Tagung der Gesellschaft für Archäoastronomie in Recklinghausen 2022. Ahrensburg: tredition (Nuncius Hamburgensis; Band 58) 2023, S. 16–63.

RAPPENGLÜCK, MICHAEL A.: Finis Terrae und Weltinnenraum: Symbole, Mythen und Rituale rund um Weltenfahrten und Seelenreisen. In: HEINZ, WERNER (ed.): *Symbolon:*

Gesellschaft für wissenschaftliche Symbolforschung e. V. Frankfurt am Main: Peter Lang (Symbolon, N.F. 22) 2024.

REDHOUSE, J. W.: On the Natural Phenomenon Known in the East by the Names Sub-hi-Kazib, etc., etc. In: *Journal of the Royal Asiatic Society of Great Britain and Ireland* **10** (1878), 3, p. 344–354.

REICHE, HARALD A. T.: Heraclides' Three Soul-Gates: Plato Revised. In: *Transactions of the American Philological Association (1974–2014)* **123** (1993), p. 161–180.

REIDINGER, ERWIN: Die Tempelanlage in Jerusalem von Salomo bis Herodes aus der Sicht der bautechnischen Archäologie. In: *Biblische Notizen, Beiträge zur exegetischen Diskussion* **114/115** (2002), p. 89–150.

REIDINGER, ERWIN: The Temple Mount Platform in Jerusalem from Solomon to Herod: An Archaeological Re-Examination. In: *Assaph, Studies in Art of History* **9** (2004), p. 1–64.

REIDINGER, ERWIN: *Die Tempelanlage in Jerusalem von Salomo bis Herodes: Neuer Ansatz für Rekonstruktion durch Bauforschung und Astronomie.* Wiener Neustadt: Rudolf Hausstein OHG (Erweiterter Nachdruck) 2005.

RENGGLI, FRANZ: The Sunrise as the Birth of a Baby: The Prenatal Key to Egyptian Mythology. In: *Journal of Prenatal and Perinatal Psychology and Health* **16** (2002), 3, p. 1–18.

RICHTER, BARBARA A.: The Amduat and Its Relationship to the Architecture of Early 18th Dynasty Royal Burial Chambers. In: *Journal of the American Research Center in Egypt* **44** (2008), p. 73–104.

RIEDWEG, CHRISTOPH; BAUMBACH, MANUEL; MÄNNLEIN-ROBERT, IRMGARD; BECKER, MATTHIAS; GLEI, REINHOLD F. & BENJAMIN TOPP (ed.): *Die Seele im Kosmos.* Tübingen: Mohr Siebeck 2019.

ROBERTS, ALLEN T.: Passage Stellified: Speculation Upon Archaeoastronomy in Southeastern Zaire. In: *Archaeoastronomy – The Bulletin of the Center for Archaeoastronomy* **4** (1981), 4, p. 27–37.

ROMAIN, WILLIAM F.: Serpent Mound in Its Woodland Period Context. In: *Midcontinental Journal of Archaeology* **44** (2019), 2, p. 57–83.

ROMAIN, WILLIAM F.: Following the Milky Way Path of Souls: An Archaeometric Assessment of Cahokia's Main Site Axis and Rattlesnake Causeway. In: *Journal of Skyscape Archaeology* **7** (2021), p. 187.

RUGGLES, CLIVE L. N. (ed.): *Handbook of Archaeoastronomy and Ethnoastronomy.* New York, NY: Springer 2015.

RUSSO, JOSEPH: Penelope's Gates of Horn(s) and Ivory. In: *Recherches & Rencontres* **17** (2002), p. 223–230.

SAINT-ONGE SR, REX W.; JOHNSON, JOHN R. & JOSEPH R. TALAUGON: Archaeoastronomical Implications of a Northern Chumash Arborglyph. In: *Journal of California and Great Basin Anthropology* **29** (2009), 1, p. 29–58.

SANTILLANA, GIORGIO DE & HERTHA VON DECHEND: *Die Mühle des Hamlet: Ein Essay über Mythos und das Gerüst der Zeit.* Berlin: Kammerer & Unverzagt (Computerkultur; 8) 1993.

SATHER, CLIFFORD: Mystery and the Mundane Shifting Perspectives in a Saribas Iban Ritual Narrative. In: *Beyond the Horizon: Essays on Myth, History, Travel and Society.* Edited by CLIFFORD SATHER. Helsinki: Finnish Literature Society / SKS (Studia Fennica Anthropologica) 2008, S. 51–74.

SCHUETZ-MILLER, MARDITH K.: Spider Grandmother and Other Avatars of the Moon Goddess in New World Sacred Architecture. In: *Journal of the Southwest* **54** (2012), 2, p. 283–293, 295–303, 305–347, 349–397, 399–421, 423–435.

SCHWEIZER, ANDREAS: *The Sungod's Journey Through the Netherworld: Reading the Ancient Egyptian Amduat.* Ithaca, NY: Cornell University Press 2010.

SCHWERIN, JENNIFER VON: The Sacred Mountain in Social Context. Symbolism and History in Maya Architecture: Temple 22 at Copan. Honduras. In: *Ancient Mesoamerica* **22** (2011), 2, p. 271–300.

SEHGAL, LINDA: Climbing Jacob's Ladder: Symbolism of a Fantastic Journey Along the Milky Way. In: *Rock Art Papers; Volume 6.* Edited by KEN HEDGES. San Diego, Calif.: San Diego Museum of Man (San Diego Museum Papers; 24) 1989, S. 83–90.

SIMEK, RUDOLF: *Dictionary of Northern Mythology.* Woodbridge: Brewer 1993.

SNODGRASS, ADRIAN: *The Symbolism of the Stūpa.* Ithaca, N.Y.: Cornell University Press (Southeast Asia Program) 1985.

SOMMARSTRÖM, BO: Ethnoastronomical Perspectives on Saami Religion. In: *Scripta Instituti Donneriani Aboensis* **12** (1987), p. 211–250.

STAAL, JULIUS D. W.: *The New Patterns in the Sky: Myths and Legends of the Stars.* Blacksburg, VA: McDonald & Woodward 1988.

STEIN, ROLF ALFRED: *The World in Miniature: Container Gardens and Dwellings in Far Eastern Religious Thought.* Stanford, California: Stanford University Press 1990.

STRÄTZ, VOLKER: Materialien zu Tierkreisen in China. In: *Monumenta Serica* **44** (1996), p. 213–265.

TATE, CAROLYN E.: Patrons of Shamanic Power: La Venta's Supernatural Entities in Light of Mixe Beliefs. In: *Ancient Mesoamerica* **10** (1999), 2, p. 169–188.

TAUBE, KARL A.: Flower Mountain: Concepts of Life, Beauty, and Paradise Among the Classic Maya. In: *Anthropology and Aesthetics* **45** (2004), p. 69–98.

TAYLOR, RABUN: Watching the Skies: Janus, Auspication, and the Shrine in the Roman Forum. In: *Memoirs of the American Academy in Rome* **45** (2000), p. 1–40.

THERIK, TOM: *Wehali: The Female Land: Traditions of a Timorese Ritual Centre.* Canberra, Australia: ANU Press (Comparative Austronesian Series) 2023.

TOLMACHEVA, MARINA: On the Arab System of Nautical Orientation. In: *Arabica* **27** (1980), 2, p. 180–192.

TOPP, BENJAMIN: Σύμβολον τού Κόσμου Astronomisch-Astrologische Vorstellungen in Porphyrios' De Antro Nympharum. In: *Die Seele im Kosmos.* Edited by CHRISTOPH RIEDWEG, MANUEL BAUMBACH, IRMGARD MÄNNLEIN-ROBERT, MATTHIAS BECKER, REINHOLD F. GLEI & BENJAMIN TOPP. Tübingen: Mohr Siebeck, 2019, S. 117–139.

UHDEN, RICHARD: Die antiken Grundlagen der mittelalterlichen Seekarten. In: *Imago Mundi* **1** (1935), p. 1–19.

ULANSEY, DAVID: The Eighth Gate: The Mithraic Lion-Headed Figure and the Platonic World-Soul. In: *The Ancient World* **XXXIV** (2003), 1, p. 67–81.

VAHIA, MAYANK NALINKANT; HALKARE, GANESH & PURUSHOTTAM DAHEDAR: Astronomy of the Korku Tribe of India. In: *Journal of Astronomical History and Heritage* **19** (2016), 2, p. 216–232.

VAN BUREN, DOUGLAS E.: The Guardians of the Gate in the Akkadian Period. In: *Orientalia* **16** (1947), 3, p. 312–332.

VAN DER WAERDEN, BARTEL LEENDERT: History of the Zodiac. In: *Archiv für Orientforschung* **16** (1952–1953), p. 216–230.

VAN DOOREN, INGRID: Navajo Hooghan and Navajo Cosmos. In: *The Canadian Journal of Native Studies* **VII** (1987), 2, p. 259–266.

WALLIS BUDGE, E. A.: *The Gods of the Egyptians or Studies in Egyptian Mythology II.* London: Methuen & Co. 1904.

WATERSON, ROXANA: The Mythical Origins of Humans and Their Houses. In: *Paths and Rivers: Sa'dan Toraja Society in Transformation.* Edited by ROXANA WATERSON. Leiden: Brill, KITLV Press 2009, S. 123–158.

WEIDNER, ERNST F.: Der Tierkreis und die Wege am Himmel. In: *Archiv für Orientforschung* **7** (1931), p. 170–178.

WEINSTOCK, STEFAN: Lunar Mansions and Early Calendars. In: *The Journal of Hellenic Studies* **69** (1949), p. 48–69.

WERNING, DANIEL A.: *Das Höhlenbuch im Grab des Petamenophis (TT33): Szenen, Texte, Wandtafeln.* Berlin: Edition Topoi (Berlin studies of the ancient world; 66) 2019.

WINSTON, DAVID: The Iranian Component in the Bible, Apocrypha, and Qumran: A Review of the Evidence. In: *History of Religions* **5** (1966), p. 183–216.

WITZEL, MICHAEL: Water in Mythology. In: *Daedalus* **144** (2015), 3, p. 18–26.

WORMS, E. A. & H. PETRI: *Australian Aboriginal Religions.* Kensington: Nel-en Yubu Missiological Unit (Nelen Yubu Missiological Series; 5), (Revised edition) 1998.

WU, HUNG: Han Sarcophagi: Surface, Depth, Context. In: *Anthropology and Aesthetics* **61/62** (2012), p. 196–212.

XAVIER PENA, JOABSON: *Temple as Cosmos: The Jerusalem Temple Imagery in Josephus' Writings.* University of Groningen, Ph.D. Thesis, 2020.

YAMPOLSKY, PHILIP: The Origin of the Twenty-Eight Lunar Mansions. In: *Osiris* **9** (1950), p. 62–83.

ZERRIES, OTTO: Sternbilder als Ausdruck Jägerischer Geisteshaltung in Südamerika. In: *Paideuma – Mitteilungen zur Kulturkunde* **5** (1952), 5, p. 220–235.

Figure 2.1:
Archaeoastronomical Complex, Andes, Peru
The thirteen towers of Chankillo Observatory, as seen from the fortified temple.
June solstice sunrise viewed from the western observing point.

(© Iván Ghezzi)

Ancient Astronomies – Ancient Worlds

Clive Ruggles (University of Leicester, UK)

Abstract: Ancient Astronomies – Ancient Worlds

We know a good deal about ancient astronomical knowledge and practices in places such as ancient China and Babylonia from the evidence contained in their recorded history, but people all over the world strived to make sense of what they saw in the sky long before the written record existed. What can we ever know of this? Some of the world's most iconic ancient monuments provide tantalising glimpses of long lost beliefs and practices relating to the sky, although they often have to be interpreted with considerable caution.

There are also valuable clues in beliefs and practices that have survived among indigenous peoples right through to modern times. Trying to make sense of this type of evidence is the business of the fields of study that have become known as archaeoastronomy and ethnoastronomy. Taking in examples from many different parts of the world, including his own field projects in Europe, Peru and Hawaii, Clive will use these insights to build up a broad picture of the diverse ways in which ancient peoples perceived and understood the world – the cosmos – within which they dwelt.

Zusammenfassung: Alte Astronomien – Alte Welten

Wir wissen viel über das astronomische Wissen und die Praktiken in Ländern wie dem alten China und Babylonien aus den Zeugnissen ihrer aufgezeichneten Geschichte, aber die Menschen auf der ganzen Welt bemühten sich, dem, was sie am Himmel sahen, einen Sinn zu geben, lange bevor es schriftliche Aufzeichnungen gab. Was können wir jemals darüber wissen? Einige der berühmtesten antiken Denkmäler der Welt bieten verlockende Einblicke in längst vergessene Glaubensvorstellungen und Praktiken in Bezug auf den Himmel, auch wenn sie oft mit großer Vorsicht zu interpretieren sind.

Wertvolle Hinweise finden sich auch in Glaubensvorstellungen und Praktiken, die bei indigenen Völkern bis in die Neuzeit überlebt haben. Der Versuch, diese Art von Beweisen zu verstehen, ist die Aufgabe der Studienbereiche, die als Archäoastronomie und Ethnoastronomie bekannt geworden sind. Anhand von Beispielen aus vielen verschiedenen Teilen der Welt, einschließlich seiner eigenen Feldforschungsprojekte in Europa, Peru und Hawaii, wird der Autor diese Erkenntnisse nutzen, um ein umfassendes Bild der verschiedenen Arten zu zeichnen, in denen alte Völker die Welt – den Kosmos –, in dem sie lebten, wahrnahmen und verstanden.

2.1 Literatur

Chankillo Observatory, Peru, Astronomical Heritage: tangible immovable. Unesco List 1624 (2021), https://whc.unesco.org/en/list/1624, UNESCO-IAU *Portal to the Heritage of Astronomy,* https://web.astronomicalheritage.net/show-entity?identity=51&idsubentity=1.

Dengfeng Observatory, China – "The Centre of Heaven and Earth". Unesco List 1305 (2010), https://whc.unesco.org/en/list/1305.

RUGGLES, CLIVE L. N. (ed.): *Ancient Astronomy: An Encyclopedia of Cosmologies and Myth.* Santa Barbara, USA: ABC-CLIO 2005.

RUGGLES, CLIVE L. N. (ed.): *Handbook of Archaeoastronomy and Ethnoastronomy, 3 Vol.* New York, NY: Springer 2014, XXXVI, 2297 pages.

Figure 2.2:
Dengfeng Observatory (Gaocheng Astronomical Observatory),
Henan, China, Yuan dynasty (1276).
The Dengfeng Observatory was designed by astronomers
Guo Shoujing (1231–ca. 1215) and Wang Xun (1235–1281)
for timekeeping, to observe the Sun in the meridian
for the new Shoushi calendar (1281).

(© GAO Chenxiang)

Abbildung 3.1:
Mensch im Kosmos

(Collage: Michael A. Rappenglück,
Foto: Kykladenidol von Michael A. Rappenglück, Teilbilder CC0)

Von der Höhle zum Himmel: Kulturastronomie & Kulturelle Kosmologie

Michael A. Rappenglück (Gilching)

Abstract: From the Cave to the Sky: Cultural Astronomy & Cultural Cosmology

With the emergence and development of human self-consciousness based on the capacities of thinking, language and counting, the dichotomy of the world into complementary subjective and objectifiable realities emerged and grew. This reduced the original instinctive involvement in the ecosphere and released the Homo species into a freer interaction with the world, which at the same time led to an increasing distancing from the natural world. To this day, cultures serve to compensate for the dwindling embedding in the natural world by providing systems of order and orientation. Humans developed certain models for understanding the functioning of ecospheres as well as the position of humans in them and interaction with them: cultural cosmologies. The living worlds are perceived and understood like a multiverse: They have their own phenomena, structures, origins and developments that run parallel to each other and interact with each other. The characteristics of human existence are in "dialogue" with the essential structures and processes of the ecosystems in which humans live. This results in cosmologically holistic models that enable members of particular social groups to integrate, orient and assess themselves within changing ecospheres. Since the Palaeolithic, they have appeared worldwide in symbols, myths and rituals relating to landscapes, seascapes, celestial landscapes, living beings, monuments and objects. Systems of power, jurisprudence and social order are based on them.

Beyond the cosmological code go ideas about modes of existence, shape-shifting, states of consciousness, other realities and transcendence, the why and wherefore of people and the world. The aim of a cultural cosmology is to reconstruct prehistoric and historical cosmologies (and cosmogonies) as well as the respective current ideas about the position of humans in the cosmos. Cultural cosmology is concerned with a culture's understanding of itself in its world (cosmos). This definition also makes it possible to understand the ongoing cosmic change of perspective that began in modern times (15th/16th century) – geocentrism, heliocentrism, galactocentrism ... multiverse, or from the singularity of the position of the Earth and man in the cosmos to the plurality of worlds – into cultural cosmology.

The rapid development of earth system sciences together with the growing need for solutions to global problems, which accompanied the technical-scientific-ecological cultural development of modern times and modernity with the juxtaposition of man and the world, interestingly leads back to the connection of the different earth spheres, the interacting network of life forms and a communicative world sphere. However, the cosmos has since expanded gigantically in space and time. This makes a future cultural cosmology difficult, but not hopeless. At the same time, however, modern techniques make it possible to translate distant spaces and times, other worlds and realities into perception and, within limits, into a kind of experience. With the focus on the "blue planet" as the only ecosphere and home of humankind to date, the advance of space travel into interplanetary space and the simultaneous extreme expansion of the view into the diversity of the structures of the cosmos, the topic of a cultural cosmology is forming anew: How do humans – Earth – cosmos belong together?

Zusammenfassung

Mit der Entstehung und Entfaltung des menschlichen Selbstbewusstseins, das auf den Fähigkeiten des Denkens, der Sprache und des Zählens beruht, entstand und wuchs die Dichotomie der Welt in komplementäre subjektive und objektivierbare Realitäten. Dadurch wurde die ursprüngliche instinktive Einbindung in die Ökosphäre reduziert und die Gattung Homo in einen freieren Umgang mit der Welt entlassen, was gleichzeitig zu einer zunehmenden Distanzierung von der natürlichen Welt führte. Bis heute dienen Kulturen dazu, die schwindende Einbettung in die natürliche Welt zu kompensieren, indem sie Ordnungs- und Orientierungssysteme bereitstellen. Der Mensch entwickelte bestimmte Modelle zum Verständnis der Funktionsweise der Ökosphären sowie der Stellung des Menschen in ihnen und der Interaktion mit ihnen: kulturelle Kosmologien. Die Lebenswelten werden wie ein Multiversum wahrgenommen und verstanden: Sie besitzen ihre eigenen Phänomene, Strukturen, Ursprünge und Entwicklungen, die parallel zueinander verlaufen und miteinander interagieren. Die Merkmale der menschlichen Existenz stehen im „Dialog" mit den wesentlichen Strukturen und Prozessen der Ökosysteme, in denen der Mensch lebt. Daraus ergeben sich kosmologisch-ganzheitliche Modelle, die den Mitgliedern bestimmter sozialer Gruppen eine Integration, Orientierung und Einschätzung innerhalb sich verändernder Ökosphären ermöglichen. Seit dem Paläolithikum tauchen sie weltweit in Symbolen, Mythen und Ritualen auf, die sich auf Landschaften, Meereslandschaften, Himmelslandschaften, Lebewesen, Denkmäler und Gegenstände beziehen. Systeme der Macht, der Rechtsprechung und der sozialen Ordnung beruhen auf ihnen.

Über den kosmologischen Code hinaus gehen Vorstellungen über Existenzweisen, Gestaltwandel, Bewusstseinszustände, andere Wirklichkeiten und Transzendenz, das *Warum* und *Wozu* von Menschen und Welt. Ziel einer kulturellen Kosmologie ist es, prähistorische und historische Kosmologien (und Kosmogonien) sowie die jeweils aktuellen Vorstellungen über die Stellung des Menschen im Kosmos zu rekonstruieren. Die kulturelle Kosmologie beschäftigt sich mit dem Selbstverständnis einer Kultur in ihrer Welt (Kosmos). Diese Definition ermöglicht es auch, den laufenden kosmischen Perspektivenwechsel, der in der Neuzeit (15./16. Jahrhundert) begann – Geozentrismus, Heliozentrismus, Galaktozentrismus ... Multiversum, oder von der Singularität der Position der Erde und des Menschen im Kosmos zur Pluralität der Welten – in die kulturelle Kosmologie einzubeziehen.

Die rasante Entwicklung der Erdsystemwissenschaften zusammen mit dem wachsenden Bedarf an Lösungen für globale Probleme, die mit der technisch-wissenschaftlich-ökologischen Kulturentwicklung der Neuzeit und Moderne mit der Gegenüberstellung von Mensch und Welt einherging, führt interessanterweise auf die Verbindung der verschiedenen Erdsphären, das interagierende Netzwerk der Lebensformen und eine kommunikative Weltsphäre zurück. Allerdings hat sich der Kosmos inzwischen in Raum und Zeit gigantisch erweitert. Das macht eine zukünftige kulturelle Kosmologie schwierig, aber nicht hoffnungslos. Zugleich aber ermöglichen es die modernen Techniken, ferne Räume und Zeiten, andere Welten und Wirklichkeiten in die Wahrnehmung und in Grenzen in eine Art von Erfahrung zu übersetzen. Mit der Fokussierung auf den „blauen Planeten" als bisher einzige Ökosphäre und Heimat der Menschheit, dem Vordringen der Raumfahrt in den interplanetaren Raum und der gleichzeitigen extremen Erweiterung des Blicks in die Vielfalt der Strukturen des Kosmos formiert sich das Thema einer kulturellen Kosmologie neu: Wie gehören Mensch – Erde – Kosmos zusammen?

RAPPENGLÜCK, MICHAEL A.: Human Beings in Cosmic Lifeworlds: Anthropology, Ecospheres and Cultural Cosmologies. In: SILVA, FABIO & LIZ HENTY (ed.): *Solarizing the Moon: Essays in Honour of Lionel Sims*. Oxford: Archeopress 2022.

Abbildung 4.1:
Stonehenge, United Kingdom of Great Britain and Northern Ireland

Stonehenge and the Major Standstill Moon

Clive Ruggles (University of Leicester, UK)

Abstract

Despite countless speculations concerning the astronomical significance of Stonehenge, the only tangible connection between this iconic monument and the skyscape that has achieved broad consensus among archaeologists as being intentional and meaningful is the solstitial orientation of its main axis – something that is also found at the nearby, and broadly contemporary, multiple timber circles at Woodhenge and Durrington Walls.

Whether Stonehenge had intentional lunar associations remains an open question, but recent investigations have succeeded in casting some new light on the subject. There is no evidence of any high-precision lunar observations, but it remains plausible that the people who used the site were well aware of the lunar node cycle and placed great significance upon the moon's occasional appearance unusually far north or south during "windows of opportunity" lasting for two or three years that occurred once in a generation (every 18–19 years).

Planning for the approach of the next "major standstill" in January 2025 has helped to highlight issues relating to the practicalities of extreme moonrise and moonset observations at Stonehenge.

In this presentation I shall explore current "mainstream" thinking about Stonehenge and the moon, and describe recent work investigating how lunar observations may have influenced some of the practices carried out there.

Zusammenfassung: Stonehenge und die Große Mondwende

Trotz zahlloser Spekulationen über die astronomische Bedeutung von Stonehenge ist die einzige greifbare Verbindung zwischen diesem ikonischen Monument und der Himmelslandschaft, die unter Archäologen breite Zustimmung gefunden hat, die absichtliche und sinnvolle Ausrichtung seiner Hauptachse – etwas, das auch bei den nahe gelegenen und weitgehend zeitgenössischen Mehrfach-Holzkreisen von Woodhenge und Durrington Walls zu finden ist.

Ob Stonehenge absichtlich mit dem Mond in Verbindung gebracht wurde, ist nach wie vor eine offene Frage, aber jüngste Untersuchungen haben ein neues Licht auf dieses Thema geworfen. Es gibt keine Belege für hochpräzise Mondbeobachtungen, aber es bleibt plausibel, dass die Menschen, die die Stätte nutzten, den Mondknotenzyklus gut kannten und dem gelegentlichen Erscheinen des Mondes ungewöhnlich weit nördlich oder südlich während der zwei- oder dreijährigen „Gelegenheitsfenster", die einmal in einer Generation (alle 18–19 Jahre) auftreten, große Bedeutung beimaßen.

Die Planung für den nächsten „großen Stillstand" im Januar 2025 hat dazu beigetragen, die praktischen Aspekte extremer Mondaufgangs- und Monduntergangsbeobachtungen in Stonehenge zu beleuchten.

In diesem Vortrag habe ich das derzeitige „Mainstream"-Denken über Stonehenge und den Mond untersucht und neuere Arbeiten beschrieben, die untersuchen, wie Mondbeobachtungen einige der dort durchgeführten Praktiken beeinflusst haben könnten.

4.1 Literatur

RUGGLES, CLIVE L. N. (ed.): *Handbook of Archaeoastronomy and Ethnoastronomy, 3 Vol.* New York, NY: Springer 2014, XXXVI, 2297 pages.

RUGGLES, CLIVE L. N. & AMANDA CHADBURN: *Stonehenge – Sighting the Sun.* Liverpool: Historic England/ Liverpool University Press 2024, 232 pages.

Stonehenge, Avebury and Associated Sites, Astronomical Heritage: tangible immovable. Unesco List 373bis (1986, minor boundary modification 2008), https://whc.unesco.org/en/list/373, UNESCO-IAU *Portal to the Heritage of Astronomy*, https://web.astronomicalheritage.net/show-entity?identity=49&idsubentity=1.

Figure 4.2:
Stonehenge – one of the most famous ancient monuments in the world with its solar alignment

The midwinter sunset at Stonehenge taken from the Stonehenge Avenue, and showing the setting winter solstice Sun seen through the monument

(Photo: James O. Davies, © Historic England, Photo Library N030018)

Abbildung 5.1:
Laserscan der Steine von Callanish

(Foto: Alistair Carty)

Die Stehenden Steine von Callanish: Ein 3D-Datensatz für *Stellarium*

Georg Zotti (Wien, Österreich), Victor Reijs, Emma Rennie & Alistair Carty

Abstract: The Standing Stones of Callanish: A 3D Data Release for Stellarium

Callanish (in English) or Calanais (in Scottish Gaelic) on the Isle of Lewis is one of the oldest, and largest, megalithic sites in the UK. It is formed of a central stone circle and tall central stone with radiating rows of stones forming a cruciform-like shape with an "avenue" formed of parallel stone rows towards the north but slightly towards the east. There is also a chambered cairn inset into the stone circle between the ring and the centre stone.

Archaeological evidence of activities was found for 5300 BCE, and 3200 BC to the 2nd millennium BC. The site likely fell out of use between 1500–1000 BCE and, between 1000–500 BCE, the stones were covered with a layer of turf. Around 800 BCE, the site was abandoned.

In his *Bibliotheca Historica*, Diodorus Siculus (1st c. BC) describes a "spherical temple" in use by the Hyperboreans, where Apollo descends every 19 years (Book II/47). This indicates a tale relating to the *Lunar Major Standstill* event. In fact, in the geographical latitude of Callanish, around Major Standstill, the Moon appears to move only very low over the southern horizon, and the near-southern orientation of the double stone row forming the avenue (azimuth 191°) has been interpreted in connection to the standstill Moon setting (e. g., [Ponting & Ponting 1981, Curtis & Curtis 2006]).

We have started to use a laser scan taken in 2005, converted to a 3D scenery for Stellarium, to explore the views of the lowest Moon as seen from the end of the avenue. A slight misalignment against earlier photos became apparent, caused by imperfect georeferencing of the original model (which was not intended to be used for archaeoastronomical simulation). In addition, as seen from the northern end of this avenue, the southern horizon is mostly hidden by the nearby rock outcrop called Cnoc an Tursa. It appears that the avenue points towards the highest stone in this landscape feature, and the Moon can be observed to hide behind it and later reappear between the stones before finally setting. This, and the orientation of the other stone rows, have been noted and documented by the earlier researchers. We therefore have added commercial photogrammetrical digital elevation model data to our 3D model and started to carefully align the laser scan model against this. A final round of corrections which require photographic documentation in fair weather is still outstanding, but at least a preliminary scenery can now, and hopefully our final model later in 2024, be replayed and studied in an accurate simulation in the Stellarium desktop planetarium, making this the first open-access 3D scenery of a site of archaeoastronomical relevance made available for Stellarium [Callanish3D 2023].

Zusammenfassung

Callanish, ein neolithischer Steinkreis mit kreuzförmig auseinanderlaufenden Steinreihen auf der Insel Lewis vor Schottland, wird seit langem mit systematischer Mondbeobachtung in der Vorgeschichte assoziiert. Wir haben einen 3D-Laserscan um zusätzliche Geländedaten erweitert und stellen den Datensatz als Modell für das freie Desktop-Planetarium Stellarium zur Verfügung. Mit dessen Hilfe können die Mondwenden und ihre Wirkung zwischen den Steinen von Callanish anschaulich simuliert werden.

5.1 Einleitung

Auf der Schottland im Nordwesten vorgelagerten Insel Lewis steht eine der bekanntesten prähistorischen Steinsetzungen der Britischen Inseln: Callanish.[1] Diese größte Anlage einer ganzen rituellen Landschaft besteht aus einem Steinkreis aus 13 bis zu 3,5 m hohen Steinen um einen 4,7 m hohen Monolithen und von diesem radial ausgehende Steinreihen [Scotland 2017]. Eine verläuft genau nach Süden, eine nach Westen, und eine nach Ost-Nordost, möglicherweise zum Aufgang der Plejaden um 1600 v. Chr. [Burl 2011, S. 35]. Zwei lange Reihen laufen parallel gegen Azimut 11° und bilden eine „Avenue". Blickt man von deren Nordende in Richtung Steinkreis, konnte man den Tagesgang des niedrigsten Mondes entlang des Horizonts und durch die Steine beobachten, und der Monduntergang erfolgte in Richtung der Avenue [Ponting & Ponting 1981]. Insgesamt sind 49 der ursprünglich bis zu 80 vermuteten [Scotland 2018] großen Steine mit nur minimaler Rekonstruktion erhalten.

1 Genauer: Callanish I. In der Umgebung sind 12 Steinsetzungen bekannt, teilweise mit gegenseitiger Sichtverbindung [Scotland 2017]. Bislang wird nur Callanish I in der archäoastronomischen Literatur diskutiert ($\lambda = 6°44'42,5''$, $\varphi = 58°11'51,1''$, $h = 21$ m). Seit den späten 1990er Jahren wird in behördlichen Dokumenten die Schreibweise „Calanais" bevorzugt, in diesem Beitrag wird jedoch die traditionelle Schreibung des 20. Jahrhunderts beibehalten.

Wie viele andere Steinkreise auf den Britischen Inseln datiert auch dieser in das Spätneolithikum (3300–2500 v. Chr.). Der Steinkreis datiert auf 2900–2600 v. Chr. Ein kleines, nachträglich in den Kreis gesetztes Steinkammergrab datiert auf 2500–1750 v. Chr. und enthielt Glockenbecher-Keramik. Die letzten archäologischen Funde werden auf 1200–800 v. Chr. datiert. Im rauher gewordenen Klima war kein Ackerbau mehr möglich, und die Anlage wurde vom Moor überwachsen. [Scotland 2017]

Ab 1857 wurden 1,5 m Torf abgegraben, und die moderne archäologische Erkundung begann.

Schon 1726 hatte John Toland eine mögliche Erwähnung des Steinkreises in einer geographischen Beschreibung der Welt von Diodorus Siculus (1. Jh. v. Chr.) gefunden. Nach dieser Interpretation entspricht die Insel Lewis dem legendenumwobenen antiken Land der Hyperboräer, über die Diodorus in seiner *Bibliotheca Historica* (Buch II/47) seinerseits einen Bericht des Hecataeus von Abdera, ca. 360–290 v. Chr.) nacherzählt [Diodorus Siculus, ed. Oldfather et al. 1935]:

> [...] *in einer Region jenseits des Landes der Kelten liegt eine Insel, nicht kleiner als Sizilien.* [...] *Die Insel ist fruchtbar und genießt wegen des milden Klimas 2 Ernten pro Jahr;* [...]
> *Leto* [die Mutter von Artemis und Apollo] *wurde auf dieser Insel geboren, und die Inselbewohner verehren daher Apollo über alle anderen Götter.*
> [...]
> *Sie haben einen heiligen Bezirk für Apollo und einen bemerkenswerten Tempel mit vielen Opfergaben, der von sphärischer Form ist* [...]
> *Es wird weiters berichtet, der Mond erscheine der Erde sehr nahe und zeige Erhebungen* [Berge] *wie die auf der Erde, die für das Auge sichtbar sind. Es wird berichtet, der Gott* [Apollo] *besuche die Insel alle 19 Jahre zur Erde herab, die Periode, in der die Wiederkehr der Sterne* [vermutlich: des Mondes?] *zur selben Stelle am Himmel zustande kommt; daher die 19-jährige Periode von den Griechen „Meton-Jahr" genannt wird.* [... Er spielt] *die Cithara und tanzt vom Frühlingsäquinox bis zum Aufgang der Pleiaden.*

Die Schilderung verbindet ein „sphärisches", möglicherweise auch nur kreisrundes Bauwerk mit einem festlichen Ereignis, das mit dem 19-jährigen Meton'schen Mondzyklus bzw. der 18,6-jährigen Abfolge der Großen und Kleinen Mondwenden zusammenhängt. Allerdings waren zu Diodorus Zeiten die Steine schon lange verlassen, er muß also selbst eine sehr alte Geschichte aus viel älteren Erzählungen übernommen haben. Auch fehlt im archäologischen Fundmaterial in Callanish jeglicher Bezug zu „griechischen" Votivgaben.

Die Lage im äußersten Nordwesten der Britischen Inseln macht die Anlage zu einem im Vergleich zu Stonehenge deutlich weniger besuchten Monument, dafür ist es und die umgebende Landschaft ursprünglicher geblieben.

Eine allfällige astronomisch motivierte Steinsetzung kann im Idealfall *in situ* untersucht und der dortige Himmel beobachtet werden. Seit der Errichtung hat sich die Erdachse etwas aufgerichtet, was zu Verschiebungen der Auf- bzw. Untergangspunkte von Sonne und Mond im Gradbereich geführt hat. Allfällige Sternbeobachtungen am Horizont wären heute aufgrund der Präzession gänzlich anders und müßten mit archäologisch abgesichterter Altersbestimmung rückgerechnet werden.

Derartige Rückrechnungen werden seit Jahrzehnten gemacht und fließen dann in einfache Pläne oder Skizzen ein. Astronomische Bezüge in prähistorischen Monumenten können jedoch mittlerweile in virtuellen Modellen unter der astronomisch korrekten Himmelssimulation von Stellarium dargestellt werden [Zotti 2018, Zotti 2020, Zotti et al. 2021], was die Anschaulichkeit deutlich erhöht und in der Analyse und Vermittlung dieser Erkenntnisse neue Möglichkeiten eröffnet hat.

5.2 Der Mondwendezyklus

Aufgrund der Mondbahnneigung von etwa $5,15^\circ$ und des Umlaufs der Mondbahnknoten im Verlauf von 18,61 Jahren (Abb. 5.2) erreicht der Mond in diesen Zeitintervallen größte nördliche bzw. südliche Deklinationen am Himmel dann, wenn sich Ekliptikschiefe ε und Neigung i addieren (Abb. 5.3). Oft wird angenommen, daß zur Zeit dieser „Großen Mondwende" der Vollmond in seiner südlichsten Deklination beobachtet worden wäre. Allerdings erreicht der Vollmond gar nicht die südlichstmögliche Deklination des Mondes:

Die Mondbahnneigung beträgt, etwas genauer, ca. $5^\circ 9' \pm 8'$ zur Ekliptik mit einer Periode von 173,31 Tagen. Sie erreicht ein Maximum, wenn die Knotenlinie zur Sonne zeigt [Meeus 1997, S. 27]. Zur Großen Mondwende muß der Aufsteigende Knoten Ω im Frühlingspunkt liegen, sodaß sich die Bahnneigungen gegenüber dem Himmelsäquator addieren (Abb. 5.3). Daher sind „Extreme maximale Deklinationen", bei denen die Sonne auch noch nahe Frühlings- bzw. Herbstpunkt steht, tatsächlich nur im Ersten oder Letzten Viertel möglich! Ein Vollmond zur Großen Mondwende erreicht niemals die theoretisch maximale Deklination.

Abbildung 5.4 zeigt den Verlauf der Deklinationen von Sonne und Mond über 20 Jahre, kombiniert mit den 4 Licht-Hauptphasen. Die Ausschnittvergrößerung zeigt den südlichen Ausschnitt um die Große Mondwende 2025. Die wellenförmige Kette der „Umkehrpunkte" des Mondes (also südlichste Deklinationen jedes Mondumlaufs zeigt den 173,31-tägigen Zyklus. Voll- (○) und Neumonde (●) an den Umkehrpunk-

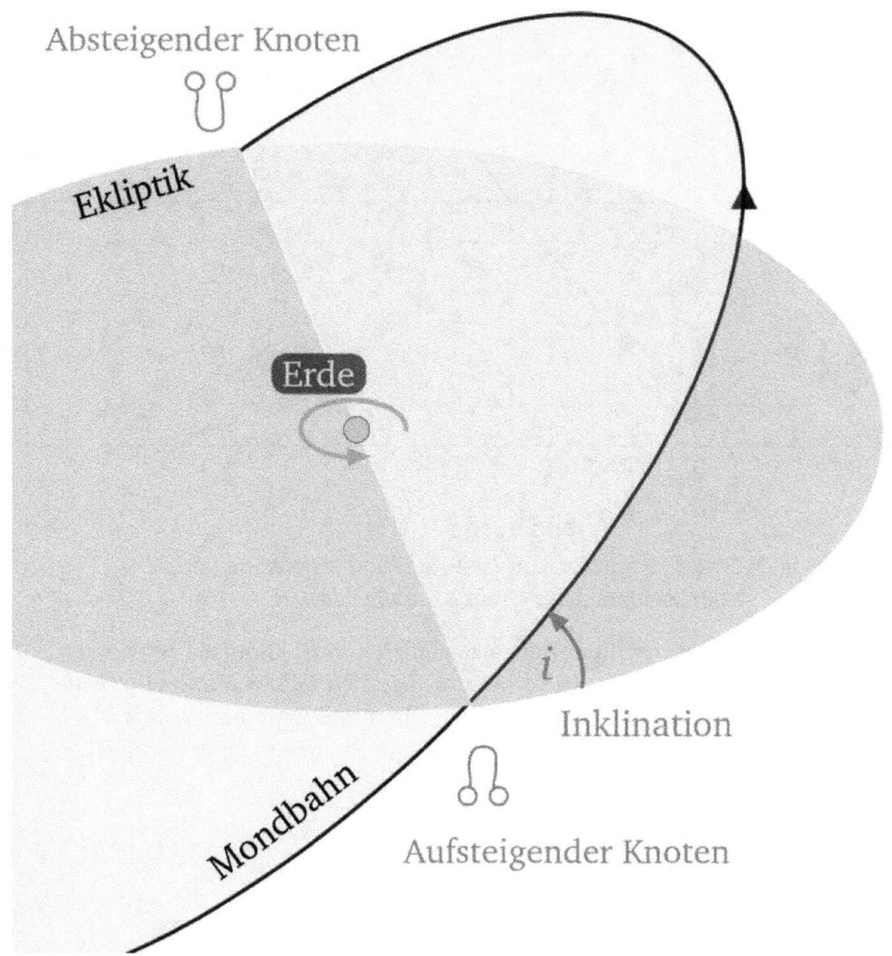

Abbildung 5.2:
Die Mondbahn ist ca. $5,15^\circ$ zur Ekliptik (scheinbare Sonnenbahn) geneigt. Die Schnittpunkte der Mondbahn mit der Ekliptik (Knotenlinie) wandern in 18,6 Jahren einmal vollständig entlang der Ekliptik retrograd herum. Daraus ergeben sich beobachtbare Extremsituationen.

(User:W!B: based on an image created by Urhixidur: Image: Orbit.png – Image:Orbit.svg, CC BY-SA 2.5, `https://commons.wikimedia.org/w/index.php?curid=1818493`)

ten sind auch hier stets abseits der südlichsten Extrema, während erste (◐) und letzte (◑) Viertelphasen oft am südlichen Ausschlag der „Welle“ liegen.

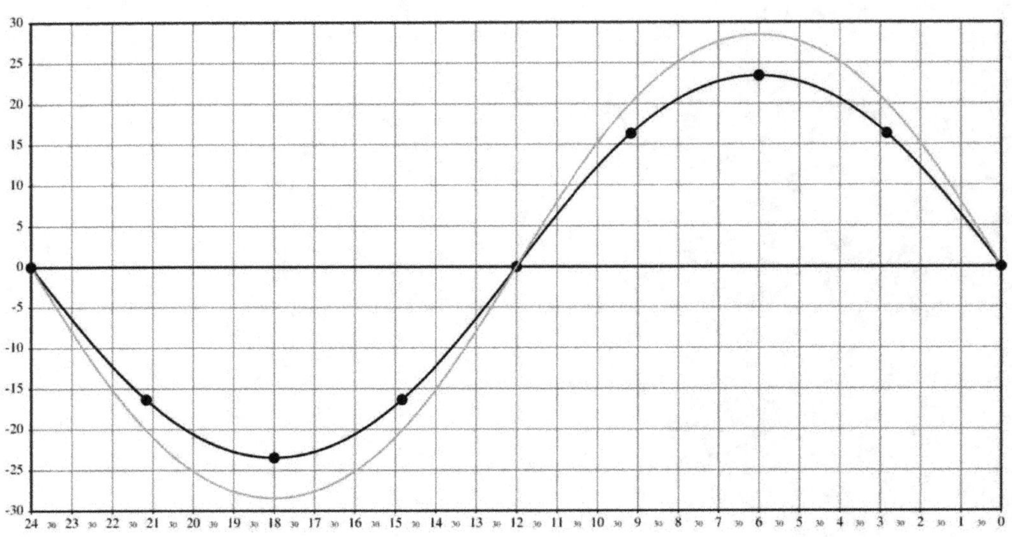

Abbildung 5.3:
Äquatoriales Gradnetz mit Sonnen- (schwarz) und Mondbahn (grün)
zum Zeitpunkt der Großen Mondwende

Der Frühlingspunkt ist rechts. In diese Bahnlage addieren sich ekliptikale Breiten des Mondes und die Schiefe der Ekliptik, der Mond erreicht größte Deklinationen. Im Laufe von 18,6 Jahren verschiebt sich die grüne Linie entlang der Ekliptik nach rechts. Nach der Hälfte dieser Zeit liegt der nördliche Teil der Mondbahn unterhalb des nördlichen Teils der Sonnenbahn: Kleine Mondwende.

(Grafik: Georg Zotti)

5.3 Laserscan

Im Jahre 2005 beauftragte Emma Alistair, einen Laserscan der Steine vorzunehmen. Dieser wurde mit einem Mensi GS200 3D Laser Scanner (Abb. 5.1) durchgeführt. Das Gerät zeichnete etwa 1000 Punkte pro Sekunde auf, und das Gelände um die Steine wurde innerhalb von 3 Tagen von 14 Standorten aus erfaßt. Die Meßpunkte werden mit hoher Winkelauflösung und $\pm 6\,mm$ Genauigkeit in Entfernung erfaßt. Leider war die Qualität der Photos aus diesem Gerät nicht geeignet, daraus ein texturiertes Modell zu erstellen.

Die Standorte des Scanners wurden damals nur mittels mobilem GPS erfaßt, die einzelnen Punktwolken wurden anhand von Referenzobjekten zu einem Gesamtmodell von 50 Millionen Punkten zusammengerechnet [Callanish3D 2023].

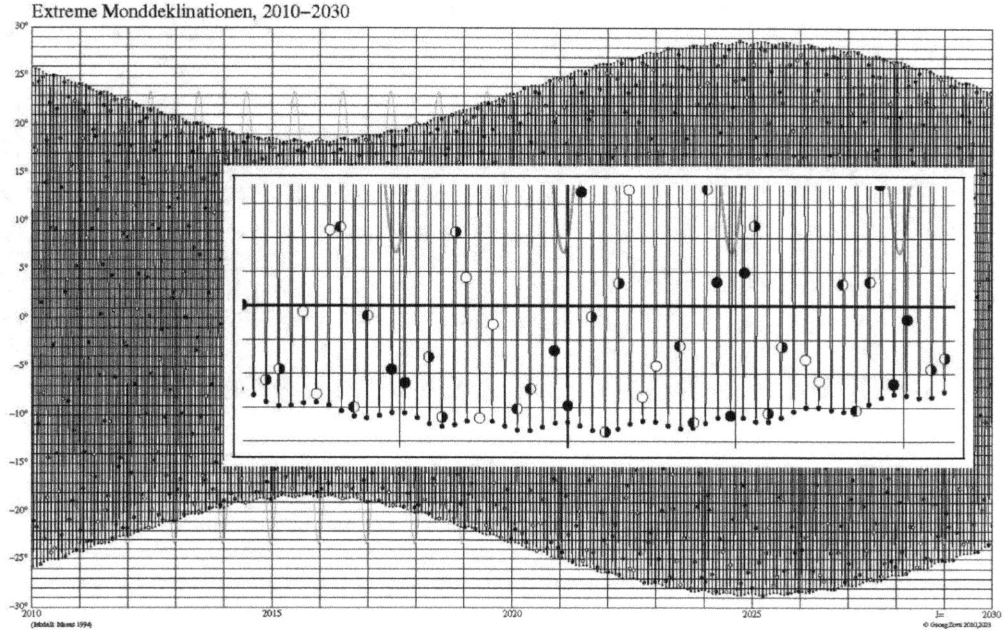

Abbildung 5.4:
Geozentrische Deklination von Sonne (rot) und Mond (schwarz) über mehr als einen Meton-Zyklus. Die Hüllkurve der maximalen Deklinationen verläuft wellenförmig. Die Phasenbildchen an den südlichsten Extrempunkten sind stets Halbphasen.

(Graphik: Georg Zotti nach [Meeus 1994])

5.4 Ein 3D-Modell für Stellarium

Im Sinne einer Nachnutzung existierender hochwertiger, aber nicht speziell für archäoastronomische Simulation erfaßter Daten standen wir nun vor dem Problem, die nicht perfekte Georeferenzierung des Datensatzes zu verbessern. Außerdem umfaßte der Datensatz nur das Gelände unmittelbar um die Steine. In der älteren Literatur wird gerne ein wichtiges Geländedetail vergessen: unmittelbar südlich der Steine ragt ein Felsausbiß, Cnoc an Tursa, aus dem Boden. Dieser versperrt vom Steinkreis aus die Sicht auf den entfernten Südhorizont und bildet einen wichtigen Teil des unmittelbar angrenzenden Geländes, sodaß er auch im 3D-Modell enthalten sein sollte. Ein weiterer Laserscan war jedoch nicht möglich.

Seit kurzem gibt es jedoch wenigstens photogrammetrisch aus der Luft erfaßte Geländedaten mit immerhin 25 *cm* Rasterauflösung [Bluesky 2023](Bluesky gestattete

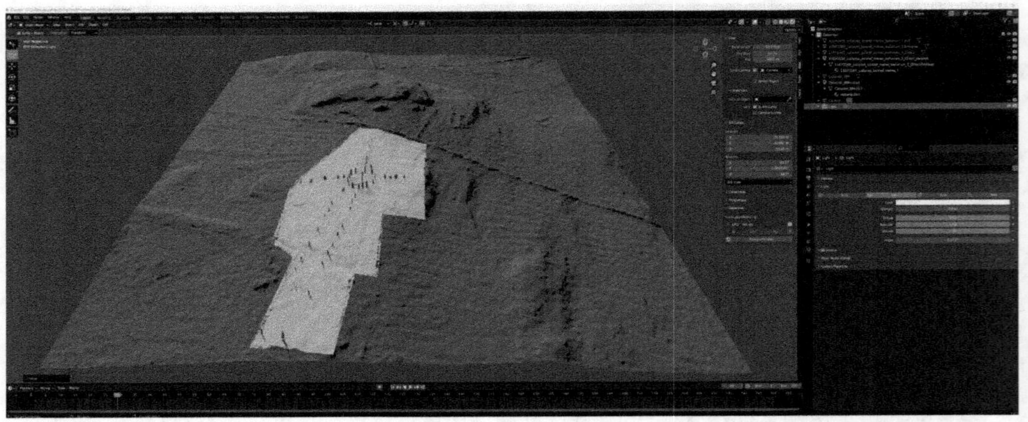

Abbildung 5.5:
Laserscan der Steine von Callanish (grau) eingebettet in das photogrammetrisch erstellte Geländemodell (grün). In Blender wurden die beiden Teilmodelle verbunden und durch wiederholte geringfügige Lagekorrektur des Laserscan-basierten Modells der photographisch ermittelte Winkelfehler allmählich ausgeglichen.

(Foto: G.Z.)

die Verwendung dieser Daten in unserem 3D-Modell). Dies sollte zumindest einen guten Eindruck der Wirkung dieser Felsen auf den Horizontverlauf bringen, und außerdem dient dieser Datensatz, da er aus einer landesweiten luftgestützten Vermessung stammt, den entsprechenden georeferenzierten Untergrund. Die Steine von Callanish sind darin auch rudimentär als geglättete „Stümpfe" zu erkennen. Jetzt galt es noch, das hochaufgelöste Modell mit dem Untergrund zu verbinden.

Im kostenfreien Progamm [CloudCompare 2022] lassen sich zwei vermaschte Modelle gegeneinander angleichen. Das (photogrammetrische) Außengelände wurde als fixe Referenz deklariert, und das innere Modell (aus dem Laserscan) so angepaßt, daß die große Fläche um die Steine möglichst parallel zur selben Fläche im Referenzmodell ist. Die genauen Steinmodelle sollten jetzt mittig in den Stümpfen des Referenzmodells stehen. Eine visuelle Kontrolle machte auch einen guten Eindruck, da aber die beiden Oberflächen eben nicht aus derselben Quelle stammen (und auch etliche Jahre zwischen den Datenaufnahmen lagen, wodurch sich Fehler durch Bewuchsunterschiede einschleichen können), wurde die in beiden Modellen enthaltene Geländeoberfläche nicht ganz zur Deckung gebracht.

Eine erste Einpassung und Verwendung als Stellarium-Modell zeigte daher ein paar kleine Unterschiede in der virtuellen Beobachtung von Stein 8 am Nordende der Avenue zum Erscheinungsbild aus früherer Literatur [Ponting & Ponting 1981]. Jetzt

Abbildung 5.6:
Horizontprofil aus [Ponting & Ponting 1981] überblendet auf einer Serie von Photos von 2023. Nur vom tatsächlichen Standort gesehen paßt das gezeichnete Profil. So konnte der Standort genau vor Stein 8 am Ende der Avenue ermittelt werden

(Graphik: Victor Reijs)

wollten wir versuchen, den Restfehler durch händisches Verschieben des Laserscan-Modells gegenüber dem als korrekt eingepaßt anzunehmenden Geländemodell im Dezimeter- oder gar Zentimeterbereich zu minimieren. Diese weitere Justierung erfolgte im kostenfreien 3D-Modellierprogramm Blender 3.4 (Abb. 5.5).

Zuerst sollten Unterschiede im Erscheinungsbild zwischen Geländehorizont (Cnoc an Tursa) und den Steinen behoben werden. Dazu mußte der genaue Standort von Pontings Profilzeichnung durch Vergleich mit Photos im Gelände ermittelt werden (Victor Reijs & Emma Rennie, Abb. 5.6).

Von diesem Standort aus sollte nun innerhalb der Simulation das Erscheinungsbild der Steine gegenüber dem Hintergrundgelände (Cnoc an Tursa) und dem für den Standort gerechneten Fernhorizont mit den Photos übereinstimmen. Dies gelang nach ein paar Verschiebungen und Verkippungen in Blender ganz gut (Victor Reijs & Georg Zotti, Abb. 5.7).

Abbildung 5.7:
Im Vergleich mit dem Photo vom Standort wurde das 3D-Modell der Steine ebenfalls iterativ angeglichen, bis der Anblick von Stein 8 aus paßte. Photo von 2023 im Vergleich mit Simulation in Stellarium.

(Graphik: Victor Reijs)

5.4.1 Callanish für alle!

Zur Weiterverteilung des Datensatzes wurde eine Website eingerichtet [Callanish3D 2023]. Das 3D-Modell kann heruntergeladen und in Stellarium installiert werden. Während man das Modell erkundet, sollte für den genauestmöglichen Fernhorizont auch je nach Standort (meist Steinkreis oder Nordende der Avenue) ein Hintergrundpanorama („Landschaft" in Stellarium) eingestellt werden. Zusammen mit dem neuen Dämmerungsmodell kann man sich damit einen guten Eindruck verschaffen, wie Steine und Himmel, Sonne, Mond, vielleicht auch Sterne miteinander wirken, sowohl für heutige Beobachter, als auch für die Zeit der Errichtung und ursprünglichen Nutzung (Abb. 5.8).

Ein wichtiger Aspekt darf nicht vergessen werden: Bei 58° nördlicher Breite sank die Sonne im Sommer nur auf etwa −8° unter den Nordhorizont, der Himmel blieb auch um Mitternacht dämmrig. Ein morgens durch die Steine wandernder Mond im Letzten Viertel, oder auch der marginal höhere Sommervollmond gegen Mitternacht, konnte da eine imposante Erscheinung bieten (Abb. 5.9). Eine Beobachtung des niedrigsten Mondes von Frühlingsäquinox bis zum (heliakischen) Aufgang der Plejaden

Abbildung 5.8:
Niedrigster Mond -2275 in Callanish. Der abnehmende Mond kurz nach Frühlingsbeginn erscheint noch einmal („regleam“), nur teilweise sichtbar, zwischen den Steinen im Süden.

(Simulation: Georg Zotti mit Stellarium 23.4)

(aufgrund der flachen Ekliptik am Frühlingsmorgenhimmel und der hohen geographischen Breite erfolgte dieser vermutlich erst etwa 2 Wochen nach die Sommersonnwende!) scheint zu den anfangs beschriebenen alten Legenden zu passen.

5.5 Ausblick: Photographische Justierung

Für den letzten Feinschliff bräuchten wir jedoch weitere astronomische Hilfe. In diesen Jahren um die Große Mondwende 2025 läuft der niedrigste Mond monatlich (Tropischer Mondzyklus, 27,32 Tage) so niedrig über den Himmel von Callanish, daß er, gesehen von Pontings Standort bei Stein 8, bereits 2023 zwischen den Steinen stehen kann. Die Idee war nun, ein zeitlich genau datiertes Bild zu machen, wenn der Mond von einem der Steine teilverdeckt war, und gleichzeitig auch Steine einander teilweise verdecken und so am sichtbaren Horizont ein markantes Identifikationsmerkmal bilden. Aus so einem Bild sollte sich im Vergleich mit dem Erscheinungsbild der im

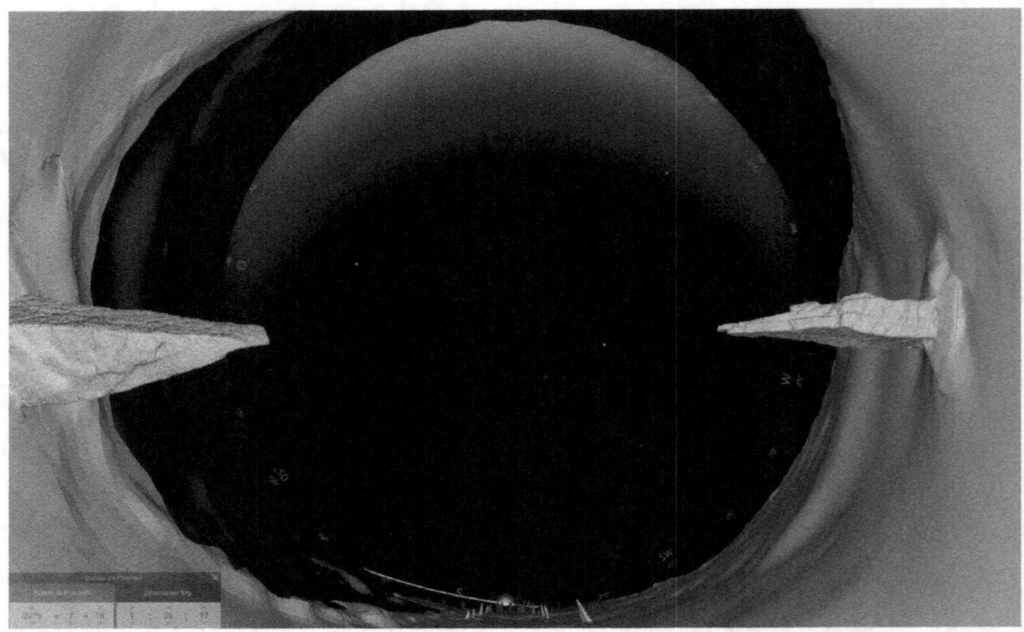

Abbildung 5.9:
Sommernacht -2275 in Callanish. Der Vollmond über dem Steinkreis im Süden, die helle Mitternachtsdämmerung im Norden.

(Simulation: Georg Zotti mit Stellarium 23.4)

Modell einander schneidenden Steine einerseits die Position des Photographen und damit auch der genaue Standort von Pontings Profil finden lassen. Im Anschluß hilft der Mond in Kontakt mit einem der Steine dann andererseits, einen verbleibenden Azimut-Fehler des inneren Steinmodells zu korrigieren. Der Prozeß ist im Detail online beschrieben [Reijs 2023].

Bedauerlicherweise gab es von Mai bis Dezember 2023 nur an einem Tag die Gelegenheit, den Mond auch zu photographieren, und auch da waren die Bilder von Wolken massiv beeinträchtigt, auch konnte leider der Zeitpunkt der Photos nicht zufriedenstellend erfaßt werden, diese Justierung ist also noch nicht abgeschlossen. Dennoch hoffen wir, im Jahre 2024 eine zweite Version der 3D-Szenerie über die Projektwebsite [Callanish3D 2023] veröffentlichen zu können.

Danksagung

Wir danken Gail Higginbottom für wertvolle Hinweise und Diskussion. Ein längerer Artikel ist in Vorbereitung.

Das Ludwig Boltzmann Institut für Archäologische Prospektion und Virtuelle Archäologie (2010–2023) beruhte auf einer internationalen Kooperation der Ludwig Boltzmann Gesellschaft (A), des Amts der Niederösterreichischen Landesregierung (A), der Universität Wien (A), der Technischen Universität Wien (A), der ZAMG – Zentralanstalt für Meteorologie und Geodynamik (A), 7reasons (A), des RGZM Mainz – Römisch-Germanisches Zentralmuseum Mainz (D), des LWL – Archäologie für Westfalen (D), NIKU – Norwegian Institute for Cultural Heritage (N) und Vestfold fylkeskommune – Kulturarv (N).

5.6 Literatur

[Bluesky 2023] *Bluesky* 2023, Website: `https://blueskymapshop.com/` (accessed 2023-11-15).

[Burl 2011] BURL, AUBREY: *Prehistoric Stone Circles, nr 9.* London: Shire Publications (Shire Archaeology) 2005, (4th edition) 2011.

[Callanish3D 2023] *Callanish3D*, 2023, Website: `https://callanish.archaeoptics.co.uk/` (accessed 2023-11-15).

[CloudCompare 2022] *CloudCompare*, 2022, Website: `https://cloudcompare.org/` (accessed 2023-03-05).

[Curtis & Curtis 2006] CURTIS, MARGARET R. & G. RONALD CURTIS: Callanish, 2006. Website: `http://www.archaeocosmology.org/eng/callanish2006.pdf` (accessed 2023-11-25).

[Diodorus Siculus, ed. Oldfather et al. 1935] DIODORUS SICULUS: *The Library of History.* Books 1–32. Translated by C. H. OLDFATHER; C. L. SHERMAN; C. BRADFORD WELLES; RUSSEL M. GEER & F. R. WALTON. LacusCurtius (Loeb Classical Library; Vol. II) 1935. Website: `https://penelope.uchicago.edu/Thayer/E/Roman/Texts/Diodorus_Siculus/2B*.html` (accessed May 8th, 2023).

[Meeus 1994] MEEUS, JEAN: *Astronomische Algorithmen.* Leipzig, Berlin, Heidelberg: Johann Ambrosius Barth (2nd edition) 1994.

[Meeus 1997] MEEUS, JEAN: *Mathematical Astronomy Morsels.* Richmond: Willmann-Bell 1997.

[Ponting & Ponting 1981] PONTING, MARGARET R. & GERALD PONTING: Decoding the Callanish complex: Some initial results. In: RUGGLES, CLIVE L. N. & A. W. R. WHITTLE (eds): *Astronomy and Society in Britain during the Period 4000–1500 BC.* Oxford: BAR Publishing (British Archaeological Reports; nr 88) 1981, p. 63–110.

[Reijs 2023] REIJS, VICTOR: *Using the Calanais I 3D scenery in Stellarium: some experiences when ground proofing.* 2023, Website:

http://www.archaeocosmology.org/eng/Calanais-3Dscenery-experiences.htm (accessed 2024-01-13).

[Scotland 2017] SCOTLAND, HISTORIC ENVIRONMENT: Calanais Standing Stones: Setting Document. Edinburgh: Historic Environment Scotland 2017, Website: https://www.historicenvironment.scot/archives-and-research/publications/publication/?publicationid=2ad491be-9e09-4e7a-ad50-a80700d5e450 (accessed 2023-11-15).

[Scotland 2018] SCOTLAND, HISTORIC ENVIRONMENT: Calanais Standing Stones. Statement of Significance Property in Care ID: PIC280, Historic Environment Scotland. Edinburgh 2018, Website: https://www.historicenvironment.scot/archives-and-research/publications/publication/?publicationid=daf72741-0f80-4d90-8034-a57000c6b0c5 (accessed 2023-11-15).

[Zotti 2018] ZOTTI, GEORG: Computerbasierte Methoden zur kulturastronomischen Landschaftsanalyse. In: WOLFSCHMIDT, GUDRUN (ed.): *Baudenkmäler des Himmels – Astronomie in gebautem Raum und gestalteter Landschaft.* Proceedings der Tagungen der Gesellschaft für Archäoastronomie 2013–2016. Hamburg: tredition (Nuncius Hamburgensis; Vol. 35) 2018, chapter 11, p. 202–231.

[Zotti 2020] ZOTTI, GEORG: GIS, Landschaft, 3D-Modelle und Himmelssimulation. In: WOLFSCHMIDT, GUDRUN (ed.): *Maß und Mythos, Zahl und Zauber: Die Vermessung von Himmel und Erde.* Proceedings der Tagung der Gesellschaft für Archäoastronomie in Dortmund 2018. Hamburg: tredition (Nuncius Hamburgensis; Vol. 48) 2020, chapter 18, p. 296–305.

[Zotti et al. 2021] ZOTTI, GEORG; HOFFMANN, SUSANNE M.; WOLF, ALEXANDER; CHÉREAU, FABIEN & GUILLAUME CHÉREAU: The simulated sky: Stellarium for cultural astronomy research. In: *Journal for Skyscape Archaeology* **6** (December 2020), 2, p. 221–258, doi:10.1558/jsa.38690.

Abbildung 6.1:
Rückblick vom Stein Nr. 41 zur Oberkaser Alm
und zum Nordpunkt am Schwarzkopf

(© Roland Gröber)

Höhenheiligtum am Pfitscher Sattel, Bedeutung und Entwicklung

Roland Gröber (Leverkusen)

Abstract: High altitude shrine on the Pfitscher Sattel, significance and development

At previous conferences of the Society for Archaeoastronomy I have reported, with different emphases, on the cup-marked stones at the Pfitscher Sattel. Especially the so-called "star plate" with numerous star bearings, which allowed a dating to 2450 BC, and the representation of 18 concrete constellations by cups were in the focus of the considerations.

This contribution will present some hitherto neglected findings. They show that the site is not "only" an astronomical observatory but also a cult site or a high altitude sanctuary. Already the access to the saddle, past some signposts and through a prominent gate point to the special nature of the site from the early Bronze Age. In addition to the special construction of the hexagonal enclosure with a standard dimension and Pythagorean triangles, a circular and an elliptical stone setting bear witness to mythical rites. A prehistoric script, a sacrificial stone with a "blood gutter", special "cult rooms" and a burnt-offering site proven by archaeological excavations complete the complex. Further observations in the immediate vicinity, some of them with astronomical references, provide clues to a possible development of the cult site. Together with the astronomical findings mentioned earlier, the cult site at the Pfitscher Sattel is an outstanding example of the culture of the Early Bronze Age and of particular importance for South Tyrol, and beyond.

Zusammenfassung

Auf früheren Tagungen der Gesellschaft für Archäoastronomie habe ich, mit verschiedenen Schwerpunkten, über die Schalensteine am Pfitscher Sattel berichtet. Vor allem die sog. „Sternplatte“ mit zahlreichen Sternpeilungen, die eine Datierung auf 2450 v. Chr. ermöglichte, und die

Darstellung von 18 konkreten Sternbildern durch Schalen standen im Zentrum der Betrachtungen.

In diesem Beitrag sollen einige bisher vernachlässigte Erkenntnisse vorgestellt werden. Sie zeigen, dass die Anlage nicht „nur" ein astronomisches Observatorium sondern auch eine Kultstätte bzw. ein Höhenheiligtum ist. Bereits der Zugang zum Sattel, an einigen Wegweisern vorbei und durch ein markantes Tor weisen auf das Besondere der Örtlichkeit aus der frühen Bronzezeit hin. Neben der speziellen Konstruktion der 6-eckigen Anlage mit einem Standardmaß und pythagoräischen Dreiecken zeugen eine kreisförmige und eine elliptische Steinsetzung von mythischen Kulthandlungen. Eine urzeitliche Schrift, ein Opferstein mit „Blutrinne", spezielle „Kulträume" und ein durch archäologische Grabungen nachgewiesener Brandopferplatz ergänzen die Anlage. Weitere Beobachtungen in umittelbarer Umgebung, z. T. mit astronomischen Bezügen, liefern Hinweise auf eine mögliche Entwicklung des Kultplatzes.

Zusammen mit den schon früher erwähnten astronomischen Erkenntnissen ist die Kultstätte am Pfitscher Sattel ein herausragendes Beispiel zur Kultur der frühen Bronzezeit und für Südtirol, und darüber hinaus, von besonderer Bedeutung.

6.1 Einleitung

Auf früheren Tagungen habe ich, mit verschiedenen Schwerpunkten, über Schalensteine am Pfitscher Sattel in der Texelgruppe bei Meran berichtet.[1] Vor allem die astronomischen Erkenntnisse auf der sog. „Sternplatte" mit zahlreichen Sternpeilungen, die eine präzise Datierung auf 2450 v. Chr. ermöglichte, und die konkrete Darstellung von 18 Sternbildern durch Schalen standen im Zentrum der Betrachtungen.[2] Im Band 42 wurde die Konstruktion von megalithischen Steinsetzungen am Beispiel des Höhenheiligtums am Pfitscher Sattel in der Texelgruppe thematisiert. Und im Band 51 wurde untersucht, ob es Zusammenhänge zwischen Schalensteinen und alpinen Brandopferplätzen gibt.

Im Folgenden sollen einige bisher vernachlässigte und neue Erkenntnisse vorgestellt werden. Sie zeigen, dass die Anlage nicht „nur" ein bedeutendes astronomisches Observatorium sondern darüber hinaus auch ein bemerkenswertes Höhenheiligtum ist.

6.2 Der Weg zum Höhenheiligtum

Schon am Fuß der Texelgruppe gibt es zahlreiche Schalensteine, und an allen Zugängen zur Kultstätte am Pfitscher Sattel auf 2150 m Höhe sind etwa auf halber Höhe ei-

1 Gröber (Nuncius Hamburgensis; Band 35) 2018, S. 26–42. Gröber (Nuncius Hamburgensis; Band 42) 2019, S. 24–35. Gröber (Nuncius Hamburgensis; Band 51) 2020, S. 96–114.

2 Gröber (Nuncius Hamburgensis; Band 35) 2018, S. 26–42.

Abbildung 6.2:
a: Panoramabild der Texelgruppe mit „Wächter-Schalensteinen"
an den Zugängen zum Pfitscher Sattel (roter Pfeil)
Oberkaser Alm und Silex-Splitter, die dort in unmittelbarer Nähe gefunden wurden.

(© Gemeinde Dorf Tirol mit Bearbeitung Roland Gröber, © Roland Gröber)

nige Schalensteine als „Wächter“ oder „Wegweiser“. Bei einigen wurde eine „Schrift“ interpretiert. Die Wegweiser sind im engen Zusammenhang mit der höher gelegenen Kultstätte zu sehen (Abb. 6.2a).

Der klassische Weg führt von Dorf Tirol durch das Spronser Tal hinauf zur Oberkaser Alm. Nach einem etwa dreistündigen anstrengenden Aufstieg über rund 1000 Höhenmeter ist die Alm ein willkommener Rastplatz zur Stärkung, zur Übernachtung und als Stützpunkt für die Forschungen am Pfitscher Sattel. Im Bereich der Alm fand man im vergangenen Jahrhundert zahlreiche Silex-Splitter, die auf einen mittelsteinzeitlichen Jägerrastplatz hindeuten. Und dort könnte man auch möglicherweise auf frühe, vielleicht nur zeitweilige Siedlungsspuren treffen (Abb. 6.2b).

Etwas oberhalb der Alm zeigt der Blick eine karge, baumlose Landschaft an den unteren Spronser Seen. Der Pfitscher Sattel, liegt nahe am kleineren See, der Pfitscher Lacke. Von der Oberkaser Alm führt ein ebener Weg direkt an der Pfitscher Lacke entlang hinüber zum Sattel. Dabei kommen wir an drei großen Felsblöcken vorbei. Zwei davon bilden ein markantes Tor. Ein stimmungsvoller Zugang zur Kultstätte (Abb. 6.3).

Aribert Egen (1995), auf dessen grundlegenden Forschungen am Sattel einige meiner Erkenntnisse beruhen, glaubt hier einen „heiligen Weg“ mit den typischen Fixpunkten aus der griechischen Mythologie zu erkennen. Und es ist in der Tat ein eigenartiges Gefühl, wenn man durch das Tor entlang des Sees hinüber zum Sattel wandert (Abb. 6.4).

6.3 Im Außenbereich der Kultstätte

Aus der Sicht von Google Earth[3] führt der Weg genau über einen Schalenstein (Punkt 5 in Abb. 6.4), vorbei zu einem markanten Steinkreis etwas außerhalb einer Steinmauer. Die Umfassungsmauer der gesamten Anlage ist im Satellitenbild noch gut zu erkennen.

Der Blick vom Schalenstein Nr. 41 zurück geht zum Schwarzkopf, über dessen rechtem Gipfelkamm genau Norden ist. Er spielte bei den astronomischen Betrachtungen zum Nordpunkt eine wichtige Rolle, da am Kamm die Verschiebung der Aufgangspunkte des Großen Wagens über die Jahrhunderte zwischen 3000 und 2200 v. Chr. durch Bedeckung erkannt werden konnte[4] (Abb. 6.1).

Der Schalenstein enthält neben einigen runden auch mondförmige Schalen. Er ist Teil einer Visiereinrichtung zum Horizont zur Beobachtung der großen nördlichen Mondwende.

3 Koordinaten Sternplatte: $(46°43'38,26''N, 11°06'10,50''O)$.

4 Gröber (Nuncius Hamburgensis; Band 35) 2018, S. 38.

Abbildung 6.3:
a: Blick auf die unteren Spronser Seen mit Pfitscher Sattel
b: Felstor am Weg zum Pfitscher Sattel

(© Roland Gröber, © Roland Gröber)

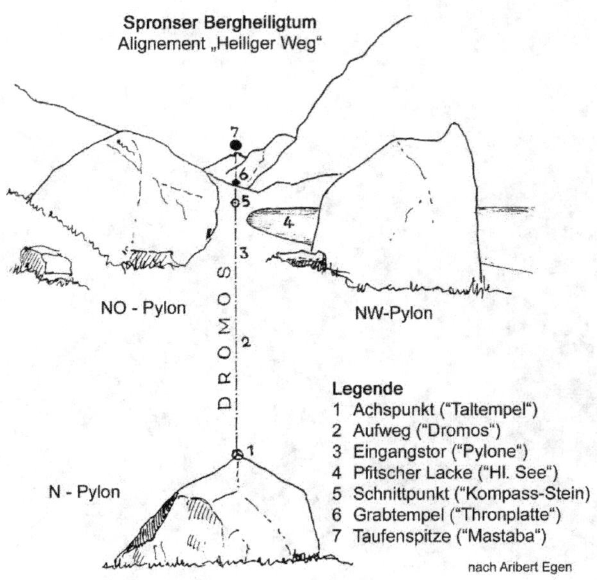

Abbildung 6.4:
„Heiliger Weg" zum Pfitscher Sattel

(© Aribert Egen)

Der englische Forscher Prof. Alexander Thom (1894–1985), hat hunderte von Steinkreisen vermessen. Er vertrat die These, dass Sonnen- und Mondfinsternisse nur in den Extremstellungen des Mondes möglich sind. Daher war es wichtig für die Vorhersage einer Finsternis diese „Gefahrenzeiten" zu kennen und deshalb die Maxima einer Lunation zu bestimmen. Burkhard Steinrücken wies in seinem ausführlichen Artikel über die Sonnen- und Mondwenden nach, dass die These von Thom zur Vorhersehbarkeit von Finsternissen auf vereinfachenden Annahmen beruht und so nicht haltbar ist.[5]

Trotzdem ist festzuhalten, dass es mehrere Hinweise auf die Beobachtung der verschiedenen Mondwenden am Pfitscher Sattel gibt.

Die Verlängerung des „heiligen Wegs" führt nach wenigen Metern zu einem Steinkreis, der außerhalb an die Umfassungsmauer grenzt (Abb. 6.5a).

Nach Thom sind megalithische Steinsetzungen häufig mit einem Einheitsmaß angelegt worden. Vermutlich auch hier. Der Steinkreis hat einen Durchmesser von 3,35 m, was 4 megalithischen Ellen (ME) von je 0,835 m entspricht. Die megalithischen Ellen

5 Steinrücken, Burkhard: Sonnenwenden und Mondwenden (2011).

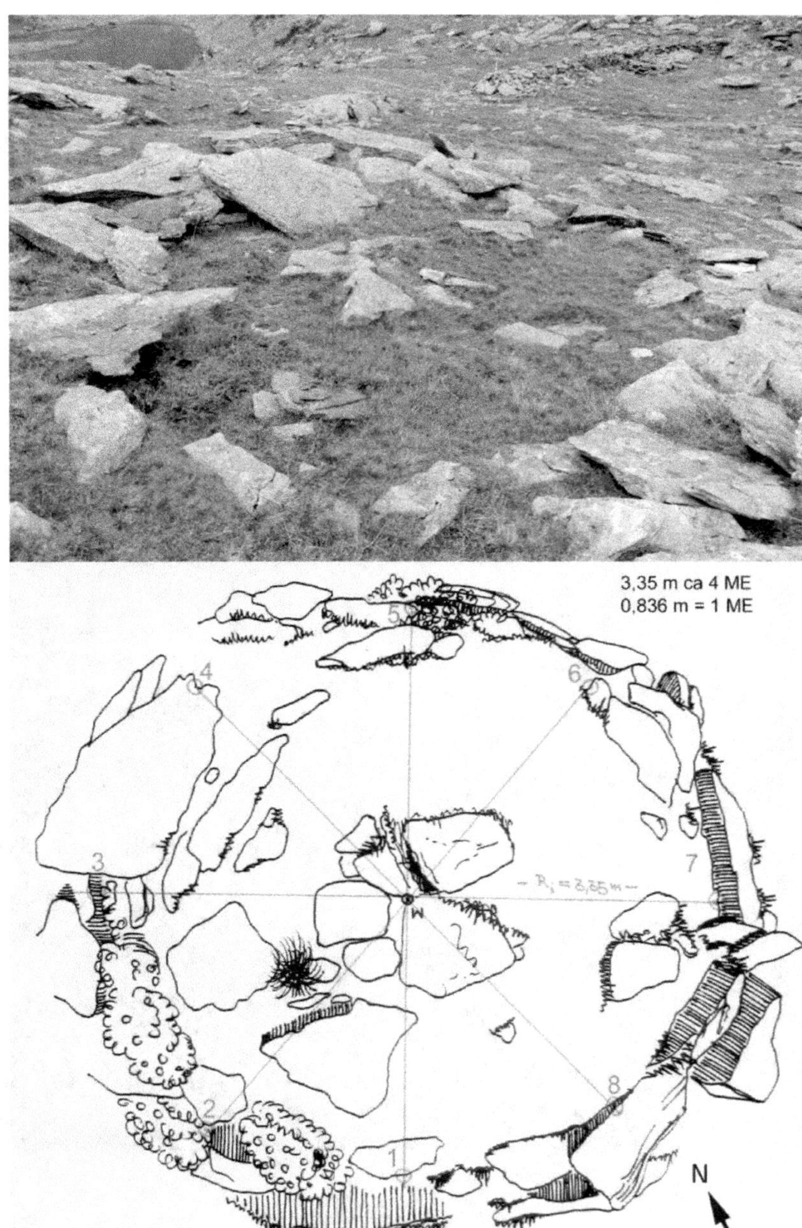

Abbildung 6.5:
a: Steinkreis außerhalb der südwestlichen Umfassungsmauer
b: Skizze des Steinkreises mit Abmessungen in Megalithischen Ellen

(© Roland Gröber, © Aribert Egen)

werden uns an anderer Stelle noch etwas ausführlicher begegnen (Abb. 6.5b). Man kann hier zu Recht einwenden, dass der Durchmesser beliebig festgelegt und auf die ME „hingebogen“ wurde. Im Zusammenhang mit anderen Steinsetzungen der Anlage, vor allem der Umfassungsmauer, erscheint die Annahme eines Einheitsmaßes zumindest möglich. Die Bedeutung des Steinkreises ist unklar. Vielleicht ein besonders privilegierter und geschützter Kultort.

Folgt man dem gedachten „heiligen Weg“ etwa 10 m weiter, erreicht man eine markante Felsnische. Auf einer nach Norden geneigten Platte ist eine mondförmige Einkerbung sichtbar. Von diesem Platz aus ist in südöstlicher Richtung eine auffallende senkrecht stehende und verkeilte Spitze zu sehen. Sie weist durch die Felskante im Vordergrund auf die kleine südliche Mondwende hin (Abb. 6.6).

Abbildung 6.6:
Peilung zur kleinen südlichen Mondwende

(Foto: Roland Gröber, Skizze: © Aribert Egen)

Da der Schalenstein mit der Peilung zur großen nördlichen Mondwende, der erwähnte Steinkreis und die vorliegende Peilung zur kleinen südlichen Mondwende noch außerhalb der Anlage liegen, könnte dies ein Hinweis sein, dass diese Beobachtungspunkte noch vor der Entstehung der Anlage, oder zumindest sehr früh entstanden sind.

6.4 Die Kultstätte, Umfassungsmauer und Innenbereich

Wenden wir uns der eigentlichen Anlage zu. Deren Konstruktion wurde bereits in Nuncius Hamburgensis; Band 42 ausführlich beschrieben.[6] Hier nur kurz die wesentlichen Ergebnisse:

Die Umfassungsmauer der Anlage ist eine Trockenmauer aus aufgeschichteten Felsplatten. Sie ist im nördlichen Teil noch sehr gut erhalten und etwa 1,50 m hoch. Im südlichen Teil ist sie durch einen Bergsturz stark gestört. Vermutlich war sie in diesem Teil wohl nie komplett geschlossen.

Die Anlage wurde nicht willkürlich sondern nach einem vorbestimmten Plan errichtet. Wichtige Konstruktionsmerkmale sind rechtwinklige Dreiecke, deren Seiten, bei Anwendung eines Einheitsmaßes (megalithische Elle 0,835 m) ganzzahlig, idealerweise pythagoreisch, sind. Im Grundriss der Anlage sind mehrere astronomische Ausrichtungen enthalten. Die Erkenntnisse erfüllen weitgehend die Regeln von Alexander Thom, der mehr als 300 Steinsetzungen genau vermessen hat. Sie können daher als Nachweis einer prähistorischen Umfassungsmauer gelten (Abb. 6.7).

Nahe der Südostecke innerhalb der Anlage fällt eine elliptische Steinsetzung auf (Abb. 6.8a). Bemerkenswert ist, dass vier der direkt in der Ellipse liegenden Steine Schalen besitzen. Stein Nr. 4 mit 186 und Stein Nr. 5 mit 158 Schalen zählen zu den vier reichhaltigsten Schalensteinen. Stein Nr. 4, der für mich der schönste Schalenstein am Pfitscher Sattel ist, wird uns noch in einem anderen Zusammenhang begegnen.

Der Versuch, die Steinsetzung durch eine Ellipse zu beschreiben, liefert etwa folgendes Bild. Wendet man die Regeln von Thom an, dann sollte die große Halbachse a, die kleine Halbachse b und die Exzentrizität e rechtwinkelige Dreiecke bilden, was sie, wie die Skizze zeigt in der Tat auch tun. Misst man jedoch die Strecken, dann ergibt sich für

a = 5,43 m = 6,49 ME; b = 3,90 m = 4,66 ME und e = 3,71 m = 4,51 ME

Also weder ganzzahlige ME noch pythagoreische Dreiecke. Nun könnte man ja versuchen, die Ellipse entsprechend „anzupassen“. Auch dieses bringt keine Lösung nach den Regeln von Thom. Seine Thesen treffen offensichtlich nicht immer auf alle Steinsetzungen zu. Umso bemerkenswerter ist es, dass die Umfassungsmauer sehr gut in das Schema passt.

6 Gröber (Nuncius Hamburgensis; Band 42) 2019, S. 24 ff.

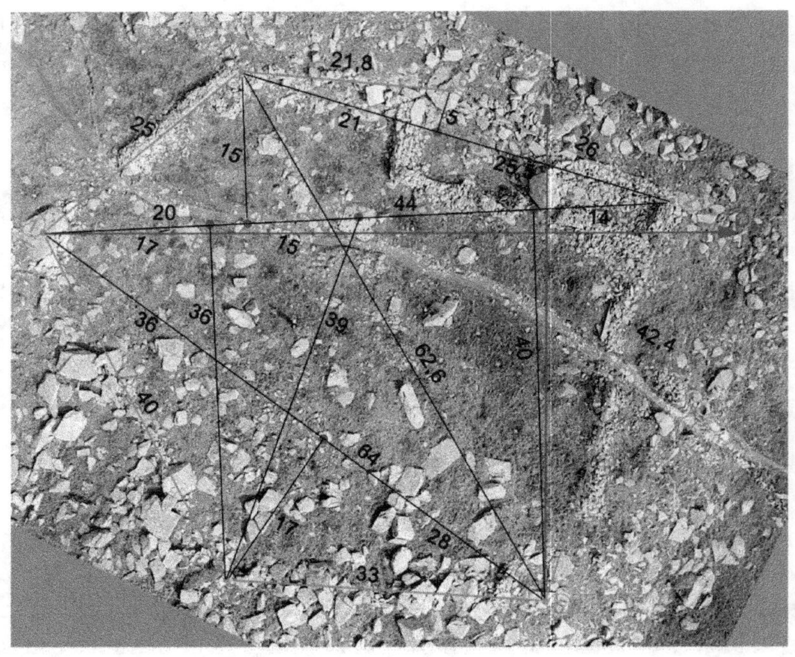

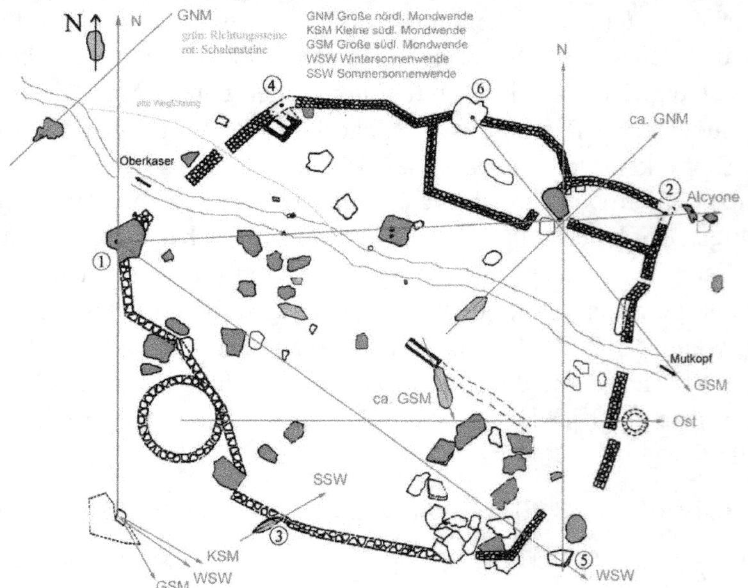

Abbildung 6.7:
a: Konstruktion der Umfassungsmauer mit geradzahligen rechtwinkligen Dreiecken
b: Gesamtübersicht der Anlage mit astronomischen Peilungen

(© Roland Gröber, © Roland Gröber)

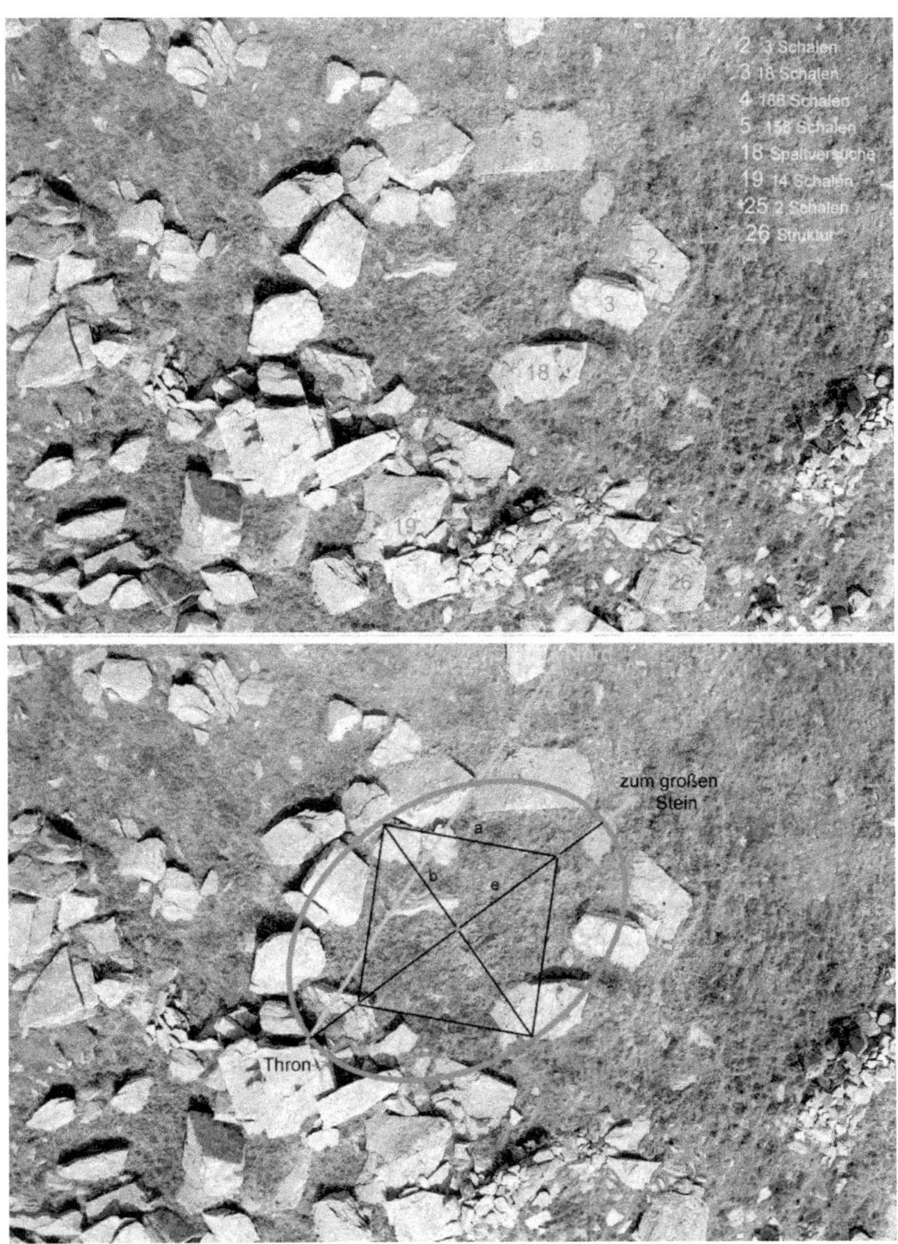

Abbildung 6.8:
a: Elliptische Steinsetzung mit mehreren Schalensteinen
b: Elliptische Steinsetzung, Versuch des Nachweises der megalithischen Regeln nach Alexander Thom

(© Roland Gröber, © Roland Gröber)

Im südlichen Bereich der elliptischen Steinsetzung ist ein vielleicht bearbeiteter Stein, der als bequemer Sitzplatz dienen konnte. Die Bezeichnung Thron suggeriert hier, dass dieser Sitzplatz eines Priesters gewesen sein könnte und das Innere der Steinsetzung innerhalb der gesamten Anlage nochmals ein besonders geschützter Raum war – sehr spekulativ, aber nicht unmöglich (Abb. 6.8b).

Die Ellipse ist mit der großen Achse genau auf den „großen Stein“ ausgerichtet. Ihm und seiner Umgebung soll das weitere Augenmerk dienen.

6.5 Der „große Stein“ und die Kammern

Der „große Stein“ ist in mehrerer Hinsicht bemerkenswert. Vom Eckpunkt 5 weist eine Verbindungslinie über die scharfe Kante am „großen Stein“ zum Kamm des Schwarzkopfgipfels, dem bereits erwähnten Nordpunkt der Anlage. Die Kante am Stein ist genau nach Süden ausgerichtet. Sie hat wahrscheinlich noch eine weitere Bedeutung (Abb. 6.9).

Etwa in der Mitte der nach Süden geneigten Dachfläche befindet sich neben zwei Schalen eine fast nach NS ausgerichtete Rinne. Gießt man im oberen Teil der Rinne in eine Vertiefung Flüssigkeit, fließt diese genau über die vorspringende Kante. Zum Eingießen der Flüssigkeit ist auf der nördlichen Seite des Steins ein speziell hergerichteter, bequemer Standplatz (rechts). Ein Flüssigkeits- Kult ist daher denkbar (Abb. 6.9b).

Möglicherweise ist dieser astronomisch ausgerichtete Kultstein nicht einzigartig. Etwa 60 km Luftlinie in nordöstlicher Richtung befindet sich im Viggartal in Tirol – einem Seitental des Wipptals nördlich zum Brenner – ein mächtiger Monolith mit ca. 60 Tonnen Gewicht und etwa 6 m Höhe. Er wird als „B'schriebener Stoa“ bezeichnet.[7] Auch bei ihm gibt es eine genau nach Süden geneigte Dachfläche die von einer Rinne durchzogen wird (in Abb. 6.9a eingeklinkt). Der „B'schriebene Stoa“ enthält Felszeichnungen und auf der Dachfläche eine Schale. Er steht im Zentrum eines frühzeitlichen Übergangs mit zahlreichen Schalensteinen als Wegweiser. Die Ähnlichkeit ist sicher nicht zufällig.

Der „große Stein“, ca. 2,50 m hoch, trennt die schmalen Eingänge in Kammer I und Kammer II. Der Zweck der Kammern ist unklar. Wenn das ermittelte Alter der Anlage von 4250 Jahren, aufgrund der Verschiebung der Sternaufgänge durch die Präzession der Erde zutrifft,[8] ist es unrealistisch eine unveränderte Situation zu erwarten. Trotzdem zeigen die ermittelten Eckpunkte der Anlage und einige Peilungen, dass die Grundstruktur der Mauer wahrscheinlich weitgehend erhalten ist.

7 Walli-Knofler: Schalensteinpfade in Tirol, 2019/21.

8 Gröber (Nuncius Hamburgensis; Band 35) 2018, S. 34.

Abbildung 6.9:
a: Der Große Stein mit nördlicher Ausrichtung zum Schwarzkopf
b: Flüssigkeits-Rinne am großen Stein

(a: © Roland Gröber, eingeklinktes Bild: Der „B'schriebene Stoa“ im Viggartal, © Thomas Walli-Knofler, b: © Roland Gröber)

Steht man am Fuß des Felsens der den Eckpunkt 6 bildet und blickt durch den schmalen Eingang zur Kammer II, dann fällt im Vordergrund der sichelförmige Felsen am Boden auf. Da diese Richtung aber genau mit der Peilung zur GSM zusammen fällt, mag man nicht nur an einen Zufall glauben (Abb. 6.7b).

Über die Bedeutung und Nutzung der beiden Kammern kann man derzeit nur spekulieren. Solange die archäologische Fachwelt an der Meinung einer neuzeitlichen Hirtenunterkunft festhält, ist die Forderung nach einer archäologischen Grabung innerhalb der Mauern zwar wünschenswert, aber illusorisch.

Die Abb. 6.7b ermöglicht eine Gesamtübersicht. Interessant ist, dass die Vorderkante des „großen Steins" offenbar ein mehrfacher Schnittpunkt ist. Peilungen zu verschiedenen astronomischen Ereignissen am Horizont, und auch die Kardinalrichtungen in der Anlage zeigen deutlich, dass der Kultplatz gezielt nach astronomischen Gesichtspunkten errichtet wurde.

Links der Kammer II liegt die sog. Archivplatte. Sie ist immer noch ein großes Geheimnis. Die 169 runden und halbmondförmigen Schalen, sind teilweise in Gruppen angeordnet und durch Rinnen verbunden. Ihre Bedeutung ist, trotz verschiedener Ansätze, nach wie vor unklar, z. B. ist eine spezielle Markierung der Nullpunkt eines Längenmaßstabes mit 0,5–1–1.5–2 ME.[9]

Zwischen der „Archivplatte" und der nördlichen Mauer wurden 1990 archäologische Grabungen von Paul Gleirscher durchgeführt.[10] In etwa 50 cm Tiefe fand man Holzkohlenreste in einer künstlichen Grube. In anderen Suchschnitten wurden eine weitere aschehaltige Schicht, eine Silexklinge und zwei Randscherben von endbronzezeitlichen Gefäßen gefunden. Einige Schalensteine waren teilweise in die Brandschicht einbezogen. Da man keine kalzinierten Knochen fand, ist eine Deutung als Brandopferplatz sehr unsicher. Die Feuerstellen könnten auch Signalfeuer gewesen sein. Damit wäre eine Verbindung des Jahreskalenders durch Sternbeobachtungen mit den von hier aus sichtbaren Brandopferplätzen denkbar.[11]

Zurück zur bereits mehrfach erwähnten „Sternplatte". Sie ist einer der reichhaltigsten und nach bisherigen Erkenntnissen der bedeutendste Schalenstein am Pfitscher Sattel.

Abb. 6.10a zeigt die wichtigsten astronomischen Ergebnisse. Die Peilungen zum Osthorizont auf Sternaufgänge einiger heller Sterne um etwa 2450 v. Chr. und die Verbindung der Peilsterne mit den zugehörigen Sternbildern sind einzigartig. Nicht dargestellt sind die ermittelten Peilungen zu den Sonnen- und Mondwenden am Osthorizont. Die im Nuncius Hamburgensis; Band 35,[12] ausführlich erläuterten astrono-

9 Gröber (Nuncius Hamburgensis; Band 35) 2018, S. 32, Abbildung.

10 Gleirscher 1993.

11 Gröber (Nuncius Hamburgensis; Band 51) 2020, S. 96 ff.

12 Gröber (Nuncius Hamburgensis; Band 35) 2018, S. 33 ff.

Teilansicht (Westen) der Sternenplatte am Pfitschersattel

Abbildung 6.10:
a: „Sternplatte“ mit den wichtigsten astronomischen Erkenntnissen
b: Schriftdeutung auf der „Sternplatte“

(© Roland Gröber, © Herbert Kirnbauer)

mischen Aussagen sind jedoch nicht das Thema. Ein anderer Grund führt uns zur „Sternplatte".

Weil die zentrale Peilschale von Jahr zu Jahr immer wieder zugewachsen war, habe ich 2014 vorsichtig die Grasnarbe entfernt und zu meiner Überraschung vier weitere Schalen gefunden. Besonders faszinierend war der kleine Felsabsatz der mitten durch die zentrale Peilschale geht und genau nach Norden weist.

Die zusätzlich gefundenen vier Schalen ermöglichten, zusammen mit den Schalen im Bereich der mäanderförmigen Linien, die nichts mit den Sternpeilungen zu tun haben, eine weitere Interpretation einiger Schalen.

6.6 Schriften auf einigen Schalensteinen

Die Bedeutung der Schalen wird oft mit einer „Urschrift" in Verbindung gebracht. Ein Vergleich moderner „Schriften", die durch ähnliche Symbole wie die Schalen gekennzeichnet sind, legt diese Verbindung nahe. In der Blindenschrift bilden sechs „Beulen" je nach Lage die Zeichen, die mit den Fingern ertastet werden. Im Fernschreibcode können durch 6 Löcher in einem Papierstreifen 64 Zeichen codiert werden. Sie sind die Vorläufer der Bits in der Informationstechnik, für Schrift, Bilder, Musik, Videos usw. Die Tifinagh-Schrift ist heute noch in Nordafrika bei den Tuaregs zu finden. Sie kam schon in der Bronzezeit von den „Nordmeerleuten" nach Nordafrika. Diese „Urschrift" wurde auf zahlreichen Schalensteinen, neben den astronomischen Informationen, von den Schriftforschern Prof. Bary Fell (1917–1994), Dr. Dietrich Knauer, Herbert Kirnbauer und anderen identifiziert und in eine „Ursprache" übersetzt.[13] Anhand der vielen in sich schlüssigen Beispiele sind die Erkenntnisse der „Schriftgelehrten" verblüffend und plausibel. Allerdings ist hier viel Erfahrung und Übung notwendig, da z. B. die „Schreiber" der Urzeit weder eine einheitliche Schreibrichtung noch „normierte Buchstaben" hatten.

Derzeit gibt es für drei von 29 Schalensteinen Erklärungen für mögliche Texte. Die übrigen warten noch auf eine Entzifferung oder einen Ausschluss als „Textsteine". Herbert Kirnbauer lieferte mir auf meine Bitte eine plausible Interpretation der „Sternplatte": Text A: *Da (geht) Weg hin* und Text B: *Da Weg hin zum Wasser* (Abb. 6.10b). Ob hier die Pfitscher Lacke oder der nahe Wasserfall gemeint ist, ist offen. Vielleicht deuten auch die Mäander-Linien auf das Wasser hin.

Ein weiteres Beispiel ist der Schalenstein Nr. 19, der direkt wie ein Pult neben dem „Thron" an der elliptischen Steinsetzung liegt (Abb. 6.11). Auch hier liefert Herbert Kirnbauer eine schlüssige Erklärung. Sie lautet: *Erwach am Weg im Grab* oder durch eine möglicherweise spätere Einfügung von kleineren Schalen im unteren Teil des Textes: *Erwach am Weg durchs Grab*. Bei einigen Zeichen (Wörtern) werden, wie in

13 Kirnbauer: Steinzeitcode, 2012. Knauer: Rätsel der Felsbilder und Schalensteine, 1987/2002.

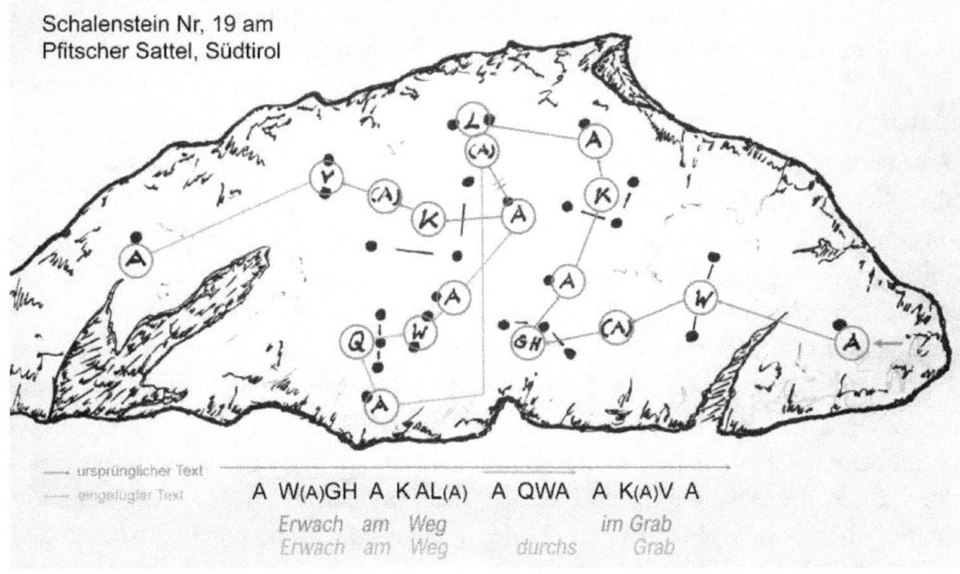

Abbildung 6.11:
a: Der Schalenstein Nr. 19, neben der elliptischen Steinsetzung
b: Die Schriftdeutung auf dem Schalenstein Nr. 19

(© Roland Gröber, © Roland Gröber, nach Herbert Kirnbauer)

alten Sprachen (Aramäisch, Arabisch, Etruskisch etc.) üblich, häufig Selbstlaute (wie hier das (A)) zwar gesprochen, aber nicht immer geschrieben.

Herbert Kirnbauer ergänzt die Interpretation: „*Dieser ‚Klartext-Wunsch‘ nach einer Reinkarnation oder Wiedergeburt nach dem Tod findet sich fast deckungsgleich in der Anwendung der diversen Schalen-Buchstabenmuster auf vielen, vielen Steinartefakten in den Alpen – und sonst wo! Nicht nur ein Schalenstein zeigt das artgleiche (unverwechselbare!) Schriftbild mit dem identen Wunsch nach einem Leben danach. Ob nun das ‚Erwachen auf dem Weg durch das Grab‘ auf dem Schalenstein vom Pfitscher Sattel tatsächlich zu einem menschlichen Totenritual gehört hat, lässt sich nicht eindeutig verifizieren. Denn es könnte ja auch die ‚Wiedergeburt‘ der Sonne nach ihrem Verschwinden und ihrem anschließenden dreitägigen Tod zur Wintersonnenwende kultisch erfleht worden sein – wie auch die ‚Inschrift‘ auf dem erwähnten ‚Bschriebenen Stoan‘ im Viggartal/Tirol vermuten lässt.*“

Die „Sonnenplatte“ ist ein weiteres umfangreiches Schriftbeispiel. Sie ist Teil der erwähnten elliptischen Steinsetzung und wird wegen der konzentrischen Kreise, die in frühen Kulturen als Sonnensymbol galten, so genannt. Daher gingen die Interpretationen meist in Richtung Sonne. Auch Peilungen nach Sternaufgängen sind nicht so zwingend wie bei der „Sternplatte“ (Abb. 6.12a).

Dietrich Knauer versuchte mit Erfolg auf dem Felsen Schriftzeichen zu finden.[14] Dazu ließ er wegen der Übersichtlichkeit die Sonnen Symbole weg und teilte die Schriftzeichen in Gruppen ein. In Abb. 6.12b sind alle Zeichen auf der Platte aufgeführt.

Die Schalen die zu einem Buchstaben gehören sind farblich gekennzeichnet. Es ergaben sich 11 Texte. In Abb. 6.13 findet sich die Übersetzung von Knauer.

Links steht die Transkription der Zeichen und rechts die Übersetzungen. Die Texte sind im Wesentlichen Wegweiser zu kultischen Punkten. Die angesprochene Höhle könnte möglicherweise unterhalb des Eckpunktes 6 in Richtung Kammer 2 sein. Derzeit ist sie sehr niedrig, wurde jedoch vielleicht im Laufe der langen Zeit durch eingeschwemmtes Erdreich aufgefüllt, bzw. der Deckstein hat sich abgesenkt.

Da auf vielen anderen Schalensteinen ähnliche Zeichen und Texte gefunden wurden, erscheinen sie plausibel. Natürlich kann manches auch anders gelesen und gedeutet werden. Berücksichtigt man jedoch, dass es auch in unserer Zeit, trotz vieler hauptamtlicher Dolmetscher, immer wieder in wichtigen mehrsprachigen Vertragstexten unterschiedliche „Lesarten“ gibt, sollte man auch Texten aus so früher Zeit eine gewisse Unsicherheit zubilligen.

Die genannten Schalenstein-Schrift-Forscher besitzen alle einen akademischen Grad und wissen daher wie wissenschaftliches Arbeiten geht. Da ihr Studium aber fach-

14 Knauer: Rätsel der Felsbilder und Schalensteine, 1987/2002, S. 84 ff.

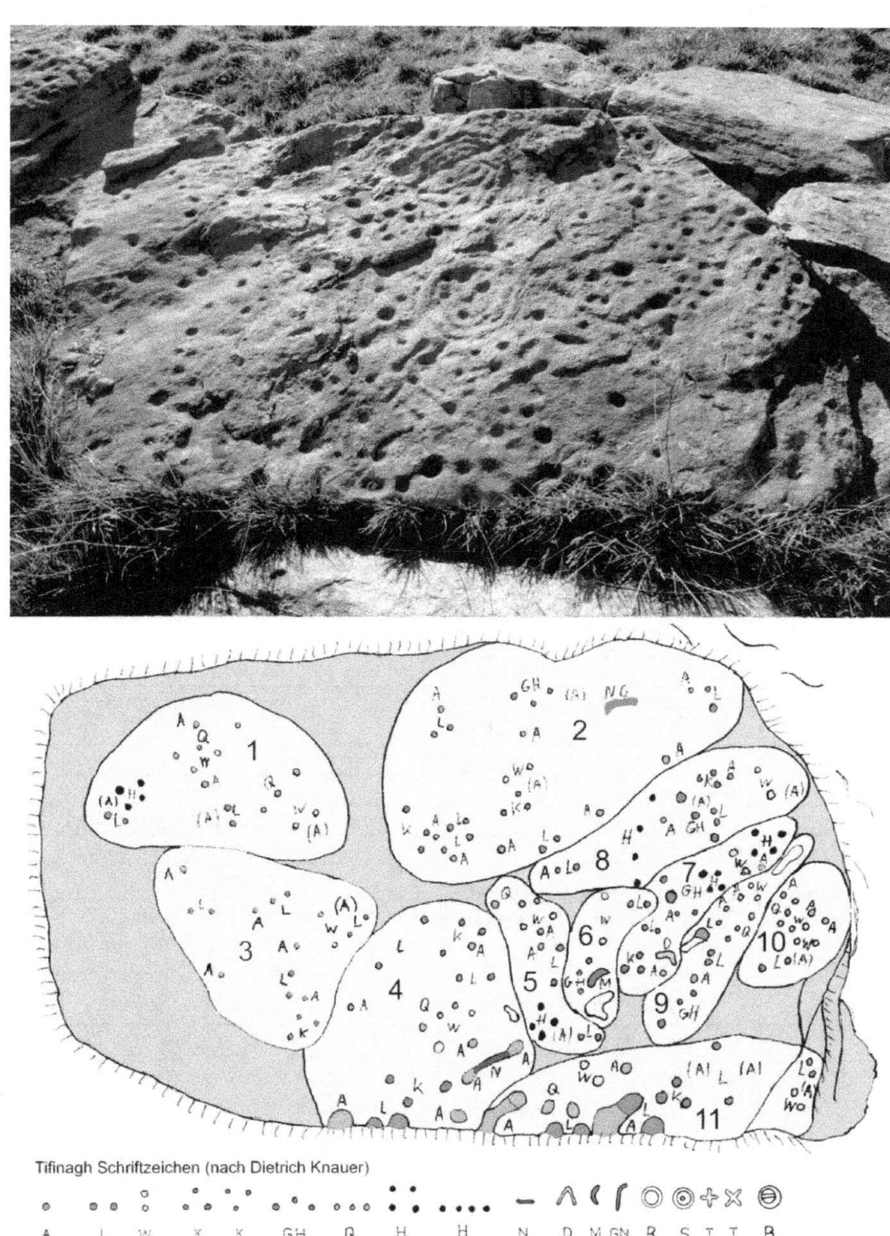

Abbildung 6.12:
a: Der Schalenstein Nr. 04, die „Sonnenplatte“
b: Die Schrift auf der „Sonnenplatte“

(© Roland Gröber, © Roland Gröber nach Dietrich Knauer)

Tabelle 6.1: Tabellarischer Überblick der Astronomie am Pfitscher Sattel

Objekt	Art	Beteiligte Elemente	Anmerkung
Mauer/Stein	Süden	Nr36 Südkante: Schatten wechselt von SO- nach SW-Fläche	
	Visur Ost	gr. SW Steinkreis - Nr5 - kl.östl. Steinkreis	
	Visur Nord	EP5 - Nr36	S-Kante Nr36
	Visur Nord	Nr49 - Nr37 (=EP1) - Nr24	Nr 24 Stein in N-Richtung
Schalenstein	Visur Nord	Nr37 (=EP1)	Kante durch zentr. Peilschale
	Visur Nord	Nr38: T-Mark.u. 4 Schalen	Längenkomparator ME
Sonne	Visur WSW	Nr49 - markante Felsformation	
	Visur WSW	EP1 - EP5.	
	Visur WSW	Nr37 - Sonnenschale	
	Visur SSW	Nr37 - Sonnenschale	
	Visur SSW	Nr32 (Menhir nahe EP3)	Längsachse nach SSW
	Visur Äuin.	Nr37 - Sonnenschale	
Mond	Visur KSM	Nr49 - markante Felsformation	
	Visur KSM	Nr37 - Mondschale	
	Visur GSM	Nr49 - mark. Fels	
	Visur GSM	EP6 - Nr58 - Nr36 (SW-Fläche)	Mond "rollt" knapp über Horiz.
	Visur GSM	Nr37 - Mondschale	
	Visur GNM	Nr30 - Nr36	SO-Fläche Nr36
	Visur GNM	Nr41 - mark. Felsformation	evtl. zur Finsterniswarnung
	Visur GNM	Nr37 - Mondschale	
	Visur KNM	Nr37 - Mondschale	
	Abbildung	Nr37	vier mondförmige Schalen
	Abbildung	Nr38	ca. 10 mondförmige Schalen
	Abbildung	Nr49	mondförmige Einkerbung
	Abbildung	Nr58	mondförmiger Felsen in K II
Sterne	Visur zum Aufgang am Osthorizont	αBoo, αLyr, αCyg, αAur, αGem, αPer, βGem, αLeo, αVir, αAqu, αCMi, ηTau, αTau, αOri, αSco, γOri, εOri, αCMa, βOri	von Stein Nr37 Visur zu 19 Sternaufgängen im Osten, etwa 2450 v.Chr. Zusätzlich αCMa ca 1800 v.Chr. (Präzession)
	Vis. Alcyone	Nr37 (=EP1) - Nr38 - Nr36 - EP2 - Nr33 - Nr34 (um 2500 v.Chr.)	
Sternbilder	Abbildung	Bootes, Leier, Schwan, Löwe, Fuhrmann, Zwillinge, Perseus, Jungfrau, Wassermann, Stier, Kleiner Hund, Orion, Skorpion, Großer Hund	auf Stein Nr.37 werden jeweils nur die hellsten Sterne der Sternbilder durch Schalen abgebildet
	Abbildung	Plejaden	Nr.38 neun Schalen
	Abbildung	Gr. Wagen	Nr.37 acht Schalen. Richtung N
	Abbildung	Kl. Wagen	Nr.37 acht Schalen. Richtung N

Abkürzungen:
GSM: Große südl. Mondwende
GNM: Große nördli. Mondwende
KSM: Kleine südl. Mondwende
KNM: Kleine nördl. Mondwnede

WSW: Wintersonnenwende
Äquin.: Äquinoktium
SSW: Sommersonnenwende
Nrxx: (Schalen-)Stein- Nummer
EP: Eckpunkt der Umfassungsmauer

1	N (A) L A Q W A (A) L Q W (A)	*zum (trinkbaren) Wasser hierhin*
2	AL KALLA A LA K(A) WA GN (A) NG A LA	*beim Ruf zur Höhle gehe dorthin*
3	A LA KALA AL W (A) L	*zum Weg ins Tal*
4	AL KAL (A) QWA KANA A LA	*zum Weg durch (die) Rinne nach da*
5	Q W A A L H (A) L	*hierdurch zum Heil (Glück)*
6	L (A) W M (A) GH	*tue recht* (ev.mit 5: *durch Recht tuen zum Heil*)
7	H A W H (A) GH A L D A K	*hab`ein heiliges Dach*
8	A L H A GH (A) L K A W (A)	*zur heiligen Höhle* oder *zur Heiligen-Höhle*
9	Q W A A L A L A GH	*hierhin zum See*
10	A Q W A A W (A) L	*hierhin zum Tal*
11	ALQ W A K A L A A W (A) L	*nach hier gehe zum Tal*

Abbildung 6.13:
Die Schriftdeutung auf der Sonnenplatte

(© Dietrich Knauer)

fremd war, werden ihre Erkenntnisse von den archäologischen „Platzhirschen" ignoriert – wie so vieles in der Schalensteinforschung.

6.7 Zusammenfassung

Im Gegensatz zu den früher vorgestellten astronomischen Erkenntnissen, die auf konkreten Beobachtungen und Messungen beruhen, sind manche der vorgestellten Befunde in ihrer Bedeutung teilweise etwas unscharf und spekulativ. Astronomische Ergebnisse lassen sich überprüfen und setzen „nur" voraus, dass die jeweiligen Astronomen das Ereignis auch gesehen haben. Bei Beobachtungen, die auf Kulthandlungen deuten, sind diese, entgegen mancher Terra-X-Sendungen, die uns deren genaue Abläufe vorgaukeln, mit großer Vorsicht und Zurückhaltung zu interpretieren. Der leider verstorbene Professor Wolfhard Schlosser, der der Archäoastronomie sehr positiv gegenüber stand, sagte immer in diesem Zusammenhang: „wir waren alle nicht dabei". Mit den gezeigten Beispielen ist die eingangs erwähnte Bezeichnung „Höhenheiligtum" für die Anlage am Pfitscher Sattel, auch ohne genaue Kenntnis irgendwelcher Rituale, sicher zulässig. Auch nach fast 20jähriger Forschung am Pfitscher Sattel, auf umfangreicher Vorarbeit von Aribert Egen (1995) beruhend, gibt es noch viele offene Fragen. Es bleibt noch viel für eine seriöse Forschung!

6.8 Literatur

EGEN, ARIBERT: Das Spronser Bergheiligtum bei Meran. Die älteste Sternwarte der Menschheit in situ? In: RICHTER, PETER (Hg.): *Sterne, Mond, Kometen. Bremen und die Astronomie.* Bremen: Hauschild 1995.

GLEIRSCHER, PAUL: Ein urzeitliches Bergheiligtum am Pfitscher Jöchl über Dorf Tirol? In: *Der Schlern – Monatszeitschrift für Südtiroler Landeskunde* (Bozen: Athesia) **67** (1993), Heft 6, S. 407–435.

GRÖBER, ROLAND: *Das Bergheiligtum am Pfitscher Sattel bei Meran. Schalensteine und astronomische Beobachtungen in der Kupferzeit.* Leverkusen: Eigenverlag 2016 (128 S.).

GRÖBER, ROLAND: Die Schalensteine am Pfitscher Sattel in der Texelgruppe. In: *Südtirol in Wort und Bild* (Thaur, Tirol) 2014, 1. Q., S. 24–33 und 2. Q., S. 21–26.

GRÖBER, ROLAND: Der Himmel über den Spronser Seen und die Himmelsscheibe von Nebra. Die „älteste konkrete Darstellung des Himmels" – ein Vergleich. In: *Der Schlern – Monatszeitschrift für Südtiroler Landeskunde* (Bozen: Athesia) **97** (Januar 2023), Heft 1, S. 70–83.

GRÖBER, ROLAND: Astronomie in Südtirol zur Zeit des Ötzi (ca. 3350–3100 v. Chr.). Bergheiligtum und älteste Sternwarte der Welt am Pfitscher Sattel!? In: WOLFSCHMIDT, GUDRUN (Hg.): *Baudenkmäler des Himmels – Astronomie in gebautem Raum und gestalteter Landschaft.* Hamburg: tredition (Nuncius Hamburgensis; Band 35) 2018, S. 26–41.

GRÖBER, ROLAND: Die Konstruktion von megalithischen Steinsetzungen am Beispiel des Höhenheiligtums am Pfitscher Sattel in der Texelgruppe. In: WOLFSCHMIDT, GUDRUN (Hg.): *Orientierung, Navigation und Zeitbestimmung – Wie der Himmel den Lebensraum des Menschen prägt.* Hamburg: tredition (Nuncius Hamburgensis; Band 42) 2019, S. 24–35.

GRÖBER, ROLAND: Gibt es Zusammenhänge zwischen Schalensteinen und alpinen Brandopferplätzen? In: WOLFSCHMIDT, GUDRUN (Hg.): *Himmelswelten und Kosmosvisionen – Imaginationen, Modelle, Weltanschauungen.* Hamburg: tredition (Nuncius Hamburgensis; Band 51) 2020, S. 96–114.

HALLER, FRANZ: *Die Welt der Felsbilder in Südtirol. Schalen- und Zeichensteine.* München: Hornung Verlag Viktor Lang 1978.

KIRNBAUER, HERBERT: *Steinzeit-Code. Die Schalensteinschrift.* Engerwitzdorf / Mittertreffling, Österreich: Freya 2012.

KNAUER, DIETRICH: *Die Rätsel der Felsbilder und Schalensteine. Die älteste Sprache und Schrift Europas.* Mailand 1987. Nachdruck: Ardagger: Verlag Dr. Michael Damböck 2002.

KOFLER, ASTRID: *Magische Plätze in Südtirol. Das Bergheiligtum am Pfitscher Sattel.* Video für den Fernsehsender Rai-Bozen (9,5 Minuten), 2018.

LUNZ, REIMO: *Archäologische Streifzüge durch Südtirol. Band 1: Pustertal und Eisacktal. Band 2: Etschtal.* Bozen: Athesia 2005.

MENARA, HANSPAUL: *Südtiroler Urwege. Ein Bildwanderbuch.* Bozen: Athesia 1984.

NIEDERWANGER, GÜNTHER: Ein bedeutender Höhenfund in der Texelgruppe. In: *Der Schlern – Monatszeitschrift für Südtiroler Landeskunde* (Bozen: Athesia) **63** (1989), Heft 7/8, S. 403–406.

STEINER, HUBERT (Hg.): *Alpine Brandopferplätze. Archäologische und naturwissenschaftliche Untersuchungen.* Trient: Editrice Temi 2010.

STEINRÜCKEN, BURKHARD: *Sonnenwenden und Mondwenden.* Vortrag auf der Tagung der Gesellschaft für Archäoastronomie vom 10.–12. März 2011 in Osnabrück, `https://sternwarte-recklinghausen.de/astronomie/forschungsprojekt-vorzeitliche-astronomie/#A01`.

WALLI-KNOFLER, THOMAS: *Schalensteinpfade in Tirol. Alpine Wegweiser und mehr und deren Übersetzung.* Innsbruck 2019/21, `https://www.raetiastone.com/`.

Figure 7.1:
Prehistoric star map of the Orion on the *Zeichenstein* of Kersbach.

Colored lines and rings pointing to the engraved marks of stars on the stone were inserted on a photo displayed in the book by Hilke Hennig. The three "girdle" stars, the "sword", the "feet", the "head triangle", comprising the "shoulders" and the "eye", can be recognized. The stars in the circles are also linked with the Orion constellation.

[Hennig 1970]

Engraved Stones from burial mounds in Franconia point to observations of the Sun, of Orion and of the Pleiades

Siegfried Hess (Berlin, Röttenbach) & Thomas Storch (Hemhofen)

Abstract:

Sand stones found in and close to prehistoric burial mounds in the area between Erlangen and Forchheim, in Franconia (Northern Bavaria), display a great variety of engraved geometric patterns. More than 120 of these stones, referred to as "Zeichensteine", were documented by Hilke Hennig [Hennig 1970]. About a dozen of the engraved stones hint to observations of the sun and of stars [Hess (2018)]. After a brief survey of the type of patterns, four of these stones are considered in more detail. One shows the sunrise over a hill about 40 days after the spring equinox (or about 40 days before the fall equinox), another one sketches a basic feature occurring when the sun is observed behind palisades.

Special emphasis of this presentation is on two stones where star maps of Orion and of the Pleiades can be identified. The comparison of the star-marks on the stones considered here with astro-photos yields:

i) The agreement of the positions of four of the five brightest stars of the Pleiades, those forming a distorted rectangle, is far beyond a random coincidence.

ii) Head triangle, girdle-stars, sword, feet, as well as six other stars of the Orion constellation can be recognized on the prehistoric star map, matching the astro-photo surprisingly well.

No reliable data are available for dating the engravings. Arguments put forward for the conjecture of an age of four to five thousand years [Nadler (2011)], are discussed.

Zusammenfassung: Gravierte Steine aus fränkischen Grabhügeln weisen auf Beobachtungen der Sonne, des Orion und der Plejaden hin

Sandsteine, gefunden in und nahe bei prähistorischen Grabhügeln in der Gegend zwischen Erlangen und Forchheim, in Franken (Nord-Bayern), zeigen eine grosse Vielfalt von eingravierten geometrischen Mustern. Mehr als 120 dieser Steine, die als „Zeichensteine" bezeichnet werden, wurden von Hilke Hennig dokumentiert [Hennig 1970]. Etwa ein Duzend der gravierten Steine weisen auf die Beobachtung der Sonne und von Sternen hin [Hess (2018)]. Nach einem kurzen Überblick über die verschiedenen Muster werden vier dieser Zeichensteine näher betrachtet. Einer zeigt einen Sonnenaufgang, etwa 40 Tage nach der Frühlings Tag-und-Nacht-Gleiche (oder 40 Tage nach der Herbst Tag-und-Nacht-Gleiche). Ein anderer skizziert ein Phänomen, das bei Auf- oder Untergang der Sonne hinter Palisaden beobachtet werden kann.

Das Hautaugenmerk dieser Präsentation ist auf zwei Steine gerichtet, auf denen Sternkarten des Orion und der Plejaden identifiziert weren können. Der Vergleich der Stern-Markierungen auf den Steinen mit Astrophotographien ergibt:

i) Die Übereinstimmung der Positionen von vier der fünf hellsten Sterne der Plejaden, nämlich jene, die ein gestauchtes Viereck bilden, ist jenseits einer zufälligen Koinzidenz.

ii) Kopfdreieck, Gürtelsterne, Schwert, Füsse, sowie sechs andere Sterne der Orion-Konstellation entsprechen dem Astrophoto überrachend gut.

Für die Datierung der Gravuren gibt es keine verlässlichen Daten. Argumente für die Vermutung [Nadler (2011)] eine Alters von vier bis fünf Tausend Jahren werden diskutiert.

7.1 Introduction and Preliminary Remarks

Since early times man felt that they were embedded in the *cosmos*, in a *world with order*, an order evident in the apparent motions of celestial bodies: sun, moon, planets and stars. Here we demonstrate that prehistoric engravings on stones from burial mounds in Franconia hint to observations of the sun, of the Pleiades and of Orion, likely four to five thousand years ago.

7.1.1 Typical Patterns of the Engraved Stones

Sand stones found in and close to prehistoric burial mounds in the area between Erlangen and Forchheim, about halfway between Nuremberg and Bamberg, in Franconia (Northern Bavaria), display a great variety of engraved patterns, some rather intriguing. These stone plates, typically 40 to 70 cm long, 30 to 50 cm wide, about 15 to 25 cm thick, are referred to as "Zeichensteine", meaning "stones with signs". Photographs and drawings of the patterns of more than 120 stones, as well as of other prehistoric objets found close by are documented in a book by Hilke Hennig (*1938), published in 1970 [Hennig 1970]. The book is based on her 1965 dissertation at the University Erlangen-Nürnberg. Some details about the sites where the stones were discovered and attempts to date the engravings, are presented later. Typical patterns as compiled by H. Hennig [Hennig 1970], are shown in Fig. 7.2.

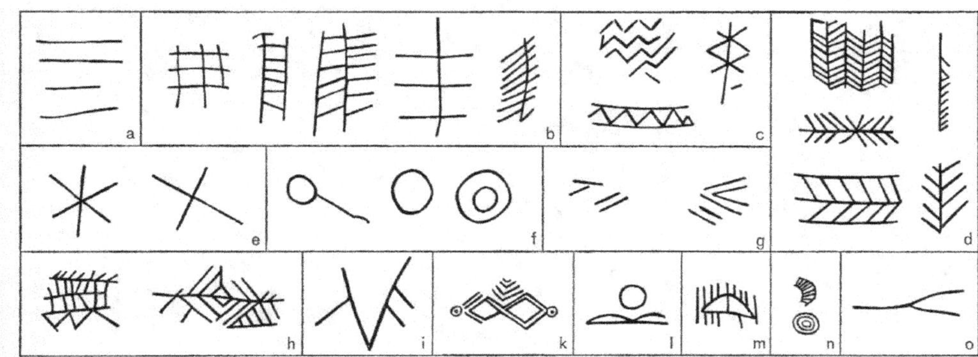

Abbildung 7.2:
Motives of the engraved patterns, found at sites "A" and "B", as presented by Hilke Hennig: a) parallel lines, b) ladders and grids, c) zigzag, d) fishbones or grains, e) crosses, f) circles, g–o) combined patterns and special forms.

([Hennig 1970], p. 27)

Where the burial mounds were not destroyed prior to the archeological excavations, the engravings were found on the outside of the stones which, in turn, were arranged in circles with three to more than ten meter diameter [Hennig 1970]. An example of a stone ring is seen in Fig. 7.3, as documented during a 1922 excavation and reproduced in [Hennig 1970].

Therefore we know that the engravings were supposed to be seen. Originally, just the interior of the stone circle might have been covered with sand. Could the "drawings" on stones have been used for teaching, instructions and for exercises? Hints at observations of the sun and of stars are discussed next.

7.2 Observations of the Sun

7.2.1 Sunrise over a Hill

The engraved pattern "ℓ" of Fig. 7.2 shows the sun or a full moon over a double hill. From a fixed point of observation, the sun regularly appears twice a year at the

Abbildung 7.3:
Stone ring of a burial mound, excavated 1922

([Hennig 1970])

position shown, whereas the moon's motion is more complex. So it is likely that the sun is depicted. The short line on the right hand side of the sun, seen on the top of Fig. 7.4 may hint to a sunrise!

The photograph showing the sunrise over the "Rodenstein" summit of the "Ehrenbürg", was taken about 200 meters north from the site where the stone was found since nowadays, trees prevent the direct sight to the hill. The date was April 30th, 2017. On May 1st the sun would have been closer to the middle of the double hill. The situation should be similar around August 15. These dates, about halfway between the equinoxes and the longest day, played an important role in prehistoric "calendars" and for festivities. A medieval chapel on the Ehrenbürg is dedicated to St. Walburga whose Saint's day is April 30th.

Abbildung 7.4:
Top: Sunrise over a hill, reproduced from [Hennig 1970]. The sketched lines were made from a plastic replica of the stone which, according to [Hennig 1970], reveals some lines better than the photo of the original.
Bottom: Sunrise on April 30, 2017, over the "Rodenstein" summit of the "Ehrenbürg". The photo, by one of the authors (SH) was taken close to the site where the sunrise stone was found.

([Hennig 1970], Fig. 2 of "Tafel" 100; photo: Siegfried Hess)

7.2.2 Remarks on sites and landscape

Brief remarks are in order on the sites where the Zeichensteine were found and on the surrounding landscape. Please consult your favorite map.

Most of the “Zeichensteine” in the region Erlangen-Forchheim stem from two sites, here referred to as “A” and “B”. In the literature, Site A, close to an area now used as a “Deponie” (garbage disposal) which, in turn, is located close to the villages Kersbach, Sigritzau, Pinzberg and Gosberg, is mainly listed under the name “Gosberg”. The first engraved stones were found over 100 years ago, a dozen burial mounds were excavated between 1922 and 1951 [Hennig 1970]. Only three of the mounds contained engraved stones in their original setting, as in Fig. 7.3. About 25 engraved stones were found in such a setting. A similar number of engraved stones, discovered close by, belonged to destroyed mounds or had found secondary use in graves from the Hallstatt period.

Site A lies 2 km east from the Regnitz river, which flows northward to the Main. Site B is about 3 km west from this river, in the “Markwald”, a state forest between Röttenbach and Baiersdorf. The location is close to the road and above a small lake called “Schwarzer Weiher”. Just as at site A, no traces of the former excavation can be found at site B. The distance between both sites is about 7 km. In 1937, a burial mound with a diameter of 13 m was excavated. at site B. It contained 58 engraved stones, set in an open circle and 6 scattered engraved stones, which might originally have stood in the gap to complete the circle, and furthermore, 3 other engraved stones, whose positions in the mound is not known. Many of the patterns of the engravings at both sites A and B are quite similar.

A few engraved stones with similar patterns have been found at five other places in the region, less than about 20 km away from sites A and B. The same applies to reported Zeichensteine which were lost or even destroyed.

The highest hill that can be seen from the site “A”, where the “sunrise” stone was found is the “Ehrenbürg”, about 7 km away. It is an elongated table mountain (about 1300 m long and 300 m wide) with two summits. The narrow northern side of the mountain, locally known as “Walberla”, (referring to “Walburga”), is depicted in the background of Albrecht Dürer’s iron etching “The Big Canon” of 1518. The southern summit of the Ehrenbürg, called “Rodenstein”, is a double hill. The oldest traces of settlements on the Ehrenbürg date back for 6000 years. On the table mountain, fortified walls were built and destroyed three times, during the Bronze Age and the Iron Age, between 3400 and 2300 years ago.

Nowadays trees prevent the direct sight from the original site “A” to the Ehrenbürg, however, the view is free at about 200 m north from site “A”. There, the picture with the sunrise over the Rodenstein was taken in the morning of April 30th, 2017. At the same date, four to five thousand years ago, the sunrise might well have been in the middle of the double hill [Ruggles 2015]. Nowadays, the situation depicted on the stone is reached on May first or second, and again around August fifteenth.

It is remarkable that the position where the "Orion-stone" was found, see below, not only lies on the line connecting sites A and B but this line also points to the Rodenstein summit, which was mentioned above in connection with the sunrise-stone. Seems unlikely that such an alignment occurred by chance.

A side remark: site *B* is just about 3 km away from the twin villages Röttenbach and Hemhofen, where the authors live.

7.2.3 Solar Observatory

The engraved pattern "*m*" of Fig. 7.2, displayed in more detail in Fig. 7.5, is a sketch of a solar observatory where sunrise and sunset were watched behind wooden posts or palisades. The stone was found at site "A". The drawing explains an effect which is even surprising today: the sunlight which passes through gaps between the posts, or between trees as on the bottom of the photo Fig. 7.5, is so bright that the sun seems to be in front the obstacles. Only the stick close to the observer, needed to determine the direction of sight, is seen in front of the sun in the prehistoric sketch seen on the top of Fig. 7.5.

For thousands of years, wooden posts arranged in a circle were used for the observation of the changing positions of sunrise and sunset, throughout the year. The reconstructed solar observatory Goseck [Goseck Circle], dated at around 6800 years BP is about 200 km north from the place where the stone Fig. 7.5 was found. Recent excavations [Tschuch 2017], just about 2 km away from site A, revealed traces of extended prehistoric settlements from 4000 to 2500 BP. At two places, where one has a free 360 degree view of the horizon, remains have been found of circular fences with openings on the northern side. These could very well have been solar observatories for which the drawing on the stone was meant to give some explanations. Of course, the palisades can also be used for the observation of stars, cf. [Steinrücken (2004)], and at the same time, are handy to fence in sheep, cattle or horses.

7.2.4 Further hints for solar observations

The angle between the most northern sunrise in June and the most southern sunrise in December is referred to as "*angle between the solstices*", or as "*Pendelbogen*". On two of the engraved stones, where the sketched lines indicate broken circles, parts of the circle are engraved considerably deeper. One example, from site "B", is Fig. 2 of "Tafel" 105 in [Hennig 1970]. Though the center of the circle is not precisely marked, the arc of about $75^o \pm 5^o$ corresponds to the actual angle between the earliest and latest sunrise or sunset of Forchheim which is 76°, for a flat horizon. Of course, the actual

Abbildung 7.5:
Top: Sunrise or sunset as seen in a Solar Observatory, sketched on stone.
Bottom: The sun seems to shine through the stems of the trees.

(Copy of the photo Fig. 1 of "Tafel" 95 in [Hennig 1970]; photo: Siegfried Hess)

solar arc which can be observed from a certain position is affected by the hight of the hills on the horizon, above the observatory. In the case considered, these heights are about 100 m for sunset and about 200 m for sunrise, both about 6 to 10 km away.

One of the many stones with an engraved grain pattern is Fig. 4 of "Tafel" 105 in [Hennig 1970]. Could the stones, prior to being set up in circles surrounding graves, not have been laid flat on the ground, with the long line directed North-South? Then the directions of sunrise and sunset could have been marked, maybe over several weeks. The situation depicted on the stone corresponds to summer time, when the engraver looks northward. During winter time, he would have to look southward. From the place the stone was found, site "B', the winter observation is possible, in particular when one takes into account, that about 5000 years ago, average temperatures might have been higher than now. For the observation during summer, the stone would have to be placed at a slightly higher position, just a few hundred meters away. That might anyway have been the place where the stone was quarried.

The angles on the stone are between 110° and 120°, the complements are 250° and 240°. Facing North, the complement angles correspond to April-May and August, at the latitude of about 50° North. Looking southward, the smaller angles occur in late November and early December, as well as in January.

7.3 Star Maps

Prehistoric sketches of Orion and of the Pleiades, from the Ice Age to the Bronze Age, have been noticed before, e. g. in [Rappenglück 2003], [Rappenglück 1999], [Rappenglück 2008], [Sparavigna 2008], [Steinrücken 2010]. Next we compare the prehistoric star maps considered here with astro-photos, taken by one of the authors (TS).

7.3.1 The Pleiades or Seven Sisters

The dots on the right hand side of Fig. 7.6, could mark the *Pleiades*, also called "Siebengestirn" or "Seven Sisters". The observation of the rise and set of the Pleiades played an important role for hunters and even more so for the Neolithic and Bronze Age farmers, for determining a calendar, cf. [Sparavigna 2008], [Steinrücken – Sternphasen der Plejaden in der Bronzezeit]. The pattern of dots on the engraved stone, at first glance, resembles the star map of the Pleiades more closely than those on the Bronze age disc of Nebra [Steinrücken 2010], from about four thousand years ago. Much older traces of the Pleiades were discovered in Ice Age cave paintings of Lascaux [Rappenglück 1999], [Rappenglück 2008].

For the Pleiades, the comparison between the marks on the stone and the actual positions of the stars, as seen in the astro-photo, see Fig. 7.6, is more subtle than for Orion. One reason is the small size of the Pleiades compared with the giant Orion. Furthermore, we do not know whether all of the stars we associate with the Pleiades and which other stars in the immediate neighborhood were included in the observation of the star cloud. In other prehistoric "pictures", a conglomerate of 6 to 9 dots seemed to be sufficient to symbolize the Pleiades.

We tested the degree of overlap between the dots on the stone and the stars of the astro-photo. The result is seen in Fig. 7.6 where the positions of the stars on the stone are marked in green and yellow circles indicate the positions of the astro-photo stars. Details of the analysis are given below. A practically perfect overlap is found for four of the five brightest stars. The random probability for such a coincidence is less than one in one million.

Abbildung 7.6:
Left: Astro-photo of the Pleiades with names of stars.
Right: Comparison of star positions on the stone, marked in green, with astro-photo positions indicated by yellow circles.

(Photo: Thomas Storch)

The quantitative comparison between the star positions on the stone with those in the astro-photo was performed as follows. First masks were made where circles, with finite diameter, indicate the distribution of the stars. The yellow circles, in Fig. 7.6, correspond to the 6 stars which can be seen, with the naked eye in a clear sky. The purple rings stand for somewhat fainter stars, the lower one has variable brightness. Green disks mark the positions on the stone. An analysis yields a rather good overlap for 4 of the yellow circles with the green disks, as indicated on the right in Fig. 7.6. These 4 star positions are actually marked on the stone with deeper and broader holes, than the other ones. What is the probability that such a coincidence occurs at random? To find this probability, the area of a star-disk is compared with the area of the stone. We find: $n = 228$ of the star-disks fit onto the surface of the stone. This then sets the length scale. One has to take into account that the position of the first star should be considered as being given. Then the probability $P(n,k)$ that $k = 4$ of these circles coincided, at random, with specific star positions, as determined by the astro-photo, is given by the binomial formula for the chance of finding $k-1$ additional stars on their $n-1$ remaining "correct" positions. The result is

$$P(n,k) = (k-1)!\,(n-k)!/(n-1)!$$

For $n = 228$ and $k = 4$, one has

$$P = \frac{1}{227}\frac{2}{226}\frac{3}{225} = \frac{1}{1923825} \approx 0.52 \cdot 10^{-6} \approx 1/200000$$

and consequently $P < 10^{-6}$, as stated above. This small probability for a random coincidence strongly suggest that the prehistoric stone was indeed engraved to show the Pleiades!

What is the meaning of the crossed lines on the stone? Their directions could indicate "rising" and "setting". A possible speculative scenario is: the "professor" teaching his students the importance of the star cluster. He points out that both the rising and the setting of the Pleiades should be observed [Steinrücken – Sternphasen der Plejaden in der Bronzezeit]. To stress his point, the teacher drew the lines on his "sand stone board". More serious, the angle between the lines is 80°, so it matches about the angle of 76° between the solstices. Could the fact that one of the marks for the stars is just on one of these lines indicate a special observation of the Pleiades on either the longest or the shortest night, about 5000 years ago?

7.3.2 Orion

The stone displayed in Fig. 7.1, measuring 125 by 70 cm, is considerably larger than the stones found in burial mounds and the engraved pattern is rather different from all the others. The dots, some of which are rather deep, could mark the stars of *Orion*. The "girdle" is clearly visible, the stars of the "sword" are replaced by a deep swordlike line. The stars of the the "feet", as well as the "head triangle" comprising "shoulders" and the "eye", can be recognized.

This stone was found in a flat field, close to the train station of the village *Kersbach*, about 2 km away from site where the Pleiade-stone was found. Close by are large prehistoric settlements, recently discovered in excavations [Tschuch 2017], initiated by the construction work for the railroad line Munich-Berlin. The Orion-stone is said to stem from a destroyed grave [Hennig 1970], no graves were found within the recently inspected area [Tschuch 2017]. So the stone could originally have stood upright, like a stele or small menhir. This stone studied here might represent one of the oldest records of the Orion star constellation in Central Europe. A much older "picture" of the Orion was identified in a bone carving [Rappenglück 2003].

For a more detailed discussion of the Orion star map have a look at Fig. 7.7, which shows, clockwise from top left, the Orion constellation inferred from "Stellarium", an astro-photo by Th. Storch, the Orion stone from H. Hennig's book [Hennig 1970], taken before 1970, and a photo of the Zeichenstein located in Kersbach, taken in 2023 by S. Hess. A comparison of the prehistoric star map, lower right of Fig. 7.7 or Fig. 7.1, with an astro-photo upper right of Fig. 7.7, reveals how accurate the positions of the stars were engraved on the stone. To guide the eye, corresponding colored lines and rings were inserted in the photos. In detail, the coincidence of about 10 features, can be recognized. These are, as numbered on the upper right of Fig. 7.7:
1) the stars of Orion's "girdle", 2) the Orion nebula also called Orion's "sword", 3) the connection *kORI – Rigel*, between the "feet" of Orion, 4) the head triangle, comprising the "shoulders", with Betelgeuze at left and the "eye", 5) *uORI*, 6) *nORI*, 7) βORI, 8) *HR*1919, and 9a), 9b) could be *HR*2039 and *HR*2065. Standard nomenclature is used for the stars referred to.

It is not so easy to identify the marks for the stars of Orion on the lower left photo of Fig. 7.7. On the stone located at a corner of a meadow close to the Kersbach train station, however, holes, three to four centimeters wide and about three centimeters deep, can be identified which correspond to those shown on the lower right photo of Fig. 7.7 or in Fig. 7.1. Except for the Orion stone of Kersbach, the more than 120 Zeichensteine recorded in [Hennig 1970] are stored in museums, there mostly in depots. Easy access to a prehistoric monument resting in open air may be appealing, however erosion takes its toll.

A minor detail could be added. The vertical and horizontal lines on the left of the stone might indicate bow and arrow of the "hunter" Orion. Rather than chasing the Seven Sisters, as the story tells, here he would be aiming in the opposite direction.

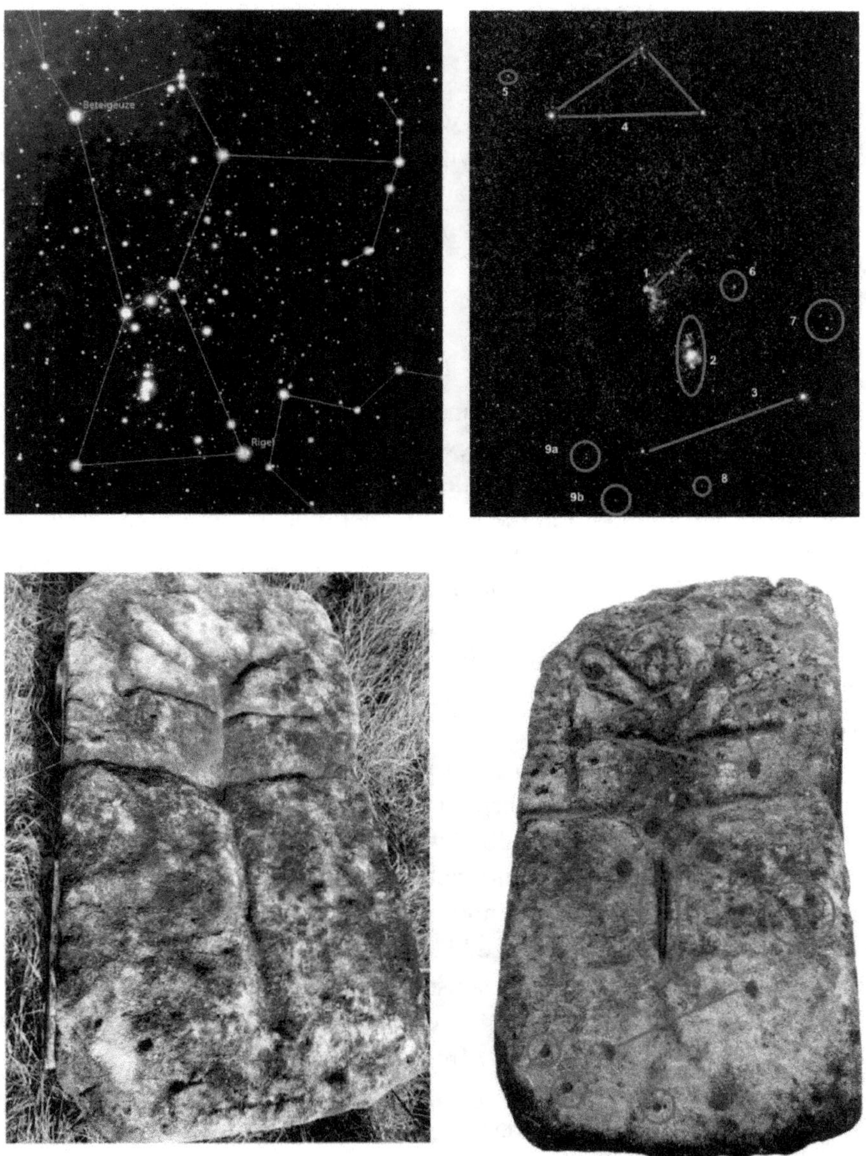

Abbildung 7.7:
The Orion constellation ("Stellarium")
and an astro-photo (Storch), top left and right;
Below are photos of the engraved stone, taken before 1970 by Hennig, right)
and in 2023 (Hess, left). The photo on the bottom right with the colored lines
and rings corresponds to Fig. 1.

(Photo: Thomas Storch, [Hennig 1970], photo: Siegfried Hess)

7.4 Concluding Remarks

The interpretations for the symbols of the prehistoric stones, as presented previously in [Hess (2018)], were plausible, but there was no proof. By comparing the distribution of the marks for the stars on the stones with the star positions on astro-photos in more detail, strong arguments are provided that the two stones considered here present the Orion and the Pleiades. There are many more patterns on the "Zeichensteine" which still await interpretation.

7.4.1 Relations to other regions

The special type of stone engravings found in a relatively small region means that the "culture" or the art to make the engravings lasted just for a short time such that it could not spread further or it was, after some time, considered as totally "useless". Still the question remains, did it pop up in the region where the engraved stones were found, or was there an influence from somewhere else, rather far away? Some of the patterns seen in Fig. 7.2, occur on neolithic stone monuments about 200 km north from our area. This was already mentioned in [Hennig 1970]. In [Nadler (2011)], similarities of some patterns with those on many petroglyphs in Western Europe were pointed out. A fair number of motives, however, seem to be rather unique for the Franconian Zeichensteine.

A close similarity of patterns also exists with stone engravings from Kamyana Mohyla, an archaeolgical site close to Terpinnya, Zaporizhzhia Oblast, Ukraine, see [Kamyana Mohyla]. This region, about $2000km$ east, is currently under Russian occupation. In the few photographs available, not only zig-zag, grain or fish bone patterns occur, but also point like holes can be noticed which might mark the positions of stars, similar to those on the stones considered here.

Distinctively different from the Zeichensteine are the "*Schalensteine*" or "cup-marked stones" with prehistoric man-made cup-like holes in solid rock. For some of Schalensteine, found in Europe rather far away from Franconia, the arrangement of cups can also be interpreted as stellar constellations, recorded three to six thousand years ago [Gröber 2018], [Gröber 2019], [Mäder 2023], [Shelley – Possible stone age star map in Kent].

7.4.2 Attempts to date the engravings

No reliable data is available for dating the engravings. However, some educated guesses are possible. Assuming the engravings were made prior to the setup of the stones in the circles of the burial mounds, and taking into account that urns were placed later into burial mounds, the engravings should have been made before the Urnfield period, which lasted from about 3300 to 2800 BP. The abbreviation BP, standing for "Before Present", sometimes interpreted as "before 1950", is meant literally here since these and the following dates may have an uncertainty of at least ±100 years.

The zig-zag and the grain or fish bone patterns, as well as a few others, occur on neolithic stone monuments [Albrecht 1999], [Hamel (2000–2001)], [Nadler (2011)], about 200 km north from the "Zeichenstein" area, as already mentioned in [Hennig 1970]. In [Nadler (2011)], it was stressed that the engravings might well stem from the late neolithic period also referred to as Copper Age, i. e. they might be 4500 to 5500 years old. This date is based on the conjecture that the Zeichensteine, set up in a ring for the burial mounds, were engraved stelae, like that one described in [Albrecht 1999] and similar to the Orion stone discussed above, which were "cut" in the middle. The original stones were assumed to be about twice as high as those found in the excavations. The argument may well apply to some stones of site A.

The shift of the number of days over millennia, between the Spring equinox and the day when the Pleiades can be seen again rising, shortly before sunrise, can provide an estimate for dating [Ruggles 2015]. Assuming this observation was made at the same time of the year as the sunrise over the double summit mentioned above, the "morning first rise" of the Pleiades should have been about 40 days after Spring equinox. Linear extrapolation (backwards) of the data presented in [Steinrücken – Sternphasen der Plejaden in der Bronzezeit] yields about 2800 BC for a flat horizon. Considering the landscape, this date might well have been two to three hundred years earlier, corresponding to about 5000 BP as a possible date when the engravings were made.

This is in accord with our previous conjecture that the star maps might have been created between 4000 and 5000 BP.

Acknowledgements

One of the authors (SH) is grateful to Hilke Hennig for the exchange of email-letters and for informative conversations where she provided friendly support for the astronomical interpretations of the patterns on some of the Zeichensteine, as put forward in [Hess (2018)], and she encouraged him to present drawings and photos of her book [Hennig 1970], in connection with studies on Archaeoastronomy. Hilke Hennig also agreed with Martin Nadler's conjecture [Nadler (2011)] that the engravings on the sand stones are considerably older than the Urnfield culture. Thanks are due to Gabriele Hess-Fernandez and to Killian Hess for helpfule comments.
Both authors thank the conference organizers of the *Gesellschaft für Archaeoastronomie* for the chance to present our findings at the meeting in Weimar and they are grateful to Gudrun Wolfschmidt for her excellent editorial work which led to this publication.

7.5 Literatur

[Albrecht 1999] ALBRECHT, KLAUS: *Die Stele von Wellen: Mondkalender – Mondsymbolik?* http://www.jungsteinsite.de/ (14. November 1999).

[Goseck Circle] *Goseck*: http://www.sonnenobservatorium-goseck.info/, https://en.wikipedia.org/wiki/Goseck_circle.

[Gröber 2018] GRÖBER, ROLAND: Astronomie in Südtirol zur Zeit des Ötzi (3350–3100 v. Chr.) – Bergheiligtum und älteste Sternwarte der Welt am Pfitscher Sattel? In: WOLFSCHMIDT, GUDRUN (ed.): *Baudenkmäler des Himmels – Astronomie in gebautem Raum und gestalteter Landschaft.* Hamburg: tredition (Nuncius Hamburgensis; Vol. 35) 2018, p. 26–41.

[Gröber 2019] GRÖBER, ROLAND: Die Konstruktionen von megalithischen Steinsetzungen am Beispiel des Höhenheiligtums am Pfitscher Sattel in der Texelgruppe. In: WOLFSCHMIDT, GUDRUN (ed.): *Orientierung, Navigation und Zeitbestimmung – Wie der Himmel den Lebensraum des Menschen prägt.* Hamburg: tredition (Nuncius Hamburgensis; Vol. 42) 2019, p. 24–35.

[Hamel (2000–2001)] HAMEL, JÜRGEN: *Anmerkungen zu Klaus Albrecht: Die Stele von Wellen: Mondkalender – Mondsymbolik?* (2000–2001), http://www.jna.uni-kiel.de/index.php/jna/article/view/50/51.

[Hennig 1970] HENNIG, HILKE: *Die Hort- und Grabfunde der Urnenfelderkultur aus Ober- und Mittelfranken.* Kallmünz: Verlag M. Lassleben, 1970.

[Hess (2018)] HESS, SIEGFRIED: *Prehistoric Engraved Stones from Northern Bavaria hint at Observations of the Sun and of Stars.* (2018), https://www.researchgate.net/publication/327106735.

[Kamyana Mohyla] *Kamyana Mohyla*: https://en.wikipedia.org/wiki/Kamyana_Mohyla.

[Mäder 2023] MÄDER, STEFAN: *Der Schalenstein von Saint Guénael, Lanester, Morbihan, Bretagne.* In: WOLFSCHMIDT, GUDRUN (ed.): *Wandlungen in Raum und Zeit.* Hamburg: tredition (Nuncius Hamburgensis; Vol. 58) 2023, p. 137–153.

[Nadler (2011)] NADLER, MARTIN: Spätneolithische Stelen und Petroglyphen? Zu einer Neubewertung der sog. Zeichensteingräber im mittleren Regnitztal. In: BEIER, HANS-JÜRGEN; EINICKE, RALPH & ERIC BIERMANN (Hg.): *Varia Neolithica* VII (2011), S. 183–210.

[Rappenglück 1999] RAPPENGLÜCK, MICHAEL A.: Palaeolithic Timekeepers Looking at the Golden Gate of the Ecliptic. The Lunar Cycle and the Pleiades in the Cave of La-Tête-du-Lion (Ardèche, France) -21,000 BP. In: *Earth, Moon, and Planets* Volumes **85/86** (1999), p. 391–404.

[Rappenglück 2003] RAPPENGLÜCK, MICHAEL A.: The Anthropoid in the Sky: Does a 32,000 Years Old Ivory Plate Show the Constellation Orion Combined with a Pregnancy Calendar. In: *Calendars, Symbols, and Orientations: Legacies of Astronomy in Culture.* Proceedings of the IXth Annual Meeting of the European Society for Astronomy in

Culture (SEAC), Stockholm, 27–30 August 2001. Uppsala: Astronomical Observatory (Report No. 59) 2003, p. 51–55.

[Rappenglück 2008] RAPPENGLÜCK, MICHAEL A.: *The Pleiades and Hyades as celestial spatiotemporal indicators in the astronomy of archaic and indigenous cultures.* In: WOLFSCHMIDT, GUDRUN (ed.): *Prähistorische Astronomie und Ethnoastronomie.* Norderstedt: BoD (Nuncius Hamburgensis; Vol. 8) 2008, p. 13–39.

[Ruggles 2015] RUGGLES, CLIVE L. N.: Long-term changes in the appearance of the sky. In: RUGGLES, CLIVE L. N. (ed.): *Handbook of Archaeoastronomy and Ethnoastronomy, vol. 1.* New York: Springer 2015, p. 473–482.

[Shelley – Possible stone age star map in Kent] SHELLEY, THOMAS: *Possible stone age star map in Kent,* `https://www.academia.edu/32812217`.

[Steinrücken (2004)] STEINRÜCKEN, BURKARD: *Archäoastronomische Untersuchung des neolithischen Palisadenrondels auf der Schalkenburg, mit einer Tabelle pähistorischer Fixstern-Äquatorialkoordinaten von 5000 bis 2500 v. Chr.* (2004), `https://sternwarte-recklinghausen.de/data/uploads/dateien/pdf/b01_schalkenburg.pdf`.

[Steinrücken – Sternphasen der Plejaden in der Bronzezeit] STEINRÜCKEN, BURKARD: *Die Sternphasen der Plejaden in der Bronzezeit. Zur theoretischen Berechnung und kalendarischen Verwendbarkeit von Sternphasen,* `https://sternwarte-recklinghausen.de/data/uploads/dateien/pdf/a05_plejadenphasen.pdf`

[Steinrücken 2008] STEINRÜCKEN, BURKARD: *A dynamical Lunar-Solar interpretation of the sky disk of Nebra.* In: WOLFSCHMIDT, GUDRUN (ed.): *Prähistorische Astronomie und Ethnoastronomie.* Hamburg: tredition (Nuncius Hamburgensis; Band 8) 2008, p. 169–185.

[Steinrücken 2010] STEINRÜCKEN, BURKARD: Die Dynamische Interpretation der Himmelsscheibe von Nebra. In: MELLER, HARALD & FRANÇOIS BERTEMES (Hg.): *Der Griff nach den Sternen. Wie Europas Eliten zu Macht und Reichtum kamen.* Internationales Symposium in Halle (Saale), 16.–21. Februar 2005. Halle an der Saale: Landesamt für Denkmalpflege und Archäologie in Sachsen-Anhalt (Tagungen des Landesmuseums für Vorgeschichte Halle; 5) 2010, S. 935–945.

[Sparavigna 2008] SPARAVIGNA, AMELIA: *The Pleiades: the celestial herd of ancient timekeepers.* 2008, `https://www.researchgate.net/publication/2209916`.

[Tschuch 2017] TSCHUCH, MATTHIAS: *Vom Endneolithikum bis zur Hallstattzeit – Ausgrabungen an der ICE-Trasse um Kersbach.* Vortrag bei der Tagung *Archäologie in Ober- und Unterfranken* des Bayerischen Landesamtes für Denkmalpflege, in Forchheim, Mai 2017.

Abbildung 8.1:
Wandstein (c) aus Grab Göhlitzsch, Landesmuseum für Vorgeschichte, Halle

(Foto: Karin Albrecht, 2020)

Im zyklischen Rhythmus des Kosmos liegt das Schicksal der Menschen – Die bildhafte Zeit im „Heydnischen“ Grab von Göhlitzsch-Halle/Leuna

Klaus Albrecht (Kassel)

Abstract: In the cyclical rhythm of the cosmos lies the fate of man – The pictorial time in "Heydnian" grave of Göhlitzsch-Halle/Leuna

The stone chamber grave of Göhlitzsch-Halle/Leuna occupies a special place among the Neolithic finds of Central Europe.[1] It is one of several burial chambers embedded in mounds in the Halle-Saale area that have been excavated over the years. Its internally ornamented stone slabs are remarkably rich and well preserved. The history of the find as well as the contemporary description of the "Heydnian Prince's Grave" by Moritz E. Hoppenhaupt in 1750 has attracted the special interest of many authors. Extensive literature has dealt with the pictorial and ornamental decoration of the tomb. Up to now, it has been assumed that the "decorations" inside the tomb are due to the decorative needs of the people of that time, similar to the corresponding decorations on the pottery. Images of weapons were also seen. The so-called "ornaments", however, are not only formal decorations but above all the fixation of time units and calendar representations. Lunar cycles are associated with cyclical processes in the solar year. In the cyclical rhythm of the cosmos lies the fate of human beings in life as well as in death. The interior decoration of the stone chamber tomb at Göhlitzsch/Leuna in comparison with similar motifs in Neolithic iconography confirms the archaic belief in selenic powers in early historical

1 Summary of a text by Klaus Albrecht: "Renewed detailed description of an ancient Heydnian tomb – In the cyclic rhythm of the cosmos lies the fate of mankind", *Archaeo-Astronomische Notizen*, Naumburg, June 13, 2023, and lecture at the conference of the *Society for Archaeo-Astronomy* in Weimar, June 23, 2023.

burial culture. The idea of a path of the eternally living individual soul through the underworld and the upper world is part of religious sentiment, which was already cultivated in the Neolithic and experienced the most diverse forms in the following period. In its pictorial form, time is to be understood as a symbol of becoming and passing away, and connected with this is the hope of eternal life.

Zusammenfassung

Das Steinkammergrab von Göhlitzsch-Halle/Leuna nimmt einen besonderen Platz unter den neolithischen Funden Mitteleuropas ein.[2] Es ist eines von mehreren in Hügeln eingebetteten Grabkammern im Halle-Saalebereich, die im Laufe der Jahre ausgegraben worden sind. Seine innenornamentierten Steinplatten sind bemerkenswert reichhaltig und gut erhalten. Die Fundgeschichte als auch die zeitgenössische Fundbeschreibung des „Heydnischen Fürstengrabes“ von Moritz Ehrenreich Hoppenhaupt 1750 hat das besondere Interesse vielen Autoren gefunden. In umfangreicher Literatur fand eine Auseinandersetzung mit den bildlichen bzw. ornamentalen Ausschmückungen des Grabes statt. Bisher wird davon ausgegangen, dass die „Verzierungen“ im Grabinneren dem Schmuckbedürfnis der damaligen Menschen, ähnlich wie auf zeitlich korrespondierenden Verzierungen auf der Keramik zu verdanken ist. Ebenso wurden Waffenabbildungen gesehen. Bei den sogenannten „Verzierungen“ handelt es sich aber nicht nur um formale Ausschmückungen sondern vor allem um die Fixierung von Zeiteinheiten und Kalenderdarstellungen. Mondzyklen werden in Zusammenhang mit zyklischen Vorgängen im Sonnenjahr gebracht. Im zyklischen Rhythmus des Kosmos liegt das Schicksal der Menschen im Leben wie im Tod. Die Innendekoration des Steinkammergrabes von Göhlitzsch/Leuna im Vergleich zu ähnlichen Motiven in neolithischer Ikonographie bestätigt den archaischen Glauben an selenischen Kräfte in der frühgeschichtlichen Bestattungskultur. Die Vorstellung von einem Weg der ewig lebendigen individuellen Seele durch Unterwelt und Oberwelt ist Teil religiösen Empfindens, welches schon im Neolithikum gepflegt wurde und in der Folgezeit die verschiedensten Ausformungen erfahren hat. In der bildhaften Form ist Zeit als Symbol für Werden und Vergehen zu verstehen, und damit verbunden ist die Hoffnung auf ein ewiges Leben.

2 Zusammenfassung eines Textes von Klaus Albrecht: „Erneute ausführliche Beschreibung eines alten heydnischen Grabes – Im zyklischen Rhythmus des Kosmos liegt das Schicksal der Menschen“, *Archäo-Astronomische Notizen*, Naumburg 13.6.2023, und Vortrag auf der Tagung der *Gesellschaft für Archäoastronomie* in Weimar, 23.06.2023.

8.1 Fundgeschichte und bisherige Interpretationen der Innenverzierungen

Das Göhlitzscher Grab wurde am 18.04.1750 vom Rittmeister Carl Leberecht von Wuthenau auf einem Plateau in der Nähe des Ortes Göhlitzsch an der Saale entdeckt. Eine Kammer von 2,30 m × 1,10 m war von einem Erdhügel bedeckt. Sie wurde ausgegraben und von Stiftsbaumeister Moritz Ehrenreich Hoppenhaupt aufgenommen. Die sechs Wandsteine sind innen bemerkenswert reichhaltig ornamentiert und gut erhalten. Sie wurden zunächst im Schlossgarten in Merseburg aufgestellt. Jetzt sind sie im Landesmuseum für Vorgeschichte in Halle ausgestellt. Hoppenhaupts wissenschaftliche ausführliche Beschreibungen und Zeichnungen vom Juli 1750 sind für die damalige Zeit einmalig. Sie sind Grundlage für weitere Untersuchungen in den folgenden Jahrhunderten. Besonders erwähnenswerte ist die Veröffentlichung von Kaufmann und Matthias mit Faksimiledrucke der Arbeit von Hoppenhaupt.[3]

Neben einigen Darstellungen von Gegenständen scheinen reine Ornamente ohne Bedeutung das Grab zu schmücken. Moritz E. Hoppenhaupt ging davon aus, laut Tacitus, dass es sich um eine Grabstätte eines Heeresführer der Chatten gehandelt hat, der im Kampf mit den Hermuduren im Jahre 59 nach Christi Geburt gefallen ist.[4] Seine Interpretationen der Gravuren an den Wandsteinen sind geprägt von der Vorstellung, dass militärisches Gerät in dem Grab dargestellt wird. Er meinte Pfeil und Bogen, Streitäxte und Schilde und sogar militärische Schlachtordnungen zu erkennen.[5] Den zeitlichen Vermutungen muss man nicht folgen, weil Funde aus dem Grab (Steinbeil) eine spätneolithische Entstehungszeit nahelegen. Die militär- und zeitbezogenen Interpretationen von Hoppenhaupt wurden nur zum Teil von späteren Forschern übernommen. Sie wurden allerdings auch nicht kritisiert.

1883/84 erfolgte durch Friedrich Klopfleisch eine Untersuchung des Göhlitzscher Grabes, der bedauerte, dass der Ostgiebel nicht im ursprünglichen Zustand belassen, sondern mit willkürlich abweichenden Ornamenten in Merseburg bemalt wurde. Er ordnete das Göhlitzscher Grab in die Zeit der „Bandkeramik“ und „Schnurkeramik“ die beiden wichtigsten neolithischen Kulturen in Mitteleuropa. In den Gräber auf der Dölauer Heide nahe Halle, die in den fünfziger Jahren des letzten Jahrhunderts ausgegraben wurden, fanden sich ähnliche Gravuren an den Wandsteinen. Die Gräber werden heute der Bernburger Kultur (ca. 3200 v.u.Z.) zugerechnet.[6]

Die Ornamente werden von den meisten Forschern als Fischgräten, Tannenbäume oder Bäume bezeichnet. Hier werden rein formal Ähnlichkeiten zu Gegenständlichem

3 Kaufmann, Matthias: M. E. Hoppenhaupt – Ausführliche Beschreibungen eines alten Heydnischen Grabes, zum hundertjährigen Bestehen des Landesmuseums für Vorgeschichte Halle (Saale). Berlin 1984.

4 Kaufmann 1984, S. 40.

5 Kaufmann 1984, S. 39.

6 Behrens, Faßhauer & Kirchner (1956), S. 13–50.

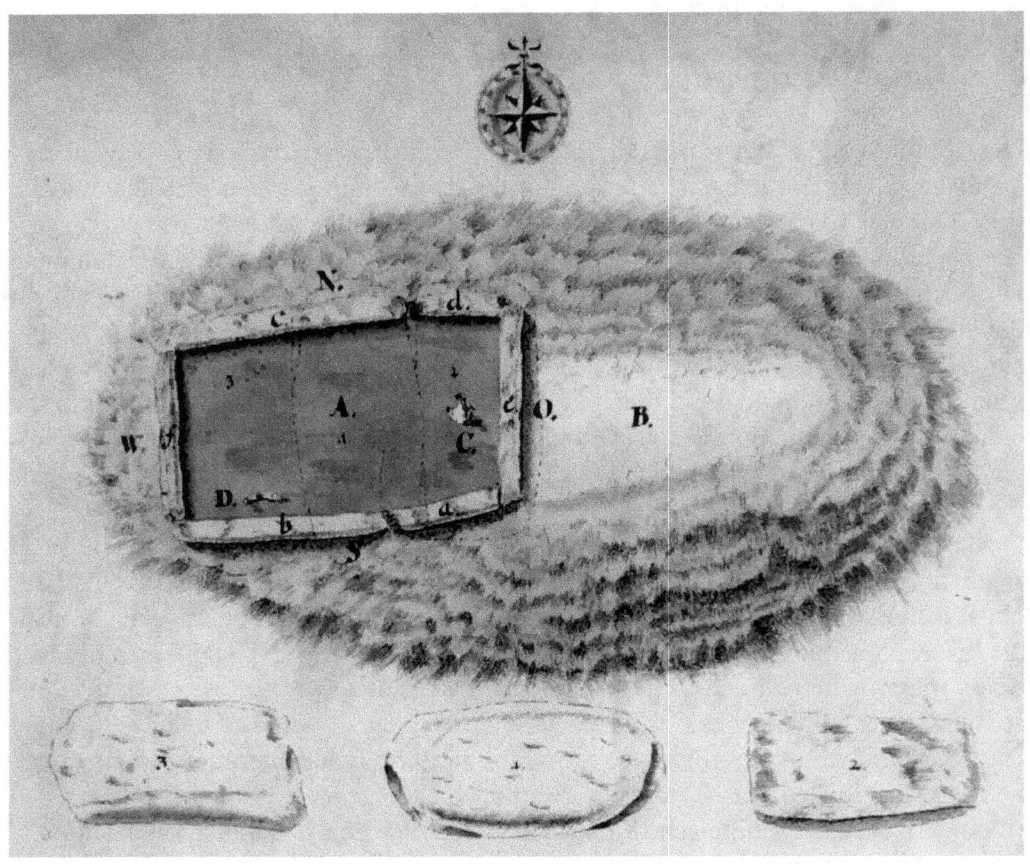

Abbildung 8.2:
Zeichnung von Hoppenhaupt 1750, Grundriss Grab Göhlitzsch: sechs Wandsteine und drei Decksteine - Ausrichtung Ost-West, Innenmaße: 2,30 m x 1,10 m x 0,75 m

(Kaufmann 1984)

gesehen. Zick-Zack-Muster sollen Wellen bzw. Wasser, ein lebensspendendes Element repräsentieren. Das Letztere führten einige auf das ägyptische Schriftzeichen für Wasser zurück. Damit zog man eine kulturgeschichtliche Verbindung in den vorderasiatischen Bereich, der sehr gewagt ist.

Im Weiteren versuchte man Beziehungen zu Textiltechniken herzustellen. Tücher, Wandbehänge, Teppiche in Web- oder Knüpftechnik sollten mögliche Vorbilder für die Wandverzierungen im Grab sein. Hier bieten flächige Ornamente Ansatzpunkte.

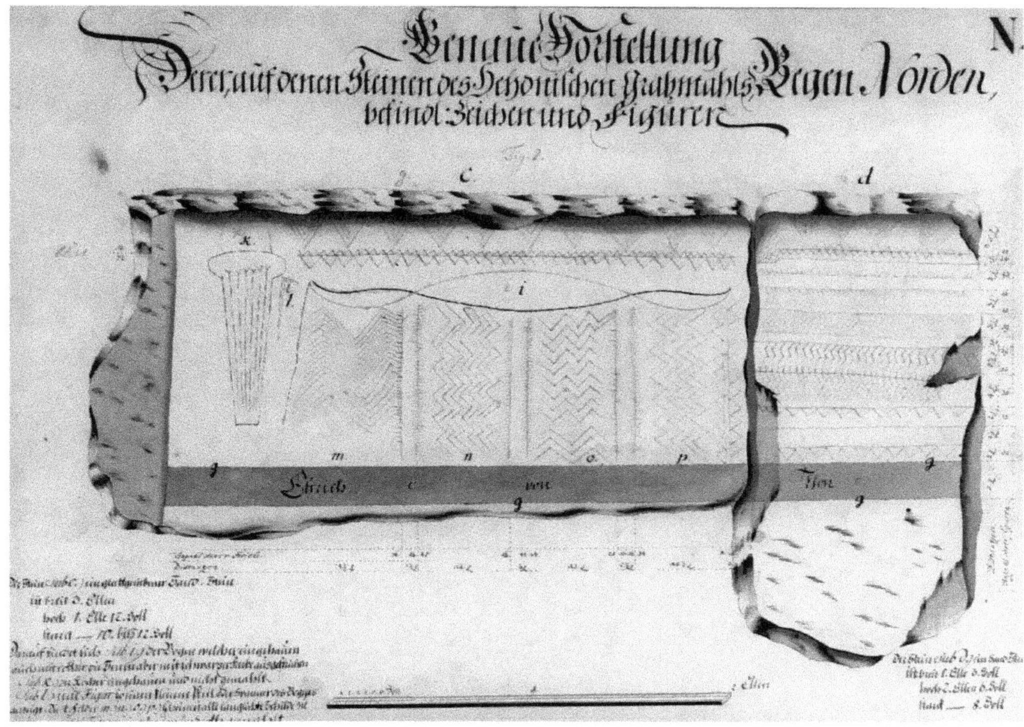

Abbildung 8.3:
Zeichnung von Hoppenhaupt 1750, Ritzungen auf der Nordseite – Stein (c) mit „Köcher, Axt und Bogen“

(Kaufmann 1984)

Eine Zeichnung von Berger 1915 zeigt den Versuch, konstruktiv die Ornamente des Grabes nachzuempfinden, wie sie in Stoffen, wie an Hauswänden verwendet wurden.[7]

Dem Vergleich mit Textilien wurde von Karl Schlabow, einem Experten für vorgeschichtliche Textiltechnik, widersprochen. Muster wie in dem Grab konnten mit steinzeitlichen Mittel nicht hergestellt werden und wurden auch nicht gefunden. Er plädiert dagegen für Kratz- und Sgraffitomalerei, die in weichen Lehmwänden von Häusern innen und außen zum Einsatz kamen.[8]

Die Dekorationen wurden auch in den dreißiger Jahren des letzten Jahrhunderts inhaltlich nicht analysiert. Zum Beispiel wurden bei den Deckenmalereien im Leunaer Rathaus Ornamente aus dem Göhlitzscher Grab nachempfunden. Der Stil korrespon-

7 Berger, Fundstellenarchiv LDA Sachsen-Anhalt,1915.
8 Kirchner 1956, S. 27.

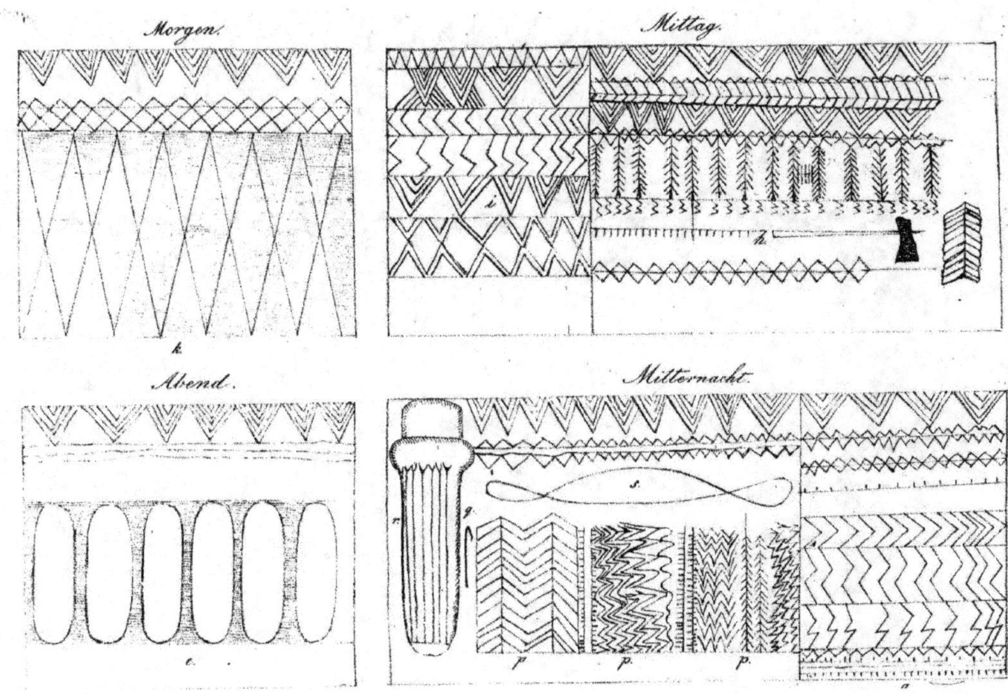

Abbildung 8.4:
Berger 1915: Ornamente wie auf Stoffen und Hauswänden

(LDA, Sachsen-Anhalt, Archiv)

diert mit Jugendstilformen bzw. Muster des Art Déco. Sie wurden im letzten Krieg größtenteils zerstört und danach zum Teil restauriert.

In einer Untersuchung von Kirchner (1956) finden sich Vergleiche mit „Schmuckelementen", in westeuropäischen Grabkammern und auf Megalithen zum Beispiel in der Bretagne oder Irland. Besonders erwähnt er das Steinkammergrab von Züschen mit dessen Einritzungen. Es gäbe ausreichend Vergleichsmaterial, womit eine neue zeitliche Einordnung der Gräber nämlich ins Neolithikum in die Bernburger Kultur 3000 v.u.Z. vorgenommen werden könnte.[9]

Nichts anderes findet sich bei Hinweisen und Beurteilungen von Detlef W. Müller für diverse Kammergräber im Saalebereich. Er unterscheidet Bilder mit Waffen und Gerät, sowie dekorativ symbolische Elemente als auch anthropomorphe Elemente. Die

9 Kirchner 1956, S. 27 ff.

Waffendarstellungen werden nicht in Frage gestellt. Bei den dekorativen Elementen sieht er die rein schmückende Komponente, findet aber auch Symbolwerte.[10]

Wieder aufgegriffen wurde die Analyse und Interpretation der neolithischen Zeichen und Symbolwelt von Torsten Schunke 2013. Sie gibt neben der Beschreibung des Grabes den aktuellen Stand der Interpretation für die Verzierung im Grab wieder.[11]

Neben der übernommenen Vorstellung, dass es sich im Göhlitzscher Grab hauptsächlich um ornamentale Verzierung handelt, versucht er die Waffenmotive fortzuführen. In einem Vergleich zu Darstellungen aus einem Grab in Klady (Kaukasus) findet er in Göhlitzsch auf dem Stein (c) neben Köcher, Pfeilen und Axt einen Bogen abgebildet. Es handele es sich um einen Reflexbogen, eine spezielle Waffenform.

Schunke beschreibt die Schwierigkeit hochkomplexe Zeichengeflechte aus Kulturen, die ihr Wissen oral nicht schriftlich bewahrt haben, zu entschlüsseln. Kammmotive, Tannenzweiglinien, ährenförmige Zeichen in Zusammenhang mit Sonnen- und Mondbildern oder gar Rinderdarstellung assoziieren seiner Meinung nach Fruchtbarkeitssymbolik. Weitergehende Interpretationen von Ankermotiven oder Zickzacklinien entzögen sich unseren Möglichkeiten. Die weitgehende Zurückhaltung bei der Interpretation der vorgefundenen steinzeitlichen Symbolwelt sieht er berechtigt aufgrund des Mangels an schriftlichen Belegen.[12]

8.2 Beschreibung und neue Interpretationen der Innenverzierungen

Im Folgenden soll der Versuch unternommen werden, einen neuen Aspekt in die Interpretationsgeschichte einzuführen, der sich auf das Weltbild vorgeschichtlicher Kulturen bezieht. Es ist von einer komplexen Vorstellungswelt der Menschen in der Frühzeit auszugehen. Diese kommt anscheinend in dem Göhlitzscher Grab zum Ausdruck.

In dieser Grabkammer findet sich eine umfangreiche, komplizierte Ornamentik, neben erkennbaren gegenständlichen Darstellungen, die aus verschiedenen Stilelementen und Ordnungen besteht. Alle Gravuren sind anscheinend einem gewissen System oder Konzept untergeordnet. Es gibt Reihungen von Dreiecken und Strichen. Gruppierung in Feldern, Symmetrien. Aber auch das Gegenteil von Ordnung und Wiederholung findet sich. Neben sauber „ordentlich“ ausgeführten Teilen verfließen manche Darstellungen vor allem in einem Ablauf von links nach rechts. (Möglicherweise der Arbeitsweise und Nachlässigkeit von rechts nach links geschuldet). Alle Formen sind

10 Müller 1994, S. 20 ff.
11 Schunke 2013, S. 152.
12 Schunke 2013, S. 262.

aus der neolithischen und auch späteren Zeit bekannt. Der Versuch in den Elementen von Kamm-, Tannenbaum-, Fischgrätenmotiven, Zickzacklinien und Dreiecken, abstrahierte Gegenstände entdecken zu wollen, scheint vergeblich zu sein. Man kann bei den abstrakten ornamental wirkenden Zeichensystemen keine einfache Interpretation vorgenehmen. Zunächst einmal ist festzuhalten, dass es sich bei den Ornamenten um umfassende Systeme von Zahlen und Reihungen handelt.

Einteilen von Flächen, Abzählen von Elementen, Linien und Punkten bis hin zur farbigen Differenzierung, erforderte ein hohes Maß an Übung und Wissen. Die Kunst Keramik, Holz, Lehmwände im Häuserbau oder Stein zu „verzieren“ ist mit geometrischen Kenntnissen verbunden bzw. zu entwickeln. Gezählt wurden Gegenstände des täglichen Umgangs. Nahrungsmittel, Arbeitsmittel, Baumaterial, Handelsgüter, Wegelängen und Orientierung im Gelände ließen sich zahlenmäßig erfassen. Daneben kamen sofort Elemente wie Zeiteinheiten für Tage, Monate und Jahre in Betracht. Hierauf soll in der vorliegenden Untersuchung besonders eingegangen werde. Das Beobachten der Gestirne am Himmel und die richtige Erfassung ihres zyklischen Verhaltens war eine wichtige Bedingung für die Orientierung in den klimatisch unterschiedlichen Jahreszeiten. Sie beeinflussten das menschliche Leben insbesondere des sesshaften, ackerbaubetreibenden Menschen der Steinzeit. Natürlich spielten die Jahreszeiten auch bei der Viehzucht oder Jagd eine Rolle, weil davon die Verfügbarkeit der tierischen Nahrungsmittel abhing. Kalender wurden erstellt um Jahresfeste, Zusammenkünfte, Markttage terminlich festzulegen. Werden und Vergehen, Leben und Tod von Mensch und Natur, das große Mysterium, versuchte man in zahlenmäßige Größen zu erfassen. Als Beispiel für grafische Verwendung von Mondphasen seien hier Schilder von Mitgliedern des Bobo Stammes im Sudan vorgestellt. Die damit verbundenen Rituale dienten der Bitte an den Wettergott um Regen. Anzahl der weißen Dreiecke auf den Schildern sind von links nach rechts: 28, 29, 30, 28.

Das Bedürfnis Gegenstände über ihren Gebrauchswert hin „ästhetisch“ zu gestalten, scheint sich aus diesen Fähigkeiten abzuleiten. Das Bedürfnis ist grundlegend in der menschlichen Kultur, wobei die Ästhetik abhängig von Inhalt, Form und Tradition ist.

Dass sich eine metaphysische Ebene bei den Zahlen einstellte, deren Herkommen, im Laufe der Zeit sich nicht einfach nachvollziehen lässt, erschwert die Interpretationen symbolhafter Zeichnungen aus der Vorgeschichte. Eine Magie der Zahlen, beruht auf Gegebenheiten, die in der realen Welt zu finden sind. Sie werden von Zahlenmagiern als Metapher angewandt und sie begründen die Kabbalistik schon in der Frühzeit der Kulturen.

Mit diesen Einschätzungen der besonderen Bedeutung von Zahlen, Wiederholungen, rhythmischer Gestaltung sind Voraussetzungen für die Interpretation der vorliegenden „Grabverzierungen“ gegeben. Wenn von eine Methodik bei der Analyse der Symbole und Zeichen der Rede ist, dann wurde von mir die Suche nach gleichartigen Darstellungen mit der Suche nach Zahlengruppen verbunden. Z. B. Zahlen wie

Abbildung 8.5:
Mondphasen auf Schildern des Bobo Stammes im Sudan

(Dolye, Robert: Das mystische Jahr - Geheimnisse des Unbekannte. Amsterdam: Time-Life Bücher 1992)

Sieben, Zwölf, Dreizehn, Neunundzwanzig und Fünfzig lassen sich insbesondere aus astronomischen, kalendarischen Zusammenhängen erklären.

Im Rahmen der vorliegenden Untersuchung kann nur auf die wichtigsten Ergebnisse eingegangen werde. Beschreibungen erfolgen hier in einer Reihenfolge, die einen Zugang zu den Inhalten erleichtern. Die drei Steine (c), (b) und (d) (entsprechend der Bezeichnung von Hoppenhaupt) werden in besonderen Zusammenhang gebracht und ausführlicher behandelt. Auf dem Stein (c) ist die Einteilung des Jahres in vier

Abschnitte festgehalten. Auf Stein (b) finden wir eine Einteilung des Jahres in zwölf Monate und auf Stein (d) ist ein Jahreskalender mit der Anzahl der Jahrestage zu erkennen. Auf allen Steinen zusammen kann man 50 Dreiecke zählen, die die Anzahl der Wochen im Jahr angeben. Die Abschlusssteine (e) und (f) sollen hier aus Platzgründen nicht weiter betrachtet werden, aber auch sie beinhalten Zahlen und Formationen, die jahreszeitlich Bezüge haben.

8.2.1 Darstellungen „Köcher, Axt, Pfeil und Bogen“ auf Stein (c)

Der Bau und die Ausgestaltung eines Grabes sind von der Absicht getragen, einen Bezug zum Tod im Allgemeinen oder im Besonderen zur Person des Bestatteten herzustellen. Stammesführer oder Häuptlinge waren durch Herrschaftsattribute öffentlich herausgehoben. Äxte oder Hämmer waren Zeichen der Macht. Bei ihrem Tode nahmen sie die Insignien mit ins Grab oder sie wurden vergraben. In dem Grab von Göhlitzsch fand man eine perfekt geschliffene Axt aus schwarzem Marmor und gleich zweimal als Zeichnungen von Äxten auf den Steinen (b) und (c). Hier scheint sich die Vermutung Hoppenhauptes zu bestätigen, dass es sich um ein Grab höhergestellter Persönlichkeit gehandelt haben muss. Daneben werden die Zeichnungen mit dem auffälligen Schleifenmotiv, dem „Bogen“, einem „Köcher mit Pfeilen“ bisher als Waffen angesehen. Diese können aber ganz anders gesehen werden.

Den von allen Autoren bezeichneten Köcher im Göhlitzscher Grab werte ich als Helm eines Stammeshäuptlings. Kleidung, Schmuck oder Kopfbedeckungen eines Herrschers waren besonders ausgeprägt. Die Abbildung des Köchers im Göhlitzscher Grab ist in der Form vergleichbar mit den Goldhüten von Priesterhäuptlingen aus der Bronzezeit. Er zeichnet sich durch eine Krempe aus, wie sie an den Goldhüten zu finden sind. Der Henkel, der bei einem Köcher unerklärlich wäre, ergibt als Kinnriemen einen Sinn. Die „Pfeile“ in seinem Inneren – der Köcher müsste durchsichtig sein, um sie zu erkennen – könnten die äußeren Verzierungen auf dem Helm darstellen. Auf den bronzezeitlichen Goldhüten sind komplexe Ornamente zu finden. Wenn auch nicht zweifelsfrei im Detail belegt, ist ein Bezug zu magisch mythischen Zahlenfolgen denkbar. Ein Ablesen von Zeiträumen in Mond- oder Sonneneinheiten wäre möglich, so Wilfried Menghin 2008.[13] Priesterhäuptlinge hätten damit ihr Wissen und Stellung demonstriert. Abbildungen astronomischer Kalenderfunktionen auf Basis eines lunisolaren Systems sind denkbar. Dass der Helm mit der Spitze nach unten gezeichnet wurde, kann damit erklärt werden, dass er die Situation in der Unterwelt repräsentiert, in der alles gegensätzlich zur oberen Welt, der realen Welt, existiert.

13 Menghin, Wilfried: Zahlensymbolitik und digitales Rechnersystem in der Ornamentik des Berliner Goldhutes. Berlin 2008, S. 157 ff.

Abbildung 8.6:
Wandstein (c), Zeichnung

(LDA Sachsen-Anhalt, Archiv, 1994)

8.2.2 Die vier Jahreszeiten auf Stein (c)

Bei dem sogenannten Bogen scheint eine vollkommen andere Deutung sinnvoll. Bei Schunke wird er als Waffe gesehen, die so ähnlich auf einem Grabstein von Klady im Kaukasus abgebildet ist. In Klady ist die Bedeutung als Waffe gegeben, weil dort in der Mitte des Bogens eine Verdickung als Handgriff angesehen werden kann. Es handele sich um sogenannten Recurvebogen (Reflexbogen), der an den beiden Ende aufgebogen ist. Die bikonvexe Form des Bogens erinnert an Bogenwaffen östlicher Reitervölker. Gegen eine Bezeichnung als Bogen in unserem Fall sprechen sowohl die Form als auch die Tradition der Bogenwaffe im vorgeschichtlichen Europa.

Recurvebögen (Reflexbögen) sind in vorgeschichtlichen Zusammenhängen technisch nicht herstellbar.[14] Um eine Form dieser Art zu erhalten sind komplizierte Verleimverfahren notwendig. Der technische Aufwand für die Herstellung eines Recurvebogens und des nicht vorhandenen höheren Spannungseffekts sprechen gegen die Nutzung von solch einem Bogen in der damaligen Zeit. Traditionell waren in der Vor-

14 Auskunft von Gerd Gerhold Bogenbauer zur Technik der Kompositbogen, Naumburg/H.

zeit bis ins Mittelalter in Europa gerade Bögen, zu meist aus Eibenholz, gebräuchlich. Diese sind durch Schnitzen herzustellen und sie waren sehr effektiv. Wenn es diese Bögen tatsächlich gegeben haben sollte, so wären sie höchst selten gewesen und hätten damit auch einen geringen Wiedererkennungswert. Auf Wandmalereien und Steingravierungen der Vorzeit wurden fast ausschließlich gerade Bögen gefunden. Wenn, dann häufig in gespanntem Zustand, was die Anwendung zeigte und die Wiedererkennbarkeit gewährleistete.

Meine Interpretation geht nicht von einer Waffe aus. Die Linie, die sich von links gesehen in einer Kurve unter, über und wieder unter einer geraden Linie bewegt, teilt diese Gerade in drei Teile. Auffallend ist, dass das unter den Kurven liegende Element eine Vierteilungen durch senkrechte Linien aufweist. Diese korrespondiert mit der oberen Dreiteilung, wenn man das Kurvenextrem in der Mitte als Teilung der mittleren Strecke hinzunimmt. Die vier gleich breiten Felder sind mit unterschiedlich gestalteten Zickzacklinien ausgefüllt. Diese senkrechten Trennungslinien sind mit einer Vielzahl kleiner Querstriche versehen. Der „Bogen“ beschreibt meiner Meinung nach einer Vierteilung des Jahres, eine Einteilung nach den Jahreszeiten, Frühling, Sommer, Herbst und Winter. Entsprechend den klimatischen Bedingungen in Mitteleuropa wurde seit alters her eine Vierteilung des Jahres vorgenommen: Erwachen der Natur im Frühling, Aufwuchs im Sommer, Reife und Absterben im Herbst, Ruhe im Winter. Diese Einteilung wurde auf den menschlichen Lebenslauf Geburt, Aufwuchs, Reife und Tod projiziert. Diese Interpretation des „Bogens“ würde gut zur Situation in einer Grabstelle passen.

Die Erfassung der Jahreszeiten kann durch die Beobachtung der Natur erfolgen und in Zusammenhang mit dem Lauf der Gestirne, Sonne, Mond und Sterne gebracht werden. Die Beobachtungen mussten über einen längeren Zeitraum erfolgen. Man brauchte speziell interessierte und ausgebildete Menschen, die die Messungen vornahmen und dokumentierten. Beim Mond waren es die Mondphasen, die das Jahr unterteilten. Dies reichte nicht aus. Ein aufzeichenbarer Ablauf des Jahres konnte erst über die Beobachtung des Sonnenverlaufes gelingen. Dazu dienten sicher Sonnenuhren, die mit ihren Schattenwürfen die Jahreszeit erkennen ließen. Hier liegt nahe, dass mit der Kurve im Grab der Sonnenstand, also die Sonnenhöhen (Deklinationen) gemeint sei könnten. Anhand einer gedachten Linie (Bogen des Himmelsäquators) am Himmel befindet sich die Sonne mal unterhalb und mal oberhalb. So zeigt sich im Lauf des Jahres ein Auf- und Absteigen der Sonne, was sich Jahr für Jahr wiederholt.

Dabei ist die Mittagslinie bei den Sonnenuhren, an dem die Schattenlängen zur Mittagszeit, dem Maximum der Sonnenhöhe, gemessen werden, wichtig. Auf ihr können unterschiedliche Sonnenhöhen einfach festgestellt und markiert werden. Bei einer Horizontalsonnenuhr gilt, je höher die Sonne im Sommer, desto kürzer der Schatten. Je tiefer die Sonne im Winter, desto länger der Schatten. Aber auch Lichteinfälle durch kleine Wandlöcher von Räumen nach Süden orientiert ergeben brauchbare Lichtpunk-

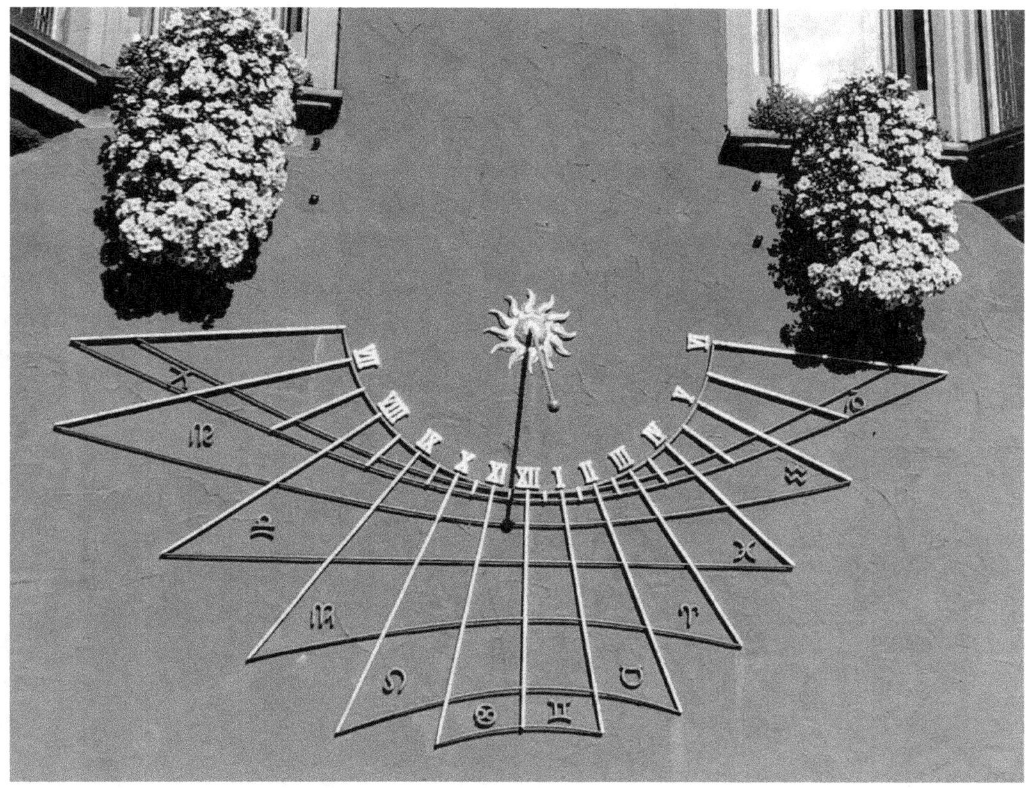

Abbildung 8.7:
Vertikalsonnenuhr am alten Rathaus von Lahr (BW) am 21. Oktober 2023 abgelesene Zeit 11.30 Uhr, Hyperbeln geben die Schattenläufe für die Monate an und die Gerade zeigt den Schattenlauf an den Tagen der Tag- und Nachtgleichen.

(Foto: Karin Albrecht)

te an Böden oder Wänden für die unterschiedlichen Sonnenhöhen im Jahr. Bei einer Vertikalsonnenuhr gilt, je höher die Sonne, desto länger der Schatten nach unten, je tiefer die Sonne desto höher der Schatten nach oben.

Diese Sonnenuhren bilden aber noch keine symmetrische Aufzeichnung der Schattenlängen für das Winter- und das Sommerhalbjahr, welche mit der Kurve im Grab dargestellt ist. Hierfür bedarf es einer weiteren „Erfindung“, die von den frühzeitlichen Astronomen erwartet werden kann: Die Äquatorialsonnenuhr. Die Schatten eines Stabes fallen auf eine schräge Ebene, die sowohl mit ihrer Mittellinie Nord-Süd ausgerichtet ist, als auch mit ihrer Schräge auf den Himmelspol (heute den Polarstern) zeigt. Dabei zeigt der Schattenstab senkrecht in der Mitte des Feldes auf den Punkt am Himmel, auf dem sich Himmelsäquator und Meridianlinie kreuzen. Die Schräge der Ebene ist je nach Breitengrad unterschiedlich und muss jeweils ermittelt werden. Durch diese Sonnenuhr entstehen wiederum Hyperbeln im Lauf der Tage, die jetzt aber symmetrisch zu einer geraden Mittellinie und dem Sonnenstand entsprechend oberhalb und unterhalb einer Linie, dem Himmelsäquator, liegen. Auf der Mittellinie liegen die „Winterschatten" oben und die „Sommerschatten" unten. Jeweils an den Extremen liegen sie dichter aneinander, in der Mitte mit gleichen Abständen.

Werden die Schattenlängen dann übertragen auf eine „Jahresleiste" ergibt sich eine Kurve, wie sie im Grab zu finden ist. Dabei sind die Sonnenhöhen anders als im Schattenwurf, entsprechend der Anschauung am Himmel im Sommer hoch, im Winter niedrig dargestellt. Das Jahr beginnt also im Grab mit den drei Wintermonaten. Interessanterweise bilden die keltischen Jahrestage die Punkte in der Kurve zwischen denen die Kurve jeweils eine Gerade bildet. Von Imbloc bis Beltane und von Lugnasa bis Samhain. Im Ergebnis wird das Jahr in vier Abschnitte geteilt, die annähernd gleich groß sind. Das heißt in unserem Fall beginnen jeweils mit dem Winter (WSW 21.12.), Frühling (TNG 21.3.), Sommer (SSW 21.6.) und Herbst (TNG 23.9.) sind jeweils drei Monate lang bzw. je ca. 91 Tage.

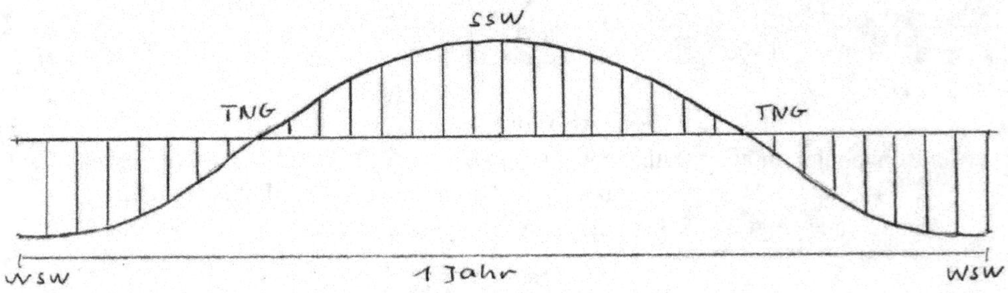

Abbildung 8.8:
Schattenlängen bzw. Sonnendeklinationen
über ein Jahr Winter, Frühling, Sommer, Herbst

(Zeichnung: Klaus Albrecht)

An den Tagen der Tages- und Nachtgleichen ist zur Mittagszeit bei der Äquatorialsonnenuhr kein Schatten zu sehen. Er fällt senkrecht zum Schattenstab. Dass hier nebenbei die Ekliptik, d. h. die Schräglage der Erde zur Sonnenbahn zu erkennen ist, kann nicht die Erkenntnis damaliger Forscher gewesen sein. Der Winkel (Deklination) der durch Schattenlänge und Schattenstab zur Mittagzeit zur Zeit der Sonnenwenden entsteht, entspricht den 23° der Ekliptik.

Für die damaligen Forscher und Kalendermacher ist eine Reihe von praktischen und theoretischen Problemen entstanden. Wahrscheinlich sind über Versuch und Irrtum die Lösungen gefunden worden, um den Verlauf eines Jahres grafisch darzustellen. Dies ist im Selbstversuch, beim Bau einer solchen Äquatorialuhr, nachvollzogen worden. Material, Werkzeuge, Messen, Zählen, Wetter (Wärme, Kälte, Wolken, Regen), genug Beobachtungszeiten, geben genug Imponderabilien ab, die einen Erfolg hinauszögern.

Abbildung 8.9:
Modell einer Äquatorialsonnenuhr; 13.6.23, ca. 13 Uhr 20 (Sommerzeit) Sonnenhöchststand, Polhöhe 52° entsprechend dem Breitengrad bei Kassel 52°, Schattenfläche für das Sommerhalbjahr

(Foto: Klaus Albrecht)

Die Felder unterhalb des Bogens sind in Breite und Länge annähernd gleich groß und entsprechen der Einteilung beim Bogen. Sie sind in ihrer inneren Gestaltung unterschiedlich, ebenso wie die Anzahl der Trennungssenkrechten mit den vielen kleinen Querstrichen. Gleichermaßen sind die inneren Felder mit Zickzacklinien ausgefüllt, allerdings in wechselnden Richtungen und mit nach rechts hin abnehmender Anzahl von Strichen und weniger sorgfältigen Ausführung. Es drängte sich die Vermutung auf, dass die Zickzacklinien, Tage bzw. Mondzyklen darstellen sollen.

Eine zahlenmäßige Erfassung von Tagen bei den schrägen Strichen lässt sich allerdings nicht feststellen, da von einer recht „freien“ Gestaltung ausgegangen werden muss. Es sind im ersten Feld 75, im zweiten Feld 75, im dritten Feld 60, im vierten Feld 49, zusammen 259 schräge Striche. Also kein ganzes Jahr. Dazu kommt, dass auf Grund des Erhaltungszustandes nicht immer die gemeinte Zahl auch gezählt werden kann, durch Störungen im Material, Ausbrüche härtere und weichere Stellen im Sandstein.

Während die Anzahl der Striche im Zickzackmuster in den vier Feldern keine Bestätigung von ca. 91 Tagen ergab, musste ich bei der Untersuchung schließlich annehmen, dass die kleinen Striche auf den trennenden Senkrechten für sich gezählt werden sollten. Sie ergeben zusammen 382 Tage. Diese geteilt durch 29,5 ergeben 13 Mondmonate. Warum 13 Monate und nicht 12? Die Eindeutigkeit lässt hier zu wünschen übrig. Möglicherweise geben die 21 Dreiecke am rechten Rand einen Hinweis auf die Differenz der Tage im Sonnen- und Mondjahr.

8.2.3 Monatszyklen auf Stein (b)

Beim zweiten großen Stein (b) ist von einer Darstellung der Mondzyklen auszugehen. Unter den zwei oben liegenden Friesen mit den großen und kleinen Dreiecken ist eine auffällige Gestaltung, ein längliches, rechteckiges Feld mit einer fast gleichmäßigen Einteilung in zwölf kleine Felder zu sehen. Darunter liegen noch ein paar Dreiecke und die große Streitaxt und ein besonderes Gebilde mit schrägen Strichen.

Während auf dem Stein (c) die Einteilung des Jahres in vier Teile dargestellt ist, kann von einer komplementären Darstellung zum Jahresablauf ausgegangen werden. Die Felder, unterteilt durch Linien mit beidseitiger „Fiederung“, stellen die zwölf Mondphasen eines Jahres dar. Links hätte es noch Platz für einen dreizehnten Monat gegeben. Dort ist der Stein zwar stark beschädigt, es kann am Original aber keine klare Zeichnung gefunden werden.

Die Einteilung eines Jahres mit Hilfe der Mondphasen ist ein archaisches Element. Das Problem aller antiken Kalendermacher war, dass die fehlenden 11 Tage im Mondjahr bis zum Ablauf des tropischen Jahres irgendwie ergänzt werden mussten, wobei man möglichst mit ganzen Tagen bzw. Nächten rechnete. Das konnte durch Beobachtung der Sonnenstände und synodischen Mondphasen geschehen. Jahresfeste und Lostage, wenn sie nicht durch die Jahreszeiten wandern sollten, mussten sich am Sonnenjahr orientieren. Es ergab sich das Problem, dass an bestimmten Jahresfesten die

Abbildung 8.10:
Wandstein (c), Zeichnung

(LDA Sachsen-Anhalt, Archiv, 1994)

Monderscheinung nicht die gleiche war wie ein Jahr zuvor. Z. B. wenn man am 21.12. einen Vollmond sah, war es ein Jahr später nur der abnehmende Mond. Im zweiten Jahr war es der zunehmende Mond und erst im dritten Jahr stimmten Jahrestag und Mondphase wieder einigermaßen überein. 10–11 Tage, ergänzten in jedem Jahr das Mondjahr zum Sonnenjahr und so ergab sich im Laufe von drei Jahren etwa einen dreizehnten Monat, der eingefügt werden musste.

Es finden sich Hinweise auf den Steinen, dass von den Erbauern des Grabes von Göhlitzsch so gerechnet wurde. Zwei waagerechte Reihen mit zwölf und zehn Dreiecken zeigen Korrekturmöglichkeiten auf dem Stein b an. Dass es sich um nur zwei von drei notwendigen Ergänzungen von Drittelmondphasen handelt, ist erstmal nicht erklärlich. Auf Stein (c) finden sich auf der rechten Seite je 10 bzw. 11 Dreiecke. Vermuten kann man, dass zwei Jahre je 10 Tage hinzugefügt wurden. Im dritten Jahr stellt sich dann von selbst die Übereinstimmung von Jahrestag und Mondbild mit einer kleinen Abweichung von einem Tag wieder ein.

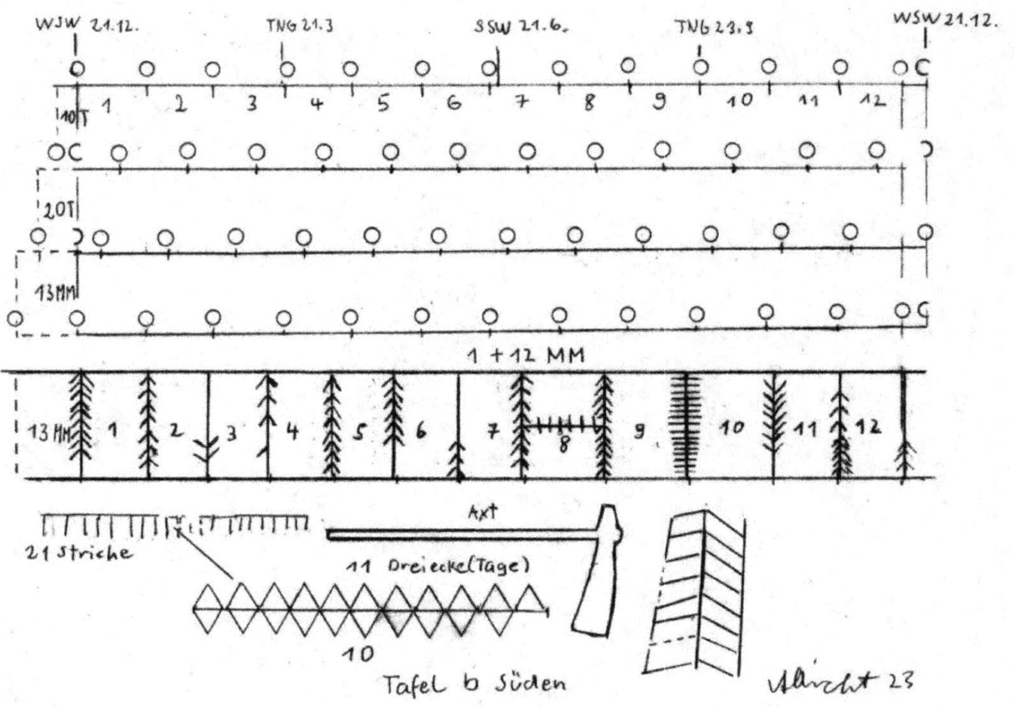

Abbildung 8.11:
Wandstein (b): 3×12 Mondmonate + 1 Mondmonat = Dreijahreszyklus mit 37 MM, jedes Jahr werden 10 Tage zum Mondjahr ergänzt, um ein Sonnenjahr mit ca. 365 Tagen zu erreichen

(Zeichnung: Klaus Albrecht)

Bei der Betrachtung der Details fallen drei Sachen auf. Bei den senkrechten Unterteilungsstrichen sind fast alle mit einer Fiederung versehen, deren Schrägstriche nach untern zeigen. Nur bei dem Dritten von links und dem Dritten von rechts sind die Schrägstriche nach oben gerichtet. Bei dem Vierten von rechts sind es waagerechte Querstriche, die diese Teilungsstriche besonders betonen. Im achten Feld von links findet man einen mittleren Querstrich mit sieben kleinen senkrechten Strichen. Darunter ist ein Element zu sehen, das sich anscheinend auf das direkt darüber liegende dritte Feld von rechts bezieht. Es bietet sich folgende Interpretation an.

Hier werden besondere Lostage oder Wochen im Jahr angezeigt. Leider kann man nicht erkennen, welche es seien könnten. Doch wenn man wie auf Stein (c) annehmen kann, dass das Jahr nach der Wintersonnenwende beginnt, wäre man auf dem Stein (b) im achten und neunten Monat im heutigen August und September mit der besonderen Kennzeichnung. Das wäre die Zeit von Ernte und den entsprechenden Festen in Mitteleuropa, die eine wichtige Rolle spielten.

8.2.4 Jahreskalender auf Steinen (d)

Der Stein (d) ist interessant, weil er als Ergänzung zu den anderen Steinen einen kompletten Tageskalender eines Jahres zeigt. Unter den großen und den kleinen Dreiecken, die im ganzen Grab umlaufend gesehen werden können, befinden sich waagerecht laufende Linien, an denen schräge Striche angebracht sind. Man kann vier zusammenhängende Bereiche erkennen. Darunter finden sich wieder zwei Reihen mit Dreiecken. Die vier Bereiche sind unterschiedlich gestaltet. Bei der Interpretation sollen zunächst nur die Linien mit den anhängenden schrägen Strichen Berücksichtigung finden. Welche Rolle die einzelne Linie im oberen Bereich mit den kurzen Strichen bedeutet, siehe unten. Es lassen sich an den Rändern nicht immer alle Striche wegen den Abbrüchen im Stein eindeutig zählen. Eine bestimmte Größenordnung je Reihe wiederholt sich aber.

Es gibt 13 Reihen mit gegenläufigen schrägen Strichen, die teils miteinander verbunden sind. Die Gegenläufigkeit setzt sich, trotz den Unterbrechungen durch Linien oder Freiräumen, gleichmäßig nach unten fort. Von oben nach unten gesehen ist der erste Bereich mit 43–45 Haken versehen, die man als zwei Reihen von gegeneinander laufenden Schräglinien betrachten kann. Zusammen sind dies 86–90 Striche – Tage – also drei Mondmonate. Im zweiten Bereich liegen Zickzacklinien in drei Reihen mit je 30 Schrägstrichen. Es ergeben sich ebenfalls drei Mondmonate. Im dritten Bereich liegen wieder drei zählbare Strichreihen mit 28 Schrägstrichen, die gegenläufig sind, also ca. drei Mondmonate. Im vierten Bereich sieht man vier Reihen im Zickzack verbunden mit je 28 Schrägstrichen, also vier Mondmonate. Wenn man davon ausgeht, dass hier ein Mondjahr dargestellt werden sollte, hätte man allerdings einen Monat zu viel. Ein dreizehnter Mondmonat würde nur einen Sinn im Zusammenhang mit einem dreijährigen Zyklus ergeben.

Warum im ersten Bereich in zwei Reihen je 1,5 Monate gezählt werden und im unteren Bereich in 4 Reihen je ein Monat gezählt wird, erschließt sich nicht. Erschwerend für eine Interpretation kommen die Dreiecksreihen im unteren Bereich hinzu. Es sind zwei Reihen mit einmal 14 und einmal 11 Dreiecken, die allenfalls Hinweise auf den Dreijahreskalender geben können.

Auf der waagerechten Linie oberhalb der vier Felder sind 43 senkrechte kleine Striche, die mit den darunterliegenden Schrägstrichen korrespondieren, ebenfalls 43 (zählbar 41). Ebenso ist beim vierten Feld zu erkennen, dass auf der Begrenzungslinie oberhalb kleine senkrechte Striche zu sehen sind mit der Anzahl 28 (zählbar 27) die mit den Schrägstrichen darunter wiederum korrespondieren. Möglicherweise dienten die kleinen Striche als Zähl-und Zeichnungshilfe.

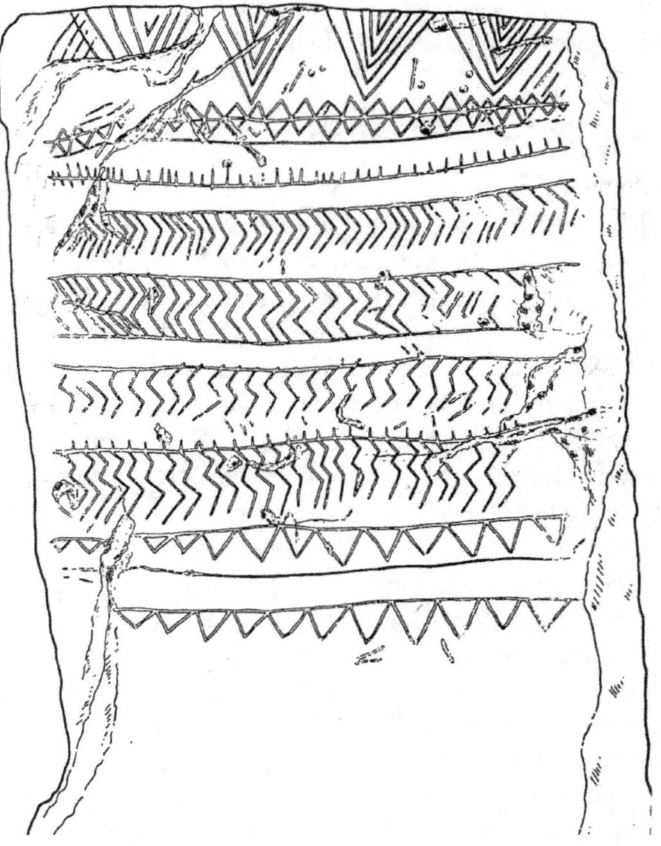

Abbildung 8.12:
Zeichnung Stein (d)

(LDA Sachsen-Anhalt, Archiv 1994)

Bei allem Zählen müssen die Schwierigkeiten bei der Festlegung der Tageszahl für die einzelnen Mondphasen benannt werden. Die siderische Mondphase, der Mondmonat gemessen nach den Positionen im Sternenhimmel beträgt 27,32 Tage, die synodische Mondphase beträgt 29,53 Tage, die Zählung der Tage von Vollmond zu Vollmond. In der Regel findet man Mondphasen von 29 oder 30 Tagen, weil die Erkennung von synodischen Mondphasen leichter war als die der siderischen. In beiden Fällen lassen sich ganze Tage nur durch Auf- oder Abrunden realisieren. 28 wird gern Zahl der Mondmonatstage genommen, weil gut durch 4 teilbar. Die Differenz der Tage des Mondjahres und des Sonnenjahres erschwerte zusätzlich die Darstellung von

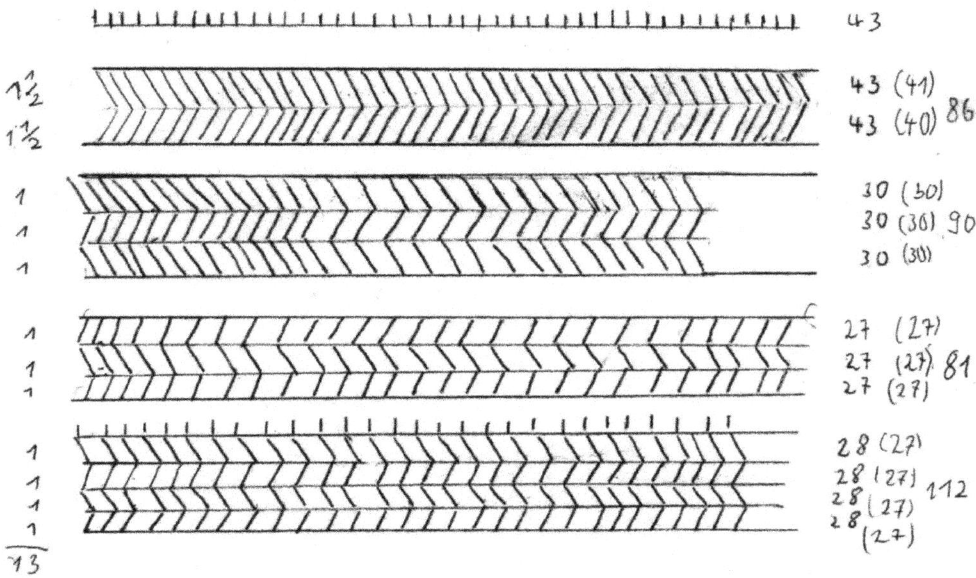

Abbildung 8.13:
Wandstein (d): $1\frac{1}{2}+1\frac{1}{2}+3+3+4=13$ Mondmonate,
1 Mondmonat zwischen 27–30 Tg, 13 MM × 28 Tg = 364 Tg

(Zeichnung: Klaus Albrecht)

Kalendern, bei der in der Kombination der Lunar- und Solarzyklen Ungenauigkeiten vorprogrammiert waren.

Aus den Zählungen am Stein (d) geht hervor, dass mit den Lunationslängen großzügig umgegangen wurde. Mal haben die Monate 27 Striche, mal 28, mal 30 Tage. Zusammengenommen (Rechnung 2) ergeben sie immerhin 368 Tage, fast ein Sonnenjahr mit 365 Tagen. Außerdem variieren innerhalb eines Jahres die Mondphasenlängen bis zu 13 Stunden, was aber damals wahrscheinlich so einfach nicht beobachtet werden konnte. Es ist z. B. schwer, den genauen Tag des Neumondes zu bestimmen, was bekanntlich auch bei dem Beginn und der Beendigung des Ramadan den islamischen Beobachtern Schwierigkeiten gemacht hat.

Interessanterweise ist auf dem gegenüberliegenden Stein a. mit einer ähnlichen Darstellung begonnen worden. Aus irgendeinem Grund ist die Einritzung nicht weiter ausgeführt worden. Ein paar Farbspuren weisen auf dem glatten Stein auf eine Bemalung hin, die ebenfalls Zickzacklinien andeuten. Die vorhandenen waagerechten Linien mit den dazwischenliegenden Haken ergeben zwei Reihen von entgegengesetzt laufenden

30 Schrägstrichen. Also ähnlich wie gegenüber $2 \times 30 = 2$ Mondmonate. Auf diesem Stein wurden aus welchem Grund auch immer die umlaufenden Linien mit den kleinen Dreiecken weggelassen.

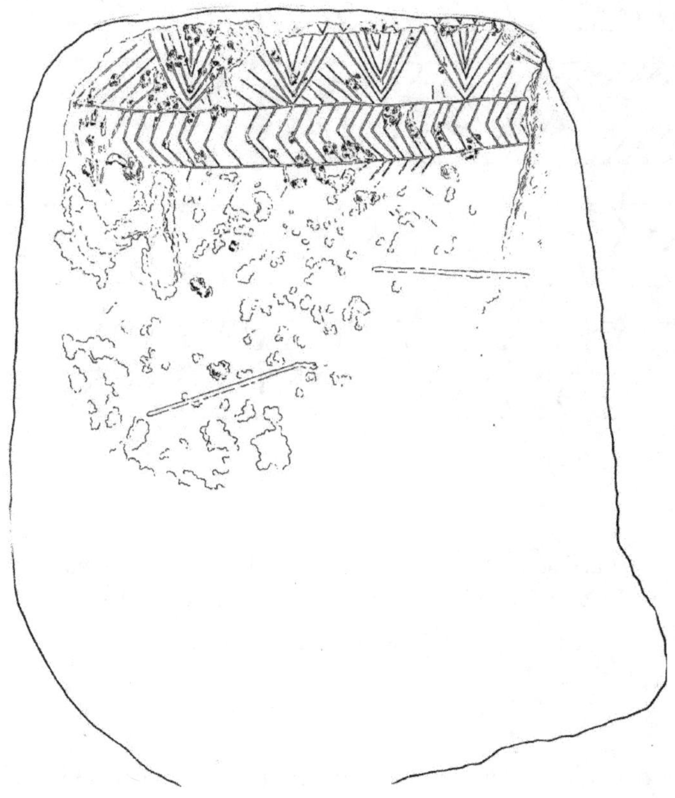

Abbildung 8.14:
Stein (a); 2 Mondmonate mit jeweils 30 und 29 Tagen

(LDA Sachsen-Anhalt, Archiv 1994)

8.2.5 Wandstein (e) im Osten, Wandstein (f) im Westen

Die beiden Kopfsteine (e) und (f) werfen in der Ausdeutung weitere Fragen auf. Es sind flächige Gestaltungen, deren Elemente auf den anderen Steinen nicht vorkommen, sich aber in ihrer Struktur ähneln. Es gibt z. B. keine Zickzacklinien. Von Hoppenhaupt sah dort Schilde oder Harnische, was nicht ernsthaft nachvollzogen werden kann.

Die Gestaltung der Ritzungen bei Stein (e) im Osten, ist nicht sehr sorgfältig ausgeführt. Die Tiefe der Ritzungen ist nicht ausreichend, um insbesondere auf der rechten Seite eine klare Strichführung zu erkennen. Dies liegt wohl daran, dass der Stein aus härterem Sandstein besteht und eine Reihe von tiefen Löchern aufweist, die schon bei der Herstellung vorhanden gewesen sein müssen. In der zeichnerischen Dokumentation kann man das Wesentliche noch erkennen.

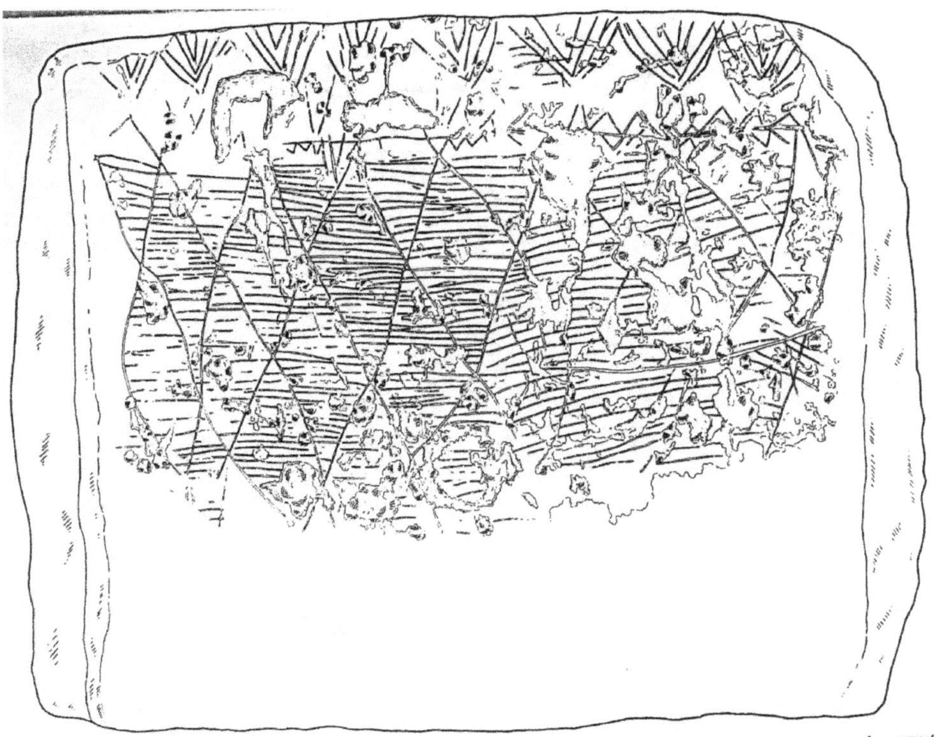

Abbildung 8.15:
Wandstein (e), Zeichnung

(LDA Sachsen-Anhalt, Archiv)

Mit schräg über das ganze Feld laufenden Linien, die sich überkreuzen, entstehen ca. 15 Rauten. In einer Reihe liegen sechs Rauten, wobei wegen der Unklarheiten am rechten Rand es nicht genau gezählt werden kann. Am oberen und unteren Rand des Feldes bilden sich halbe Rauten also 13 Dreiecke. Alle Rauten und Dreiecke sind wiederum mit waagerechten parallelen Linien ausgefüllt. Teilweise sind die parallelen

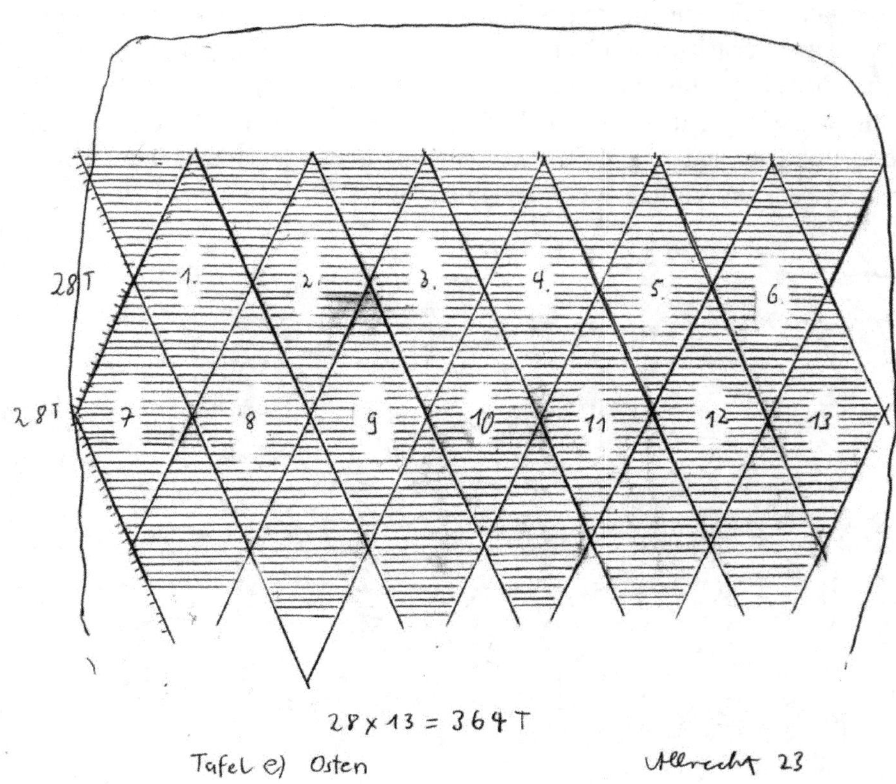

Abbildung 8.16:
Wandstein (e)

(Umzeichnung: Klaus Albrecht)

Linien durchgehend oder setzen an den schrägen Linien an. Die Linien sind alle nicht geradlinig, sondern eher „freihändig" gezogen.

Wenn man versucht die waagerechten Linien zahlenmäßig zu erfassen, ergibt sich überraschenderweise doch eine gewisse Regelmäßigkeit. In den besser erkennbaren 10 Rauten finden sich jeweils 27 bis 30 Striche mit natürlich unterschiedlichen Längen. Fehlen welche, kann man aus den Nachbarfelder Linien verlängern. Ein Kalender ist aufgrund des mangelhaften Zustandes des Steines und der Zeichnung schwerlich auszumachen. Trotzdem gibt es den Bezug auf Mondphasen.

Es bietet sich folgende Interpretation an. Die einzelne Raute bildet die abgeschlossene Mondphase an, in der der Mond sich mit zunehmender und abnehmender Helligkeit zeigt. Ineinander rhythmisch „verwoben" bezeugen die Rauten eine ganze Reihe von Mondphasen. Es entsteht der Eindruck, die Rauten weisen über die Ränder hinaus, es sei eine unerschöpfliche, unendliche Menge. Diese Darstellungsmethode, eines sich unbegrenzt in die Unendlichkeit ausdehnenden Musters, findet sich beispielsweise häufig bei orientalischen Teppichmustern. Dort werden durch die Randbordüren Motive am Rand angeschnitten, um den Eindruck einer unendlichen Fortsetzung zu erwecken, wie auch bei diesem Stein, die angeschnittenen Rauten am Rand. Mondphasen wiederholen sich jetzt und in alle Ewigkeit. Bei den vorliegenden Darstellungen könnte die Ewigkeitsvorstellung mit dem Umstand der Wiedergeburt verbunden werden.

Auf dem Stein (f), der gegenüber im Westen liegt, findet man ein ähnliches Motiv wie auf Stein (e). Unter den auf jedem Stein zu findenden gefüllten Dreiecken und Trennungslinien, befinden sich „eieruhrförmige“ Gebilde, die oben und unten weit und in der Mitte schmaler sind. Drei nebeneinander liegende Gebilde ergeben ovale Figuren, die an den Enden spitz zulaufen. Diese Motive, links beginnend, wurde nicht über die ganze Fläche fortgesetzt. Es hätten von der gleichen Größe noch mal drei Motive hingepasst, allerdings ist die Oberfläche rechts uneben.

In den „eieruhrförmigen“ Gebilden liegen parallel waagerechte Linien, die durch die äußeren Linien begrenzt sind. Zählt man jetzt die Linien in den Figuren von links gesehen, ergeben sich 2 mal 26 und dann 28 Linien. Bei den ersten zwei Figuren kann man aufgrund der unten verwischten Zeichnung noch ein oder zwei Linien hinzuzählen, sodass man auf 28 Mondtage kommt. Mit den unterschiedlichen Längen der Linien könnte das Abnehmen und Zunehmen der Helligkeit des Mondlichtes gemeint sein. Bei den Rauten auf Stein (e) sind die kürzeren Linien oben und unten, während bei den „eieruhrfömigen“ Motiven die längeren Linien oben und unten sind.

8.2.6 Durchlaufende Friese mit Dreiecken auf den Steinen

Auf fünf von sechs Steinen befinden sich am oberen Rand auf der Spitze stehende große Dreiecke. Sie sind ausgefüllt mit Strichen, die die dreieckige Außenform nachvollziehen. Wenn es, wie angenommen, bei der Gestaltung der Steine um die Manifestation von Zeit ging, könnte sie sich in dieser Dekoration wiederfinden. Alle Dreiecke an den oberen Rändern im Grab zusammengezählt, ergibt eine Zahl von 41. Auf dem Stein (b) gibt es eine zusätzliche zweite Reihe, möglicherweise hat der Platz oben nicht ausgereicht, von weiteren 9 gleichartigen Dreiecken. Also finden sich auf allen Steinen zusammen 50 Dreiecke. Fünfzig ist die Wochenzahl für ein Mondjahr bei einer Siebentagewoche mit einem Rest von 4 Tagen. Die Füllung der Dreiecke ist nicht bei allen gleich, unterschiedliche Anzahl von Strichen von 6 und 10 sind festzustellen. Auch zwischen den Dreiecken sind wie bei Stein (a) noch Striche angebracht. Die Füllstriche haben keine bemerkenswerte Zahl, also eher eine Schmuckfunktion.

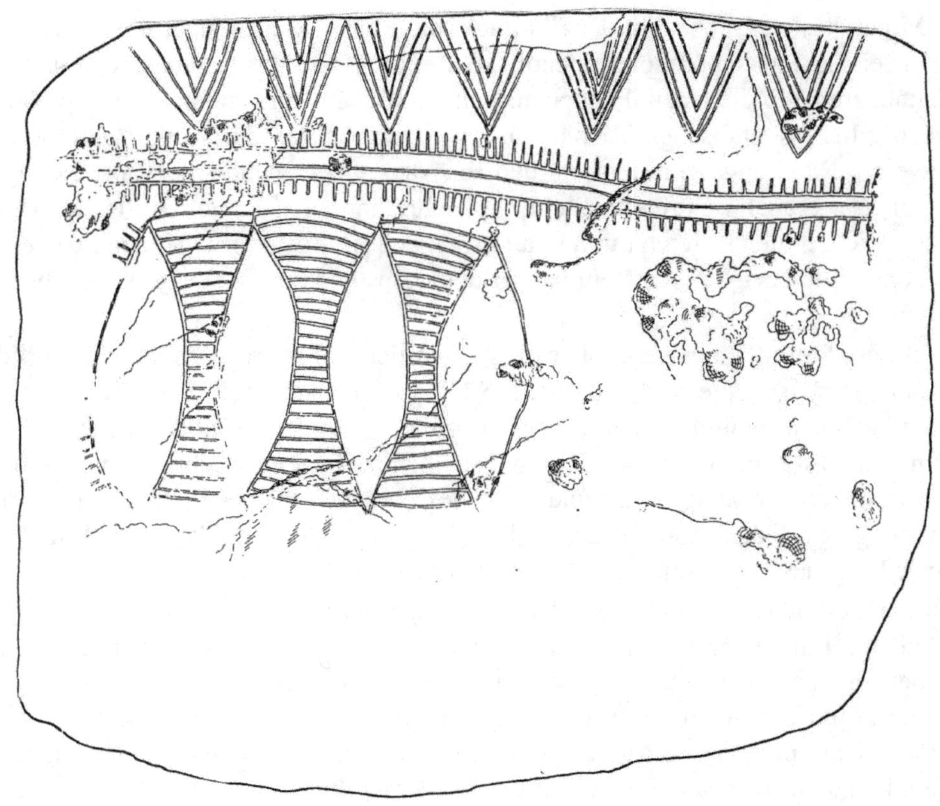

Abbildung 8.17:
Wandstein (f)

(LDA Sachsen-Anhalt, Archiv 1994)

Die Reihe der kleineren Dreiecke, die sich auf einer Linie unter der oberen Reihe befinden, erfahren auch eine Fortsetzung auf anderen Steinen in derselben Höhe bzw. unterhalb der Großen Dreiecke. Ausnahmen sind der Stein (f) und (a). Was bei (f) an kleinen Dreiecken fehlt, wird durch Linien mit Strichen ersetzt und auf dem Stein a. durch ein Feld mit Hacken.

Es handelt sich auf dem Stein (c), z. B. um 82 kleine Dreiecke unterschiedlicher Größe. Links oberhalb der durchgehenden Linie befinden sich zunächst größere Dreiecke, die nach rechts kleiner werden, um ab dem 15. Dreieck wieder an Größe zu zunehmen. Bei dem 30. Dreieck haben sie wieder eine Größe wie am Anfang. Diese Größenverwandlung in der Abfolge wiederholte sich. In der darunterliegenden Drei-

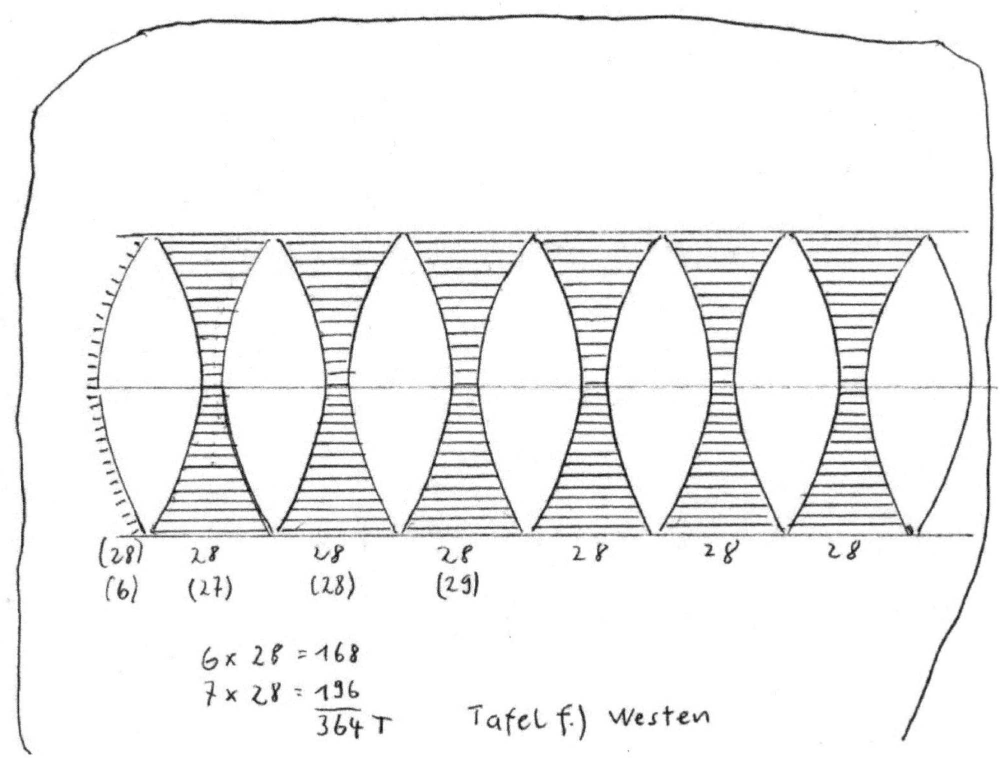

Abbildung 8.18:
Wandstein (f)

(Umzeichnung: Klaus Albrecht)

ecksreihe setzt sich diese Abfolge fort bis ca. 60. Bis 75 werden sie wieder kleiner, um dann bis 82 wieder größer zu werden. Wir haben es hier möglicherweise mit eine Beschreibung des zunehmenden und abnehmenden Lichtes des Mondes zu tun.

Dieses Auf und Ab findet sich auch auf den anderen Steinen, nur nicht so ausgeprägt. Leider ist die Gesamtanzahl aller Dreiecke nicht feststellbar, weil der Erhaltungszustand dies nicht hergibt. Ebenso scheint die Ornamentierung mit den kleinen Dreiecken auf dem Stein b. zwischen den beiden Reihen mit den großen Dreiecken nicht vernünftig auswertbar zu sein, weil nicht komplett. Eine schlüssige Interpretation der Gesamtzahl ist in diesem Falle nicht möglich.

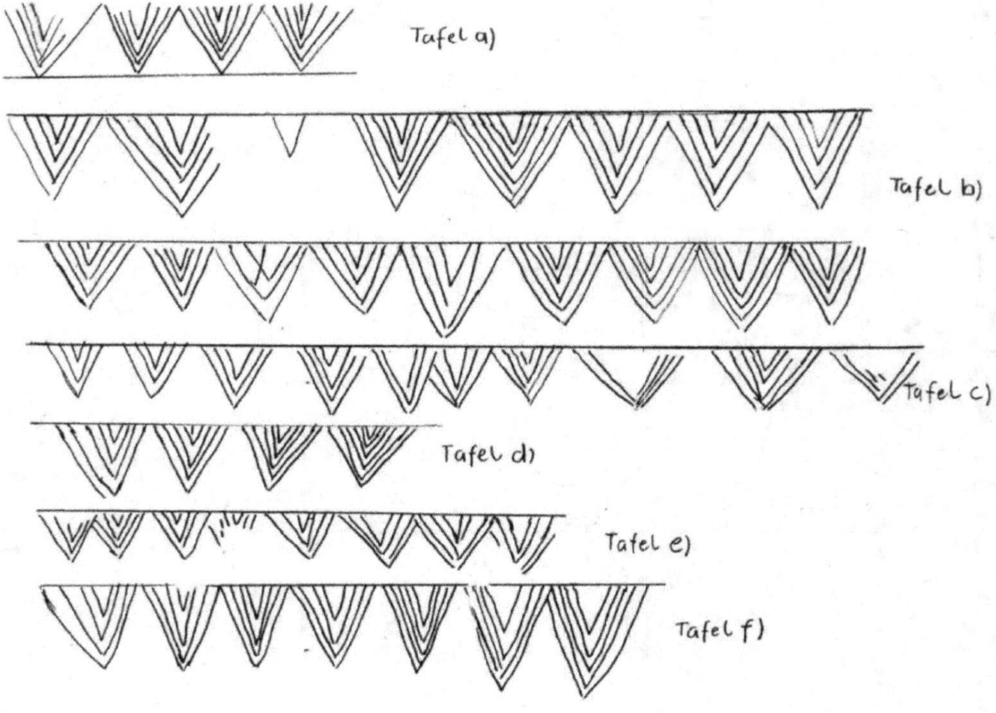

Abbildung 8.19:
Tafeln (a), (b), (c), (d), (e), (f), umlaufende Dreiecke, Jahreswochenzahl 50:
$50W \times 7 = 350Tg$, $50W \times 7,3 = 365Tg$

(Zeichnung: Klaus Albrecht)

8.3 Vergleichbare Beispiele für Ornamentik mit astronomischem Bezug

Im Vergleich zur Ornamentik im Grab von Göhlitzsch sollen ein paar wenige Beispiele aus anderen Zusammenhängen herangezogen werden, die typische Zickzacklinien oder Dreiecksmotive aufweisen. Damit können die oben beschriebenen astronomischen Interpretationen weiter untermauern werden. Es kann hier nicht weiter auf die Vielzahl von Beispielen aus dem Neolithikum eingegangen werden, in denen sich Mondphasen in Artefakten manifestieren. Auf Knochen, Felsen, einzelnen Steinen und auch auf Keramik finden sich europaweit Ritzungen und Markierungen, die von kalendarischem Wissen der damaligen Menschen zeugen.[15]

Eine mit „Zickzacklinien“ verzierte Stele wurde bei Edertal-Wellen, östlich von Fritzlar, 1961 bei Niedrigwasser als Einzelstück im Bereich der Eder gefunden. In der mir bekannten Literatur wurde noch nicht auf die Ähnlichkeiten der Verzierung im Göhlitzscher und Dölauer Grab mit dieser Stele in Nordhessen hingewiesen. Erwähnt wurde bisher nur von Kirchner, die Schrägstrichornamentik im Züschener Grab[16] und u. a. auch von Müller allgemein die vergleichbaren „Schmuckelemente“ in westeuropäischen Grabkammern und Megalithen z. B. in der Bretagne und Irland.

Abbildung 8.20:
Innenansicht Grab auf der Dölauer Heide 1956

(LDA Sachsen-Anhalt, Archiv)

Zwei Steine im Dölauer Grab sind mit dem Wellener Stein vergleichbar. Beim ersten Stein im Dölauer Grab – zweiter vom Eingang auf der linken Seite – finden sich zwölf Spalten plus einer schmalen dreizehnten Spalte in denen Schrägstriche mit abwechselnder Richtung eingeritzt wurden. Auch wenn eine Zählung der Striche nicht immer eindeutig festzustellen ist, ist anscheinend auch hier die Absicht gewesen, Mondjahreskalender zu dokumentieren. Auf dem zweiten Stein im Dölauer Grab auf der nördlichen Seite (siehe oben), befinden sich, ähnlicher dem Wellener Stein, wieder zwölf Spalten mit Schrägstrichen. Auf der rechten Seite am Stein gibt es ein paar Unklarheiten. Waagrechte parallele Linien im oberen Bereich ergeben im unbeschädigten Bereich 7 Linien, dann folgen in den Spalten 13 Schrägstriche. Darunter wieder eine waagerechte Linie, die die Spalten unterbrechen. Dort wo keine Beschädigung vorhanden ist, folgen noch 9 Schrägstriche. Das ergibt 29, die Tage einer Mondphase.

15 Albrecht; Kassel 2000.
16 Kirchner 1956, S. 26.

Auch hier gibt es eine gewisse Ungenauigkeit, was der etwas unsauberen Ausführung und dem schlechten Erhaltungszustand geschuldet ist.

Die Interpretation der Wellener Stele als lunarer Jahreskalender wurde von mir 2000 veröffentlicht.[17] Die Stele galt bis dahin als abstrahierende anthropomorphe Darstellung. Ansatzpunkte dafür waren lediglich die mittleren, parallelen Rillen, die einen Gürtel darstellen sollten. Es fanden sich aber kein Kopf und keine Füße, weshalb eine figürliche Darstellung auszuschließen ist. Die Analyse, die von einer zahlenmäßigen Erfassung ausging, ergab eine gleichmäßige Einteilung in zwölf Spalten. Diese waren wieder mit Schrägstrichen gefüllt, die in der Mitte des Steins die Richtung wechselten. In jeder Spalte befanden sich oben 14 unten 11 Schrägstriche. Wenn man die waagerechten Linien mittig und unten dazu zählt, bekommt man eine Summe von 29 Strichen. Die mittigen und unteren waagrechten Linien könnte man als die Phasen des Vollmondes bzw. Neumondes rechnen, die besonders herausgehoben worden sind. Diese Interpretation unterlag einer gewissen Kritik, weil sie anscheinend zu einzigartig war und eine Unsicherheit erfuhr durch die Interpretation der mittleren und unteren Linien.

In der Abteilung Vor-und Frühgeschichte des hessischen Landesmuseums in Kassel finden sich zwei verzierte Steinstelen. Die Stelen fanden sich in der Gemarkung von Guxhagen-Ellenberg, 15 km südlich von Kassel, im Zusammenhang mit einer Grabstelle. Im Vergleich der Ornamentik auf der einen Stele im Landesmuseum in Kassel und der Stele von Wellen fiel eine formale Ähnlichkeit mit Ritzungen im Steinkammergrab von Züschen-Fritzlar (Wartbergkultur 3500–2800 v.u.Z.) auf. Es handelt sich um gegenläufige Schrägstriche, die oft als Tannenzweigmuster bezeichnet werden. Leider sind sie im Grab und auf der Stele teils nicht vollständig, weil im oberen Bereich Beschädigungen durch Pflugscharen anzunehmen sind.

Allerdings finden sich im Grab an der Innenseite des Eingangssteins, dem Stein mit dem „Seelenloch", Zickzacklinien, die einen vollständig erhaltenen Eindruck machen. Sie ergeben eine Anzahl von dreißig klaren Strichen, die eine Mondsynode darstellen und kommen deshalb für einen Vergleich in Frage.[18] Gleichzeitig befinden sich im Grab von Züschen Darstellungen von Rinderpaaren, die einen Wagen ziehen. Die Rinderhörner in Zusammenhang mit den Zickzacklinien zu sehen, erhöht die Wahrscheinlichkeit, dass mit den zählbaren Linien Mondphasen assoziiert wurden.

Weil diese Ritzungen im Zusammenhang mit Grabanlagen gefunden wurden, liegt der Symbolwert in der auffälligen Betonung der gekrümmten paarweise angeordneten Hörner. Immer wieder, sowohl an Gräbern oder Kultstätten in Europa, z. B. Sardinien, sind Rinderköpfe mit Hörnern zu finden. *„Das Horn ist nichts anderes als Abbild des Neumondes; sicher ist es zum Mondsymbol geworden, weil es an die Mondsichel*

17 Albrecht, Klaus: Die Stele von Wellen - ein neolithischer Mondkalender, 2000.
18 Albrecht, Klaus: Morgenstund hat Gold im Mund, 1998, S. 106 ff.

Abbildung 8.21:
Wandstein nördliche Seite Zwölf Spalten mit Schrägstrichen,
Grab Dölauer Heide (1956)

(LDA Sachsen-Anhalt, Archiv)

Abbildung 8.22:
Wellener Stele (Nordhessen)

(Kappel 2000)

erinnert; das doppelte Horn repräsentiert also zwei Sichel und damit alle Phasen des Gestirns.“[19] Die Wagendarstellungen im Züschener Grab deuten nicht nur auf die technischen Möglichkeiten beim Einsatz von Rindern in der Landwirtschaft hin, sondern sie befördern auch die Vorstellung einer Reise der Toten durch die Unterwelt

19 Eliade, Mircea: Die Religionen und das Heilige, 1998, S. 194.

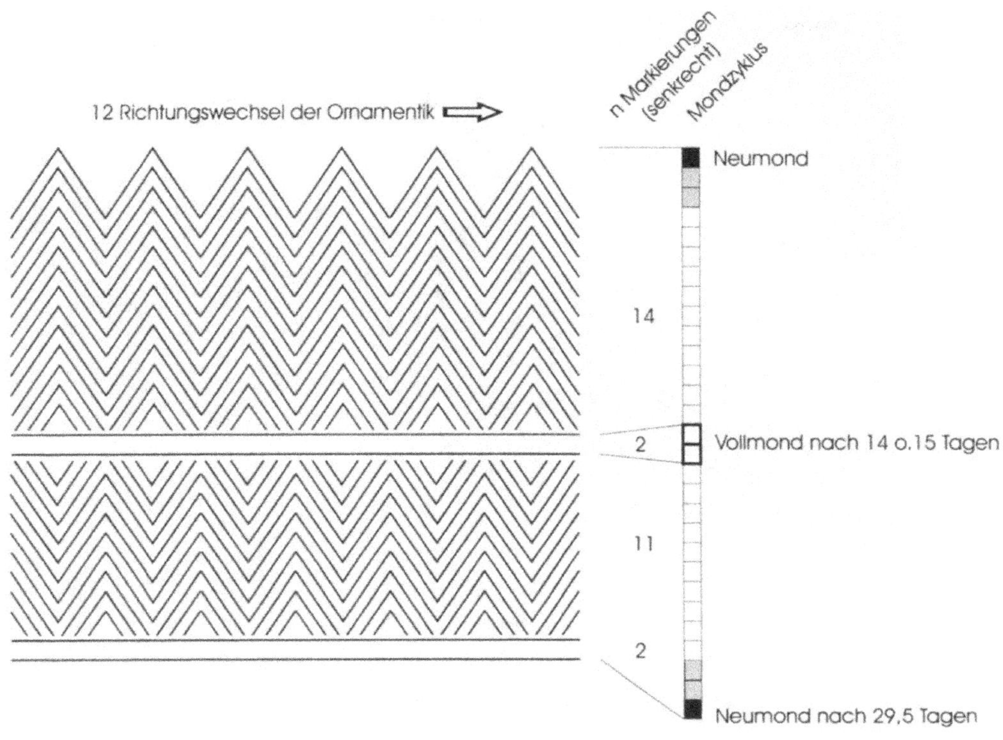

Abbildung 8.23:
Schematische Erfassung der Stele von Wellen
(Albrecht 2000)

auf einem Wagen. Die Kuh ist Attribut der großen Erdmutter. Mit ihrem lunarischen Aspekt des Werdens und Vergehens, gilt sie weltweit als Fruchtbarkeitssymbol und tellurische Kraft in der neolithischen Kultur.

Um noch mal auf die Bedeutung von Dreiecken in der neolithischen Zeichensprache einzugehen, sollen die zweite Ellenberger Stele und Schieferidole aus Portugal erwähnt werden.

Die Ellenberger Stele mit den Dreiecken, lässt an Eindeutigkeit wenig zu wünschen übrig. In sechs Zeilen entstehen durch Zickzacklinien 29 ganze und halbe Dreiecke, die sich vom Grund abheben und mit der Spitze nach oben zeigen. Dazwischen liegen 28 Dreiecke, die durch Herausschlagen aus dem Stein mit der Spitze nach unten zeigen. Durch einen entsprechenden Lichteinfall entsteht eine schachbrettartige Hell-Dunkel-Abfolge. Dies entspricht der synodischen Mondphase von 29 bzw. 29,5 Tagen.

Abbildung 8.24:
Abschluss Stein im Westen, Grab Züschen

(Abreibung Klaus Albrecht 1998)

Portugiesische Schieferplattenidole aus Megalithgräbern führen einen großen Variantenreichtum mit Dreiecksmotiven und Zickzack vor. Einige Beispiele verblüffen durch die dort abgebildeten Dreiecksreihen und ihren Zahlen von 28, 29 oder 30. Mal tauchen auch die Reihungen von 14, 15, 13 oder 7 auf, die wieder auf die bekannten Teiler der Mondphasen, Halbmond und Viertelmonde hinweisen. Es finden sich auch „Fischgrätenmuster“ bei denen die Zahlen 28 oder 29 oder ein Vielfaches zu finden sind.

Die Schieferplattenidole fanden sich in großer Menge in den portugiesischen Steinkammergräbern.[20] Ihre Gestaltungsvielfalt lässt nicht immer eine eindeutige Interpretation zu. Trotzdem sind ihre Funktionen als Amulett und ihre Mondsymbolik mit den Vorstellungen von Werden und Vergehen in der Natur, im Geborenwerden und Sterben der Menschen zu sehen.

20 Müller-Karpe, Hermann: Handbuch der Vorgeschichte, 1974, Tafel 563.

Abbildung 8.25:
Ritzungen am Seelenlochstein, Grab Züschen

(Abreibung Klaus Albrecht 1998)

8.4 Religiöse Bezüge

Bei der Ausschmückung im Göhlitzscher Grab ist von einem umfassenden Ideenkomplex auszugehen, der sich in langer Traditionslinie verfolgen lässt. Grundlegende Anschauungen über Leben und Tod, über eine kosmische Ordnung spiegeln sich wider. Der Tod war nicht das Ende, sondern die Voraussetzung für den Neubeginn. Dies in einem Grab zu illustrieren war naheliegend.

Die vier Jahreszeiten waren Phasen im zyklischen Ablauf der Zeit. Das Leben der Menschen sowie die Natur insgesamt waren abhängig von dem Jahreslauf mit den Phasen vom Erwachen, dem Aufwuchs, dem Reifen und dem Absterben. Das erfüllte

Abbildung 8.26:
Ellenberger Stele (Landesmuseum Kassel)

(Kappel 2000)

Leben eines Menschen bestand in der Abfolge von vier Lebensphasen: Kindheit, Jugend, Reife, Alter. In ewiger Schleife wiederholten sich Zeit und Leben. Im ewigen Sterben und Wiedergeboren liegt das Schicksal des Menschen. Der Zyklus wird zum Garanten der Lebensordnung schlechthin. Diese Vorstellung durchzieht alle Zeitalter, vom Neolithikum über die Hochkulturen in Vorderasien bis in die Antike. In Ägypten, im Mythos um Osiris, dem Gott der Unterwelt, der nach dem Tod wiedergeboren wurde und selbst den Toten zur Wiedergeburt verhalf. In griechischer Zeit verehrte man Dionysos den Gott des Todes und der Wiedergeburt. Bei den Etruskern gab es

Abbildung 8.27:
Schieferplattenidole aus Portugal
(Müller-Karpe 1974)

den „Vertumnus“, den Gott der reifen Früchte, der auch als Gott der vier Jahreszeiten galt. Verehrt wurde er am 15. August. In Rom wurde aus Dionysos der Vegetationsgott Bacchus. Auf römischen Grabmonumenten fanden sich öfter Bezüge zu den Vierjahreszeiten. Beispielhaft ist die Sarkophagplatte aus der Antikensammlung von Kassel mit den Vierjahreszeiten in Gestalt junger Männer – In der Mitte Bacchus der dionysische Gott der Vegetation, der auf einem Panther reitend die Jahreszeiten begleitet, links Winter und Frühling, rechts Sommer und Herbst,

Ausgehend von dem archaischen Glauben an die Seelenwanderung, in dem die Seele in ihrem Weg durch die „Untere Welt“ zur irdischen Wiedergeburt findet, ist der

Abbildung 8.28:
Sarkophagplatte mit den vier Jahreszeiten Röm. Reich Mitte 3. Jahrhundert n.u.Z.

(Staatliche Museen Kassel, Antikensammlung; Foto: Karin Albrecht)

Seele das Wissen um die kosmischen Zusammenhänge nützlich. Die Seele, die sich vom Körper gelöst hatte, bedurfte der gleichen Kenntnisse im Tode wie im Leben. Deshalb war eine bildhafte Darstellung der Zeit in einem Grab nützlich.

Die hier vorgestellte Analyse und Interpretation der Ausschmückungen des Grabes von Göhlitzsch und anderer Artefakte ist als Angebot zu verstehen, neu auf die neolithische Ikonographie zu schauen. Eine Abstraktion von realen Verhältnissen, z. B. der kosmischen Erscheinungen, gehört schon immer zur intellektuellen Grundausstattung der Menschen. In Bildern, Zeichen und Zahlen sind uns mehr Informationen überliefert als man vermuten möchte. Die Komplexität der Aussagen überrascht häufig. Es wurde der Versuch unternommen, Mensch und Kosmos mit dem gleichen Maß zu fassen. Die vermutete, ewige, gleichmäßige Wiederholung entsprach einer eher zirkularen Geschichtsauffassung der damaligen Menschen. Dazu kam die Vorstellung von göttlicher Einflussnahme. Das göttliche Maß offenbarte sich in den Zyklen der himmlischen Erscheinungen.

Auf der Stele von Sion findet man die „Kombination“ von anthropomorpher und kalendarischer Darstellung. Hier sind im unteren Teil neun Reihen mit je 29 oder 30 kleinen Dreiecken. Darüber sind zwei Reihen mit Halbkreisen auszumachen, eine Reihe mit zwölf und eine Reihe mit dreizehn Halbmonden. Wieder darüber ein Feld mit zwölf Reihen von Rauten, die allerdings eine nicht ganz vollständige 29er-Reihe aufweisen. Wenn man das Raster im unteren Teil zugrunde legt, ergibt sich eine gleiche Anzahl von Rauten wie die Dreiecke in einer Reihe. Das Feld mit den Rauten, wird von den beiden senkrechten Leisten rechts und links am Rand eingefasst. Dies sind wahrscheinlich die Arme. Es sind abstrahierte Hände in der Mitte der Stele zu erkennen, Pfeil und Bogen diagonal über dem oberen Feld und eine Halskette. Ein fehlendes Gesicht könnte man sich im oberen Teil auf einer glatten Fläche vorstellen. Bei dieser Stele wird eine gegliederte Anzahl von Mondphasen dargestellt, die auf einen mythologischen oder religiösen Zusammenhang mit den Mondphasen verweist. Der Gedanke an eine göttliche Instanz, die im Kosmos für Ordnung sorgt, lag bei vorgeschichtlichen Völkern nahe. Die Abbildung eines Himmelsgottes, der hier in Zusammenhang mit den Mondphasen präsentiert wird, ist nicht abwegig.

Genauere Beobachtungen des Sternenhimmels ergaben aber auch für sie bald Ungereimtheiten. Wurden Mondphasen und Sonnenphasen zahlenmäßig erfasst, passten sie nicht zusammen. In den Lunisolarkalendern suchte man den Ausgleich, die himmlische Harmonie, herzustellen. Dabei wurden im Laufe der Zeit die Zeiträume länger und die Modelle komplizierter, in denen man den Gleichklang von Sonne und Mond herstellen wollte. Die Aufgabe der Schamanen, der Priesterastronomen war es, diese Modelle zu kreieren und anzupassen.

Die Befangenheit herkömmlicher archäologischer Wissenschaft das Ganze auszuschöpfen, liegt wohl häufig in der Spezifizierung und Abschottung wissenschaftlicher Disziplinen untereinander. Während der Versuch Hoppenhaupts, eine Interpretation der Grabdekorationen auf Grund von militärtechnischem Hintergrund, verständlich ist, kamen Forscher später mit religionswissenschaftlichen Erkenntnissen weiter. Ethnologische Vergleichsforschungen ließen weitgehende Schlüsse auch auf die archaischen Gesellschaften zu. In diesem Fall hilft uns die Archäoastronomie, uns der Vorstellungswelt unserer Vorfahren zu nähern. Das Leben von vorgeschichtlichen Gesellschaften lässt sich nicht auf Reproduktionsfaktoren und ihre Entwicklung oder auf ihre Stammes- und Herkunftsbezüge, ästhetische, kulturelle oder religiöse Äußerungen beschränken, sondern es ergibt sich erst in der Zusammenschau ein realistisches Bild. Die Beobachtung des Himmels und die Erfahrung in der materiellen Welt auf Erden wurden von unseren Vorfahren in Zusammenhang und Abhängigkeit gesehen und dann in abstrahierte Form, in Bildern und Worten wiedergegeben.

Danksagung – Acknowledgement

Besonderer Dank gilt Frau Bettina Stoll-Trucker, Abteilungsleiterin des Landesmuseums für Vorgeschichte in Halle, die mir nützliche Hinweise und Material zur Verfügung gestellt hat und die freundliche Genehmigung für Bilder und Zeichnungen aus dem Landesamt für Denkmalpflege und Archäologie Sachsen-Anhalt-Archiv gegeben hat.

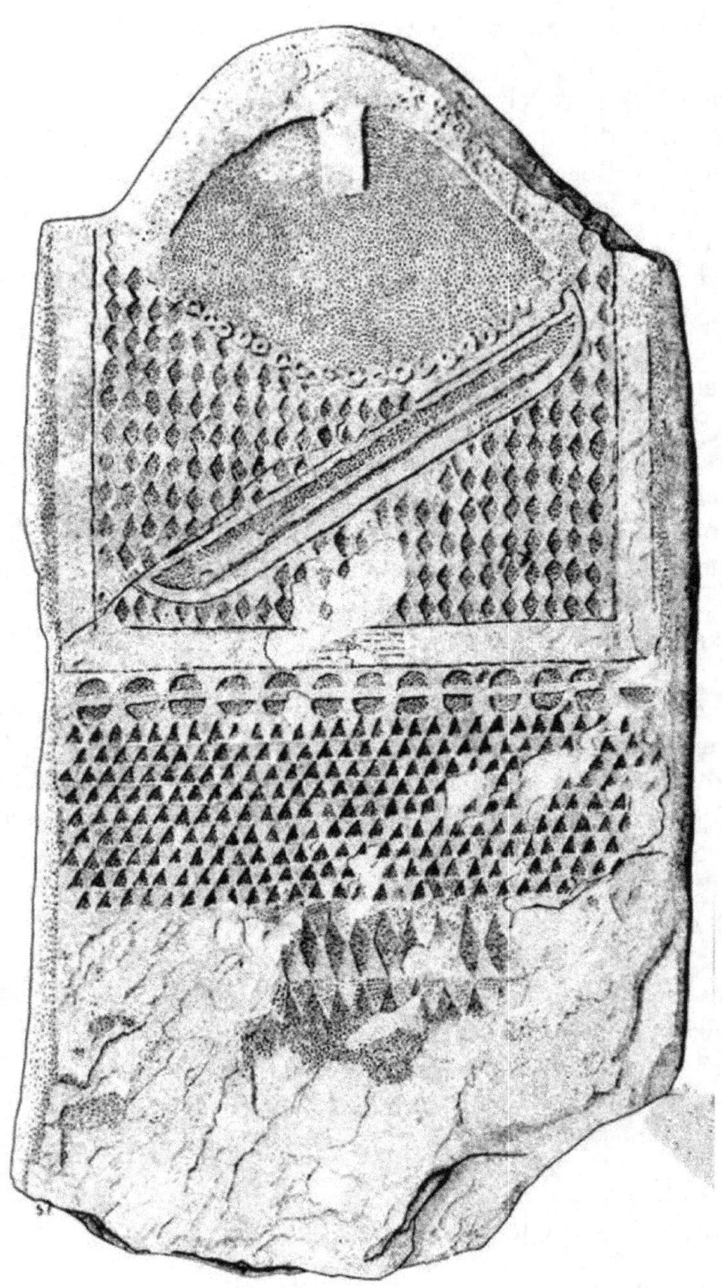

Abbildung 8.29:
Stele von Sion „Der Kleine Jäger“

(Zeichnung von S. Favre. Quelle: Louboutin: Steinzeitmenschen, 1992, S. 126.)

8.5 Literatur

ALBRECHT, KLAUS: *Morgenstund hat Gold im Mund.* Naumburg (Geschichtsverein Naumburg; Sonderband 5) 1998.

ALBRECHT, KLAUS: Mondkalender auf nordhessischen Sandsteinstelen. In: *Korona – Zeitschrift Astronomischer Arbeitskreis Kassel e.V.* **28** (2000), Nr. 83; S. 15–31.

ALBRECHT, KLAUS: Die Stele von Wellen (Gde. Edertal, Schwalm-Eder-Kreis) – ein neolithischer Mondkalender? In: *Archäologisches Korrepondenzblatt* **30** (2000), Heft 1. Mainz: Verlag des Römisch-Germanischen Zentralmuseums.

BÄCHTOLD-STÄUBLI, HANNS: *Handwörterbuch des deutschen Aberglaubens, Band 6.* Stichwort Mond. Berlin: Walter de Gruyter 1987.

BEHRENS, HERMAN; FASSHAUER, PAUL & HORST KIRCHNER: Ein neues innenverziertes Steinkammergrab der Schurkeramik aus der Dölauer Heide bei Halle (Saale). In: *Jahresschrift Mitteldeutsche Vorgeschichte* **40** (1956), S. 13–50

BRENNAN, MARTIN: *The Stones of Time – Calendar, Sundials, and Stone Chambers of Ancient Ireland.* New York: Thames und Hudson 1994.

CALVIN, WILLIAM H.: *Wie der Schamane den Mond stahl. Auf der Suche nach dem Wissen der Steinzeit.* München: Hanser Verlag 1996.

DOLYE, ROBERT (Redaktion): *Das mystische Jahr – Geheimnisse des Unbekannte.* Amsterdam: Time-Life Bücher 1992.

ELIADE, MIRCEA: *Die Religionen und das Heilige- Elemente der Religionsgeschichte.* Darmstadt: Wissenschaftliche Buchgesellschaft 1976, Frankfurt am Main: Insel-Verlag 1998.

FANSA, MAMOUN (Hg.): *Wohin die Toten gehen – Kult und Religion in der Steinzeit. Ausstellungkatalog.* Oldenburg: Staatliches Museum für Natur und Vorgeschichte Oldenburg, Isensee Verlag 2000.

GIMBUTAS, MARIJA: *The Language of the Goddess.* New York: Thames & Hudson 2001.

HAFFNER, ALFRED (Hg.): *Heiligtümer und Opferkulte der Kelten.* Stuttgart: Konrad Theis Verlag 1995.

HOCK, HANS-PETER (Hg.): *Der Tod in der Steinzeit – Gräber früher Bauern aus dem Ried.* Ausstellungkatalog. Darmstadt: Hessisches Landesmuseum Darmstadt 1991.

KAPPEL, IRENE: *Steinkammergräber und Menhire in Nordhessen.* Kassel: Staatliche Kunstsammlungen Kassel 1989.

KAUFMANN, DIETER; WALDEMAR MATTHIAS: *M. E. Hoppenhaupt – Ausführliche Beschreibung eines alten Heydnischen Grabes. Faksimiledruck zum hundertjährigen Bestehen des Landesmuseums für Vorgeschichte Halle (Saale).* Berlin: VEB Deutscher Verlag der Wissenschaften 1984.

LEVI-STRAUSS, CLAUDE: *Das wilde Denken.* Frankfurt am Main: Suhrkamp 1973.

MAHLSTEDT, INA: *Die religiöse Welt der Jungsteinzeit.* Stuttgart: Theiss Verlag 2004.

MAUERSBERGER, ARNO: *Tacitus Germania.* Frankfurt am Main: Insel Verlag 1980.

MELLER, HARALD (Hg.): *3300 BC – Mysteriöse Steinzeittote und ihrer Welt.* Ausstellung im Landesmuseum für Vorgeschichte. Halle (Saale) 2013.

MENGHIN, WILFRIED: Zahlensymbolik und digitales Rechnersystem in der Ornamentik der bonzezeitlichen Goldhüte. In: *Acta Praehistorica et Archaeologica*, Band **40** (2008). Berlin: Staatliche Museen zu Berlin.

MÜLLER, DETLEF W.: *Große Steine, alte Zeichen – Jungsteinzeitliches Bildgut in Grabbrauch und Religion.* Halle (Saale) 1994.

MÜLLER, DETLEF W.: Die Berburger Kultur Mitteldeutschlands im Spiegel ihrer nichtmegalithischen Kollektivgräber.In: *Jahresschrift Mitteldeutschland, Vorgeschichte*, Band 76 (1994).

MÜLLER, DETLEF W.: Ornamente, Symbole, Bilder – Zum megalithischen Totenbrauchtum in Mitteldeutschland. In: *Revue archéologique de l'Ouest*, Supplement, Nr. 8 (1996), S. 163–176.

MÜLLER, ROLF: *Der Himmel über dem Menschen der Steinzeit – Astronomie und Mathematik in den Bauten der Megalithkulturen.* Berlin: Springer 1970.

MÜLLER-KARPE, HERMANN: *Handbuch der Vorgeschichte, Band 2: Jungsteinzeit, Band 3: Kupferzeit.* München 1968, 1974.

PROBST, ERNST: *Deutschland in der Steinzeit – Jäger, Fischer und Bauern zwischen Nordseeküste und Alpenraum.* München: Orbis Verlag 1999.

RAETZEL-FABIAN, DIRK: *Die ersten Bauernkulturen – Jungsteinzeit in Nordhessen.* Kassel: Staatliche Museen Kassel 2000.

SCHLAG, HANNES E.: *Ein Tag zuviel – Aus der Geschichte des Kalenders.* Würzburg: Königshausen & Neumann 1998.

SCHMIDT-KALER, THEODOR: Die Entwicklung des Kalender-Denkens in Mitteleuropa vom Paläolithikum bis zur Eisenszeit. In: *Acta Praehistorica et Archaeologica*, Band **40** (2008), S. 11.

SCHUNKE, TORSTEN: Die befestigte Siedlung Bischofswiese, Halle-Dölauer Heide. Bilderflut im Dunkeln – Grabhügel in der Dölauer Heide und die innenverzierten Steinkammern. In: MELLER, HARALD (Hg.): *3300 BC – mysteriöse Steinzeittote und ihre Welt.* Halle (Saale) 2013, S. 139 und S. 143.

KLADY-GÖHLITZSCH. Vom Kaukasus nach Mitteldeutschland oder umgekehrt? Die Welt der Zeichen – Symbolik in der Salzmündener Kultur. In: MELLER, HARALD (Hg.): *3300 BC – mysteriöse Steinzeittote und ihre Welt.* Halle (Saale) 2013, S. 151 und S. 262.

STEINRÜCKEN, BURKHARD: *Lunisolarkalender und Kalenderzahlen am Beispiel des Kalenders von Coligny.* Recklinghausen 2012.

STUDZISKAJA, SWETLANA W.: *Vorgeschichte der Zeit – Zeit der Jäger in Geburt der Zeit eine Geschichte der Bilder und der Begriffe.* Katalog zur Ausstellung der Staatlichen Museen. Kassel: Edition Minerva 2000, S. 25.

REDEN, SIBYLLE VON: *Die Megalith-Kulturen – Zeugnisse einer verschollenen Urreligion.* Köln: DuMont 1989.

WESTRHEIM, MARGO: *Kalender der Welt – Ein Reise durch Zeiten und Kulturen.* Freiburg: Herder Verlag 1999.

WUNN, INA: *Götter, Mütter, Ahnenkult – Religionsentwicklung in der Jungsteinzeit.* Oldenburg: Verlag Marie Leidorf (Beiheft der Archäologischen Mitteilungen aus Norddeutschland, Band 36) 2001.

Abbildung 9.1:
Sun-God Tablet from Sippar, Iraq (860–850 BC), Mus. No. BM 91000

(© The Trustees of the British Museum, London)

Man, Myth, Cosmos – The Sun and Solar Eclipses in Ancient Mesopotamia (Second and First Millennia BC)

Anna Paule (Linz, Österreich)

Abstract

The present article examines different sources on the perception of the Sun and its eclipses in ancient Mesopotamia (i. e., during the Middle to Late Bronze and Early Iron Age periods). In particular, the increasing studies on solar iconography and mythology not only provide valuable information for decoding ancient hypotheses about the nature of the Sun, but also contain clues for actual eclipse observations. Taking a closer look at these studies reveals that the knowledge gathered by scholars in these fields presents an interesting yet contradictory picture: on the one hand, the Sun figures as a divine person with anthropomorphic traits. On the other hand, not only this mythological concept but also abstract ones are used in visual art, while different textual sources explicitly relate to the Sun's natural appearance. Despite the potential for studies on ancient, pre-scientific forms of cosmology manifested in these sources, two main questions have long been neglected will be dealt with here in more detail: first, the question of how the Sun's nature was understood and mastered in ancient Mesopotamia will be discussed by means of examples. Second, the question of whether evidence for actual astronomical observations can be found in the archaeological record will be examined. Thus, the present article can be deemed as a first step towards a better understanding of the perception of the Sun in ancient Mesopotamia, with the aim of making the latest research available to a wider public.

Zusammenfassung: Mensch, Mythos, Kosmos – Sonne und Sonnenfinsternisse im Alten Mesopotamien (2. und 1. Jt. v. Chr.)

Im vorliegenden Artikel werden verschiedene nahöstliche Quellen zur Wahrnehmung der Sonne und der Sonnenfinsternis im alten Mesopotamien untersucht (hier: während der mittleren bis späten Bronzezeit und der Früheisenzeit). Besonders die zunehmenden Studien zur Sonnenikonographie und -mythologie bieten nicht nur wertvolle Informationen zur Entschlüsselung antiker Hypothesen über die Natur der Sonne, sondern liefern auch Hinweise auf mögliche Beobachtungen von Sonnenfinsternissen. Ein genauerer Blick auf diese Studien ergibt ein interessantes, aber widersprüchliches Bild: einerseits tritt die Sonne als göttliche Person mit menschenähnlichen Wesenszügen in Erscheinung. Andererseits werden in der bildenden Kunst nicht nur dieses mythologische Konzept, sondern auch abstrakte Konzepte verwendet, während sich verschiedene Textquellen ausdrücklich auf das natürliche Erscheinungsbild der Sonne beziehen. Trotz des Potentials für Studien über antike, vorwissenschaftliche Formen von Kosmologie, welches aus diesen Quellen hervorgeht, sind zwei Hauptfragen bislang von der Forschung vernachlässigt worden. Diesen Fragen soll nun im vorliegenden Artikel genauer nachgegangen werden. Zunächst wird daher die Frage, wie das Wesen der Sonne im alten Mesopotamien verstanden und gehandhabt wurde, anhand von Beispielen diskutiert. Danach wird die Frage, ob sich in ikonographischen und mythologischen Quellen zur Sonne Beweise für tatsächliche astronomische Beobachtungen finden lassen, näher untersucht. Dementsprechend kann dieser Artikel als erster Schritt zu einem besseren Verständnis der Wahrnehmung der Sonne im antiken Nahen Osten angesehen werden, der das Ziel hat, neueste Forschungsergebnisse einem breiteren Publikum bekannt zu machen.

9.1 Introduction

Over the last few decades, archaeoastronomy (also known as cultural astronomy) or research on early, pre-scientific forms of astronomy has attracted an increasing interest. This scholarly interest is not only limited to visual astronomical alignments between landmarks, architecture, and celestial bodies or patterns in their apparent motion across the sky – such as is well-known, for instance, from pre-literate Europe – but also includes research on how astronomical phenomena were understood and anchored in prehistoric societies.[1] Against this backdrop, it seems surprising that the implementation of comparable studies on pre-classical Mesopotamia[2] has to some extent been neglected by research. The country, however, does not only provide a rich

1 Cf. the definitions by Ruggles et al. 1993, 1–31 and Sinclair 2006, 13–26.

2 Today, the surface area of ancient Mesopotamia corresponds to southeastern Anatolia, northeastern Syria, Iraq, Western Iran, and parts of Kuwait. Aruz, Benzel & Evans 2008, xxi.

and sophisticated material culture but also a longstanding tradition of cuneiform writing and record-keeping. In particular the latter constitutes fertile ground for studies on the perception of the Sun's nature. Despite this, relevant aspects including cosmology have often been dismissed or only been studied in isolation (e. g., omen texts), but not in their entirety.

9.1.1 Methodology

The present article adopts the perspective of archaeoastronomy to consider the changing nature of how the Sun's behavior was understood and mastered by major civilizations of ancient Mesopotamia. All of the data is based on methods currently applied in archaeoastronomy and has been derived by adopting material stemming from research in archaeology and the history of astronomy. The focus here is on a conceptual-comparative analysis of iconographic and written; i. e., cuneiform primary sources. The available material is in the form of archaeological artefacts bearing solar iconography (artworks, sculpture, ritual implements, etc.) and of textual sources that directly address to the Sun, such as omen texts (esp. the *Enūma Anu Enlil* series), prayers, or certain *Letters and Reports* sent to the Assyrian kings. Today, most of the material is an integral part of important museum collections worldwide.

The collected material has been combined here with clues for early sky observations that come from studies in astronomical, rather than archaeological, research. The English translations cuneiform texts used in this article are based on those by leading experts in ancient Mesopotamia.[3] In this context, mention should be made of problems relating to text interpretation: not only the nature of cuneiform writing, but also the ancient understanding of the universe may lead to the blurring of boundaries between early forms of astronomy and astrology. This is particularly true to the writings of early astronomers (the so-called *Non-Mathematical Astronomical Texts or NMAT*) that predate the start of mathematical astronomy. The latter is documented in Astronomical Cuneiform Texts or ACT that are no earlier in date than the fourth c. BC.[4] In response to these problems, extensive attention has been drawn here on careful re-evaluation and comparison of our textual sources.

3 For references, see below.
4 Neugebauer 1955.

9.2 The Imagery of the Sun. An Introduction to Solar Imagery from Ancient Mesopotamia

Although the focus here is on clues for early, pre-scientific forms of astronomy, research on solar imagery should be part of an article dealing with the perception of the Sun in ancient Mesopotamia. In the visual art and mythology of this civilization, approaches to depicting the Sun tend to result in anthropomorphic, theriomorphic, and symbolic forms rather than naturalistic representations of the celestial body visible in the sky.

Among these representations, the theriomorphic type has for long been neglected by research. One reason for this may be that hybrid creatures, such as human-faced bisons and bull-men, are difficult to understand from a current perspective. Despite this, their role as adjuncts of the Sun god is well-known from glyptic art (second half of the third and second mill. BC; cf. Fig. 9.2, p. 194).[5]

Figure 9.2:
Akkadian cylinder seal, Mus. No. VA04269 (ca. 2350–2150 BC)

(© Staatliche Museen zu Berlin, Vorderasiatisches Museum / Olaf M. Teßmer)

Aside from that, a lack of epigraphic evidence often makes it difficult to identify artwork related to the Sun god. In addition, the consultation of different written sources can provide conflicting results.

A good example in this context is the identification of two Akkadian terms that relate to the bison/bull and its mythological variants: *kusarikku* and *alû/elû (lû)*. The first

5 E. g., Ławecka 2022.

term, *kusarikku*, has been equated with the Sumerian logograms ALIM, GUD.ALIM (bison; bull), GUD.DUMU.^{D}UTU (bull, son of Sun god), and GUD.DUMU.AN.NA. (bull, son of heaven).[6] *Kusarikku* is also used to describe the human-faced bison or the bull-man in mythological contexts from the Early Dynasty (ca. 1894–1595 BC) onwards.[7] The second term, *alû/elû (lû)*, designates the *Bull of Heaven (Bull)* but is better known under the Sumerian term/logogram MUL.GUD.AN.NA.[8] Furthermore, it has been observed that the *Bull of Heaven* refers to the winged human-faced bull rather than the human faced-bull in mythological contexts of the late second and first mill. BC.[9]

Contrary to the conclusions drawn from iconographic sources, however, their relationship with the Sun god is less obvious in written sources. Instead, their astral aspect is documented in individual star catalogues, which have given rise for their supposed association with different constellations in the sky. More precisely, *kusarikku* (MUL.GUD.ALIM) has tentatively been assigned to the constellations *Ophiuchus / Serpens* or *Centaurus*.[10] By contrast, the identification of *alû/elû* (MUL.GUD.AN.NA) as the constellation *Taurus* and/or the *Hyades* (*is lê* or the *bull's jaw*) has been generally accepted.[11]

In the *Epic of Gilgamesh*, the *Bull of Heaven* (a mythological monster) does not necessarily relate to the Sun god but explicitly serves as the vehicle of the rising Sun god in daylight sky in both versions. After the slaughtering of this monster, Gilgamesh and Enkidu *"pulled out its innards, set them before Shamash* [the Sun god], *backed away and prostrated themselves before Shamash."*[12]

A last but striking example of potential links between the Sun god and the bull (of heaven) is an elaborate bull-head protome made of sheet gold and lapis lazuli. It comes

6 Sołtysiak 2001, 4–6. The logograms are also spelled using "GU$_4$" instead of "GUD" (cf. GU$_4$ALIM, ^{D}GU$_4$–DUMU–^{D}UTU; Labat & Malbran-Labat 1976, 138–139; Schramm 2010, 16 and 62). Furthermore, as the author does not add her own translations to the existing corpus of Sumerian texts, she avoids the use of small capitals, etc. characterizing the transliteration of Sumerian logograms. Instead, and for quicker readability, she focuses on the more general spelling of logograms encountered in studies on early astronomy, such as Hunger and Steele 2019 (e. g., MUL.APIN).

7 Wiggermann 1992, esp. 174–179.

8 This logogram is also spelled GU$_4$–AN–NA; Labat & Malbran-Labat 1976, 138–139 and Schramm 2010, 61.

9 Wiggermann 1992, 176.

10 See Foster 1996, 148–149 (translation); Oeslner and Horowitz 1998, 182–183 (original texts); and Sołtysiak 2001, 4 pointing to Gössmann 1950, 23 and Lewy 1965, 278 (identification).

11 Sołtysiak 2001, 4–6. The Akkadian term *is lê* is also equated with the Sumerian logograms MUL.GUD.AN.[NA]or MUL.GIŠ.DA (Schramm 2010, 59 and 61). Early evidence for the constellation *Taurus* is enumerated in Oelsner & Horowitz 1997/98, 180, s.v. Ea 3 and dates to the late second mill. BC.; this evidence can include information from earlier (hitherto unknown) sources.

12 Sołtysiak 2001, 6. The motif of hunting the *Bull of Heaven* is an ancient one and dates back to the third mill. BC (Sołtysiak 2001, 7–8).

from Tomb 789 of the Royal Cemetery in Ur in Iraq and remains from a wooden lyre instrument (cf. Fig. 9.3, bull-head protome, B17694, ca. 2450 BC).

Figure 9.3:
Bull-head protome from Ur, Iraq. Object No. B17694 (ca. 2450 BC),

(Courtesy of the Penn Museum, Philadelphia, USA, image no. 295563)

9.2.1 The Imagery of the Sun as a Divine Person. Evidence from Ancient Mesopotamia

Among the ancient Near Eastern solar deities, the Sumerian Sun god Utu or the Assyro-Babylonian Shamash (also written ^{D}UTU) is best studied and is referred to as the god of justice, morality, and truth.[13] The earliest, anthropomorphic depictions of Utu/Shamash can be found in Akkadian (i. e., early Mesopotamian) glyptic art. Here, the Sun god is shown using a saw to cut his way through the mountains, as he does every morning: this type of imagery has reasonably been identified as the metaphoric depiction of the sunrise.[14] A fine example is a small-sized depiction of the bearded Sun god who is characterized by a bundle of light rays arising from his shoulders (cf. Fig. 9.2, p. 194, Akkadian cylinder seal, Mus. No. VA04269; ca. 2350–2150 BC; without exact provenance). Crowned with the horned headgear of divinity, he appears between the *great mountains*[15] or the *Twin Mountains* (*Mashu*)[16] and is lifting his saw. Two doorkeepers, who open the *doors of heaven* for him, flank the mountainous scene.

Another popular representation relates to Shamash sitting on a throne and keeping the rod-and-ring symbol in his hand. This symbol has often been identified as measuring tools and refers to his role as the all-seeing and divine judge.[17] Although this pictorial motif dates back to the Late Akkadian or Old Babylonian periods (i. e., the late third mill. or the first half of the second mill. BC),[18] it is best known from an elaborately executed stone tablet that today is referred to as *Sun-God Tablet* (cf. Fig. 9.1, p. 190, *Sun-God Tablet* from Sippar, Iraq, Mus. No. BM 91000; 860–850 BC).[19] Chr. E. Wood's investigation of its iconography is comprehensive[20] but the scene on the *Sun-God Tablet* finds best parallels in another, much more ancient depiction of the Shamash, the judge. This refers to the relief that accompanies the Law-Codex incised on the stone *Stele of Hammurabi*,[21] King of Babylon (Susa; ca. 1792–1750 BC).

13 Black & Green 1992, 182–184.

14 Kurmangaliev 2012, 286–287, figures. 1–3.

15 E. g., Heimpel 1986, 143 and Foster 1996, 644.

16 Cf. *The Epic of Gilgamesh*, tablet 9, 38–41 (standard version).

17 Written evidence comes, for instance, from the Law-Codices of Ur-Namma (prologue; ca. 2100 BC, Roth 1995, 13) and Hammurabi, Columns XL–XLII (epilogue; ca. 1792–1750 BC). For later examples, see Foster 1996, esp. 632–664 (ca. 1500–1000 BC).

18 Ascalone and Peyronel 2001, 8–9, figures. 8–9 and Rohn 2011, 74, no. 564, pl. 46.

19 For details on the rediscovery of this and related artefacts, see Finkel & Fletcher 2016.

20 Woods 2004, 45–76.

21 Elsen-Novák & Novák 2016.

9.2.2 The Imagery of the Sun as a Divine Person. Family Ties and Astral Character

Further anthropomorphic traits are documented in the Sun god's family ties. Accordingly, Shamash is the husband of his wife Aya (Akkadian; her Sumerian equivalent was Sherda /Sherida), the goddess of the dawn, who cares for him and prepares dinner for him every evening.[22] They have several children, among them Mamu, the goddess of dreams,[23] and Kittum, another goddess of truth (justice). The twin sister[24] of Utu/Shamash is Inanna (Sumerian; her Akkadian name was Ishtar), later worshipped as the East/Northwest Semitic Astarte. She is known as goddess of love, fertility, and war and is associated with the morning/evening star (Venus).[25] Father of the divine twins is Nanna/Sin, the Moon god, whose wife is Ningal, his female counterpart.[26]

An interesting issue is his integration among the gods of astral character and, consequently, in the cosmological view of ancient Mesopotamia. It is noteworthy that the Sun (Shamash) and the much smaller planet Venus (Ishtar) were considered as twin brother and sister. By contrast, Sun and Moon do not form an antithesis; i. e., opposite but interconnected forces, as it may be thought in modern day; but were considered as a son and his father. Together with Ishtar, Shamash and Nanna formed an important divine trinity,[27] which is also rendered in the form of astral symbols: an eight-pedaled star of rosette (Ishtar), a symbol combining full moon and crescent moon (Nanna), and is a four-pointed star completed by four diagonal wavy lines that are bundled in the center; this star is inserted into a plain solar disk that is often completed by a boat-like crescent moon and/or splayed wings[28] (Shamash; cf. the interior and exterior of the throne on Fig. 9.1, p. 190).

The question if Shamash was considered the supreme god of the ancient Mesopotamian pantheon can only be given a general answer, as this depends on the period[29] and the site under examination. Renowned cult centres for the worship of Shamash were located in Sippar (present-day Tell Abu Habbah), and Larsa (Tell-as-Senkereh; Iraq). In addition, his high rank can be exemplified by the mid-third millennium BC cuneiform list SF1 unearthed at Shuruppak (Tell Fara), Iraq.[30] On this text, he is

22 Krebernik 2009/10, 394–395; Heimpel 1986, 129–130; Foster 1996, 660 ("Sunset Prayer"), and Black & Green 1992, 173.

23 Black & Green 1992, 128.

24 Black & Green 1992, 182 and 184. Another, Old Babylonian reference to "UTU, my twin" comes from a mythic narrative about Inanna (Van Dijk 1998).

25 Black & Green 1992, 108.

26 Black & Green 1992, 135 (Nanna) and 133 (Ningal).

27 Beaulieu et al. 2017, 55–56, suggest that Aya may be associated with the (unidentified) stellar constellation "Ewe" (MUL.APIN .... MUL.APIN I. i. 18).

28 Paule, forthcoming.

29 Cf. Foster 1996, 654–656, n, (prayer about) *The Supremacy of Shamash.*

30 Krebernik 1986.

listed among the highest gods of ancient Mesopotamia, such as An/Anu, Enlil/Ellil (Grandfather of Shamash), and Enki/Ea.[31]

9.3 The Imagery of the Sun in Cosmology. Important Sources of Research

From an archaeoastronomical perspective, the following details are most striking and provide interesting insights into ancient Mesopotamian cosmology, although they seem at first sight to go slightly beyond the main topic of this article. However, as these details relate to hidden knowledge of important stages of the solar cycle, they are well worth covering.

As has been indicated in the previous section, Anu, the sky god, Enlil, the storm god, and Ea, the god of wisdom,[32] form the supreme trinity of ancient Mesopotamian mythology. Interestingly, the night sky and its stars are divided into the three realms of Anu, Enlil, and Ea, as is documented by at least two major sources: the omen text series *Enūma Anu Enlil* (prologue) and the MUL.APIN compendium, tablets I–II. The *Enūma Anu Enlil* series (ca. 70 tablets)[33] is based on Old Babylonian or second millennium BC text sources but was repeatedly copied until the second century AD. The MUL.APIN compendium[34] is best known for its star catalogues and is documented by seventh century BC text copies, but may in part relate to much earlier celestial events. Inter alia, the *Enūma Anu Enlil* text passage about the division of the night sky among these three gods mentions that *"they.* [...], *divided the* [stellar] *paths, the gods, the likeness* [of them they dr]*rew, the constellations, night (and) day, as equa*[ls? they measure]*"d, month and year they created."*[35] The second text passage taken from the introductory part of MUL.APIN I provides further information about their location in the sky, which is bound to the Sun's apparent movement through these paths and their zodiacal/stellar constellations during the daylight sky (cf. Fig. 9.4, MUL.APIN tablet I, Mus. No. BM 86378; 1000–500 BC or ca. 7th c. BC).

In addition, this text passage allows for determining the seasonal calendar of ancient Mesopotamia (as follows): "[the Sun stands in] the *Path of Anu* from the first of Addaru (XII [no. of month]) to the 30th of Ajjaru, [period of] wind and storm [spring] (II); the *Path of Enlil* from the 1st of Simanun (III) to the 30th of Abu (V), [period of] harvest and hot season [summer] ; the *Path of Anu* from the first of Ululu (VI) to the 30th of Arahsamnu (VIII), [period of] wind and storm [autumn]; and the *Path of Ea*

31 Black & Green 1992, 30, 76, and 75 (s. v. Anu, Ellil, and Ea).
32 The original range of their specific characteristics has been simplified here for better reading.
33 See Weidner 1968/69, Hunger & Pingree 1999, and Rochberg 2018 for an introduction.
34 See Hunger & Pingree 1989 as well as Hunger & Steele 2019 for an introduction.
35 Beaulieu et al. 2017, 2–3, pointing to Horowitz 2011, 147 (full text, references, and later variants).

from the 1st of Kislimu (IX) to the 30th of Šabatu (XI), cold [season; winter]."[36] Hypothetically, the provided data also allow for establishing a more detailed division of the ancient Mesopotamian solar cycle (zodiacal scheme) and for determining today's ecliptic longitude λ of the different paths by using vernal equinox as a starting point (see, however, below).[37] This knowledge can be combined with further data originating from the identification of individual stars that are assigned to Enlil, Anu, and Ea in the MUL.APIN (I, i, 1–I, ii, 35).

9.3.1 The Imagery of the Sun in Cosmology. Astronomical Interpretations

So far, this research has resulted in two main hypotheses; aiming to answer to the question of how the paths' boundaries should be determined. These hypotheses, that have attracted renewed attention following the publication by J. Koch,[38] can be summarized as follows: first, the hypothesis that the *Paths of Enlil, Anu, and Ea* are based on northern, middle, and southern (imaginary) belts around the celestial equator. This interpretation is based on a first mathematical approach to this problem, resulting in a declination value (δ) of ca. $\pm 17°$ for the boundaries of the double *Path of Anu* when measured north and south of the celestial equator. Following this approach that was initiated by A. Kopff & J. Schaumberger,[39] the following formulas can be used to illustrate the different ranges of declination (Tab. 9.1; cf. today's equatorial coordinate system and Neumann 1991/92, 111):

This can be completed by Schaumberger's formula (Tab. 9.2; sine theorem/spherical triangles):[40]

As can be seen from our tables above, the *Path of Anu* measures around 34° in width when Kopff's and Schaumberger's approach is applied (i. e., $2 \times 16°6' = 33°2'$).[41] However, this result does not answer to the question of how the boundaries of the upper (*Path of Enlil*) and lower (*Path of Ea*) paths can be determined. Therefore, one might assume that these paths include all the stars visible in the sky from ca. $\pm 17°$ (δ). A related approach by B. L. Van der Waerden is based on the identification of the

36 Text passage based on Weidner 1931/1932, 170 and Koch-Westenholz 1995, 24. The additional information in square brackets has been provided by the author.

37 Schaumberger 1935, 321 still is convincing.

38 Koch 1989 and Neumann 1991/92.

39 See Bezold et al. 1913, 6–8, for pioneering research by A. Kopff & Schaumberger 1935, 321–322, for a refinement of Kopff's approach.

40 Taken from Schaumberger 1935, 322.

41 Schaumberger 1935, 322.

Figure 9.4:
MUL.APIN tablet I (1000–500 BC or ca. 7th c. BC), Mus. No. BM 86378

(© The Trustees of the British Museum, London)

Table 9.1:
Kopff & Schaumberger's approach; cf. today's equatorial coordinate system and Neumann 1991/92, 111.

Ancient Path in the Sky (Kopff-Schaumberger)	Modern Calculated Values of Boundaries	Hypothetical Center Point	Fundamental Plane (0° Latitude)	Hypothetical Poles
Path of Enlil (north)	$\delta = 90 - \varphi$; $\delta = \text{ca.} + 17°$	center of	celestial	celestial
Path of Anu (middle)	$\delta = \text{ca.} + 17°$; $\delta = \text{ca.} - 17°$	the Earth	equator	poles
Path of Ea (south)	$\delta = \varphi - 90°$; $\delta = \text{ca.} - 17°$	(geocentric)		
δ = declination or upper and lower boundaries of the paths; φ = geographic latitude				

Table 9.2:
Sine theorem/spherical triangles (Schaumberger's formula)

$\delta' = \varepsilon = 23.8°$	$\sin 45° : \sin\delta = \sin 90° : \sin 23.8°$
$\lambda' = 90°$	$0.7071 : \sin\delta = 1 : 0.404$
$\lambda = 45°$	$1 \times \sin\delta = 0.404 \times 0.7071$
δ = path width of *Path of Anu*	$\sin\delta = 0.28567$
at one side of the ecliptic	$\delta = 16.6°$

individual stars of Enlil, Anu, and Ea in the MUL.APIN compendium followed by the retrospective calculation of their coordinates in the ancient sky.[42] This approach is, however, problematic, as the original dating of this compendium cannot be determined with certainty due to the lack of written sources that are earlier in date than the 7th century BC text copies of the original script.[43]

A further problem is posed by the invisibility of the imaginary boundaries between the different *Paths of Enlil, Anu, and Ea* in the night sky and, thus, the lack of orientation framework.[44] This problem has given rise to the second hypothesis that today is associated with the approach by Reiner & D. Pingree, suggesting that these paths were inspired from early observations of important stages of the solar cycle, such as vernal/autumnal equinox and summer/winter solstices. Once again, this idea is related to much earlier pioneering works, such as that by B. Meissner who suggests that the

42 Van der Waerden 1949.

43 Van der Waerden 1968, 69 and Koch 1989, 15–16. See Hunger & Steele 2019, 16–17, for the dating.

44 Cf. Koch 1989, esp. 16.

boundaries of the *Paths of Enlil and Ea* were limited by the observable line left by the Sun when rising at summer/winter solstice (following Meissner, the double *Path of Anu* measured 25° in width).[45] In this case, the upper and lower limits for the boundaries were limited by the declination values of the maximum positions of the ecliptic. Following Meissner's approach, it is assumed that the Earth's equatorial plane was inclined to it by an angle of about 23.8° with the ecliptic plane in antiquity (i. e., at a reference period of ca. 1000 BC). Therefore, one might assume that the *Paths of Enlil, Anu, and Ea* include all the stars visible in the sky from ca. ±23.8° (δ). This can be illustrated as follows (Tab. 9.3):

Table 9.3:
Meissner's approach

Ancient Path in the Sky (Meissner)	Modern Calculated Values of Boundaries	Hypothetical Center Point	Fundamental Plane (0° Latitude)	Hypothetical Poles
Path of Enlil (north)	$\delta = +\varepsilon$ (summer solstice; ca. +23.8°); $\delta = +12.5°$	center of the Earth (geocentric)	celestial equator	celestial poles
Path of Anu (middle)	δ= ca. +12.5°; δ= ca. −12.5°			
Path of Ea (south)	$\delta = -\varepsilon$ (winter solstice; ca. −23.8°); $\delta = -12.5°$			
δ = declination or upper and lower boundaries of the paths; ε = obliquity of ecliptic				

At this point, mention should be made of the innovative approach by E. Reiner & D. Pingree.[46] Both researchers have abandoned the previous idea of three imaginary bands located in parallel each other along the celestial equator. Instead, their starting point is the flat plane of the eastern horizon and the celestial sky that is divided into three upright parts (today: segments of the celestial sphere) by the solar path during

45 Meissner 1925, 407.
46 Reiner & Pingree 1981, 17–18.

the days of the equinoxes and summer/winter solstices.[47] This approach fits perfectly, if one has the following segments in upright position in mind:

i) a smaller one in the south that is limited at the inner side by the solar path during winter solstice (*Path of Ea*);

ii) two larger/identical ones that are separated in the middle by the solar path during equinoxes (*Path of Anu*);

iii) and a large one in the north that is limited at the inner side by the solar path at summer solstice (Path of Enlil).[48]

The latter also includes some constellations that are mentioned as being circumpolar in the MUL.APIN compendium, which is not in contradiction with their location in nature.[49] In addition, this reconstruction is in line with the first-mentioned text passage on the different *Paths of Enlil, Anu, and Ea* (MUL.APIN tablet I; see above). The approach based on Reiner and Pingree has further been reinforced by a text passage taken from the *Enūma Anu Enlil* tablets 50–51, commentary III 24b.[50] This text passage transmits us that "the road of the Sun at the end of the cattle-pen is the *Path of Ea*, the road in the middle of the cattle-pen is the *Path of Anu*, and the road of the Sun at the beginning of the cattle-pen is the *Path of Enlil*." In this context, the term cattle-pen (TÙR)[51] has been identified by Reiner & Pingree as the eastern horizon, which appears convincing. A short summary is given in Tab. 9.4 (as follows):

The changing of the perspective brought about by the different reconstruction models (e. g., celestial equator – eastern horizon; celestial poles – zenith and nadir) allows for coming back to some key issues. In particular, they have been inspired from the discussion following the author's presentation in the 2023 meeting[52] of the *Gesellschaft für Archäoastronomie* held in Weimar, Germany. First, the question of whether the ecliptic plane was already known in ancient Mesopotamia and second, the question of whether precession was already discovered in ancient Mesopotamia.

47 In this context, another reconstruction model by van der Waerden 1949, 24, fig. 5, is very interesting: here, the author suggests a (first) celestial sphere that is divided by a zodiacal scheme into four segments – two horizontal and two oblique ones – assigned to the *Paths of Enlil, Anu, and Ea*.

48 Unfortunately, the idea of upright segments instead of horizontal bands is not included in Neumann 1991/92, 111, who assigns the declination value (δ) = 0 to the *path of Anu* instead of a double segment or a double path width, as it was made in the approach by Kopff & Schaumberger.

49 Reiner & Pingree 1981, 18.

50 Reiner & Pingree 1981, 17.

51 Apart from the approach by Reiner and Pingree, the Sumerian logogram TÙR (also spelled TUR_3; Akkadian: *tarbasu*) has been translated either as "cattle-pen" or "halo of the moon" (Labat & Malbran-Labat 1976, 2 and 78–79 and Schramm 2010, 151). Not only his problem, but also different omen series related to the halo of the Moon have more recently been dealt with in Verderame 2014.

52 This meeting was entitled *"Man within the Cosmos: Lifeworlds and Cosmologies"* (June 21–25, 2023).

Table 9.4:
Reiner & Pingree's approach

Ancient Path in the Sky (Reiner-Pingree)	Modern Calculated Values of Inner Boundaries/ Start Points	Hypothetical Center Point	Fundamental Plane (0° Latitude)	Hypothetical Poles
Path of Enlil (north)	$\delta = +\varepsilon$ (summer solstice; ca. $+23,8°$)	observer	eastern horizon	zenith, nadir
Path of Anu (middle)	$\delta = 0°$ (equinoxes)			
Path of Ea (south)	$\delta = -\varepsilon$ (winter solstice; ca. $-23,8°$)			
δ = heliacal rising of the Sun on the horizon/start points of the day arc at the solstices/equinoxes seen by an observer on Earth; ε = obliquity of ecliptic				

With regard to the former, van der Waerden is amongst the first to indicate that the MUl.Apin I tablet also includes another crucial text passage; i. e., *"that not only the Moon but also the Sun and the five planets walk along the same path."* This path is described as the *Path of the Moon* (zodiacal belt) in the same text passage and includes 18 stellar constellations. Therefore, ancient Mesopotamian stargazers must have noted the oblique circle related to the day arcs of the Sun in the sky, which was followed later by striking constellations visible in the night sky.[53] In response to the second question,[54] another further argument put forward by Van der Waerden can be used for clarification. More precisely, he complains of irregularities in the zodiacal scheme describes in MUL.APIN, which states that the vernal equinox takes place in the middle of the month (Nisannu). The latter is 15° of the sign [Aries][55] or would be correct only for a period at about 940 BC, but was for long copied without any corrections (cf. the 7th c. date assigned to the preserved MUL.APIN tablets).[56]

A closing remark relates to the question if ancient Mesopotamian stargazers knew that circumpolar stars appear to rotate around the north celestial pole at night. This scenario cannot be excluded, as some constellations in the *Path of Enlil* are classified as circumpolar (MUL.APIN). In this context, another text passage that indicates that

53 See Van der Waerden 1949, 24, and 1952/53, 218.

54 Reiner & Pingree 1981, 18.

55 This consequently also applies to the remaining equinox and solstices.

56 This is also true for much later texts that contains an error of up to five days or more (Van der Waerden 1952/53, 223–224. See also the more recent summary by Rogers 1998, 21.

the *"MAR.GİD.DA [star] stands all year and circles around"* comes to confirm this working hypothesis (*Enūma Anu Enlil* 50 commentary III, 28c).[57]

9.4 The Imagery of the Sun in Divination Literature

Before turning to clues for astronomical observations encountered in ancient Mesopotamian divination literature, some brief mention should be made of Shamash, the judge of heaven and earth.[58] In particular, he was evoked by the king and his diviners as an aid against evil magic and for passing his judgement on specific issues.[59] For this purpose, Shamash was preferably evoked together with Adad, the storm god. Both deities also offer advice on the occurrence of lunar eclipses that can be drawn from oil and liver divination (the so-called extispicy; esp. during the first mill. BC).[60]

Despite this archaic-seeming practices and beliefs, some clues for actual astronomical observations can be found in divination literate. This particularly applies to the solar eclipse omen texts of the *Enūma Anu Enlil* series, tablets 31–36. The omens are written in the form of protasis-apodosis statements and inform us about what will happen if the Sun is eclipsed in months I–XII on specific days. Over the last two decades, research on these textual sources could add new aspects. Above all, a related, late second millennium BC compendium on solar eclipse omens has been presented by M. T. Rutz, ending in better readability of the *Enūma Anu Enlil* series.[61]

On the other hand, J. C. Fincke's research reveals that not only total but also annular solar eclipses are recorded in this series.[62] This can be demonstrated by means of the following example (cf. Fig. 9.5, *Enūma Anu Enlil* Tablet 24/25, BM 38359; Neo-Late Babylonian): *"If the face of a normal solar disk* [is covered] *in the mid*[dle]."[63]

However, as the solar eclipses mentioned in this series do not only occur at the end of the month or at new moon – as it happens in nature – but also on different days of the month, one may have doubts whether these statements are based on real eclipse observations.[64] An explanation for this may be that the original *Enūma Anu Enlil* series was created at an earlier period during which it was difficult to add to the annual solar/lunar calendar by means of an intercalary month. Another explanation comes from U. Koch-Westenholz[65] who refers to the practice of "pseudo-mathematics" or

57 Reiner & Pingree 1981.
58 I. e., he was considered as the divine judge of the gods and humans, the living and dead.
59 See the series of texts mentioned in Foster 1996, esp. 641–664 (ca. 1500–1000 BC).
60 Koch-Westenholz 1995, 39, 44, and 111; Koch-Westenholz 2001, 72, 78, and 81.
61 Rutz 2006.
62 EAE tablet 24/25; Fincke 2014, 107–115.
63 Solar omens I, 1–1a; Fincke 2014, 109–111. See also Van Soldt 1995.
64 Steele 2000, 24.
65 Koch-Westenholz 1995, esp. 87 und 110.

Figure 9.5:
Enūma Anu Enlil Tablet 24/25, BM 38359; Neo-Late Babylonian (Fragmented)

(British Museum, London. Taken from Fincke, J. C. (2014), "Commentary on Enūma Anu Enlil 24(25) (CCP 3.1.24.E)", *Cuneiform Commentaries Project* (E. Frahm, E. Jiménez, M. Frazer, & K. Wagensonner), 2013–2023; `https://ccp.yale.edu/P461177` (accessed December 6, 2023). DOI: `10079/931zd4c`, © The Trustees of the British Museum, London)

the use of additional explanatory lists (*sâtu*) that complete ancient manuals. Unfortunately, the *sâtu* list of the *Enūma Anu Enlil* series is hardly known. However, another *sâtu* text explains that "the 22^{bd} day equals the 14th day (of the month) and the 25th equals the 15th day" and "the month of Abu" means "this month" (the *Šumma Sîn ina tāmartīšu*; *ACh Sin* 3, 49–50). This may also be the way to cope with the inevitable difficulties with date specifications in the *Enūma Anu Enlil* series. A further interesting problem concerns the identification of the different types of solar eclipse in research literature.

A good example for this is the description of the Sun in a 7th century BC *Report* to the Assyrian King, which *"at is rising is like a crescent and wears a crown like the Moon."* (written by *Rasil, the older, servant of the King*).[66] In this context, R. Stephenson tells us that this solar eclipse has been assigned by S. Parpola to that occurring on the 27th of May 669 BC. Later, the same solar eclipse has been assigned to a text passage stemming from another *Report* written by Akkullanu, the scribe, who mentions that *"the Sun at its rising made an eclipse of two fingers."* As the latter corresponds to the occurrence of a weak and hardly visible partial solar eclipse or an eclipse magnitude of ca. 0.17, both interpretations are somewhat contradictory.[67] This interpretation problem clearly demonstrates the potential of research using astronomy software, such as the open-source free software *Stellarium*[68] that can be used in problem-solving. This can be combined with further research based on the consultation of the *Five Millennium Catalog of Solar Eclipses*[69] and X. M. Jubier's *Five Mill. Canon of Solar Eclipses Web Tool.*[70] First trials show promising results and allow an independent opinion on the occurrence of solar eclipses in Antiquity.[71]

9.5 Conclusions

Investigations on ancient Mesopotamian cosmology usually focus on studying the astral aspect of distinctive classes of artefacts, such as solar iconography, mythology, or solar omen texts. So far, however, these studies lack of attention to whether the information drawn from these different sources can be combined. This research task is, however, needed as no distinction was made in ancient times between early, prescientific forms of astronomy and mythological beliefs related to the sky. This is also the reason why it remains difficult today to rediscover the ancient achievements of

66 Hunger 1992, 220.

67 Huber & De Meis 2004, 36–37.

68 `https://stellarium.org` (latest version 23.3). See Zotti et al. 2021 for an introduction.

69 Espenak & Meeus 2006/2009. Cf. `https://eclipse.gsfc.nasa.gov/SEcat5/SEcatalog.html`.

70 `http://xjubier.free.fr/en/site_pages/solar_eclipses/5MCSE/xSE_Five_Millennium_Canon.html`.

71 Paule, forthcoming.

sky and space, which obviously are hidden in solar imagery and written texts on solar mythology. Nevertheless, a century-long history of research on individual text passages has generated different models of the division of the ancient sky. With this in mind, this article seeks to outline some of the major sources and to present the relationship to each other. This also includes the consideration of selected solar imagery and the Sun god's family ties. In doing so, it has become obvious that most of the archaeological record allows for associations with the real Sun and the patterns of its movement in the sky. Despite there is no general consensus regarding the reconstructions of ancient models of the sky, such as the *Paths of Enlil, Anu, and Ea*, it can be indicated that the more ancient archaeological record, such as glyptic art and the *Enuma Anu Enlil* text series, seems to be based on the perspective of the ancient observer on Earth (eastern horizon). In addition, it seems plausible that the ecliptic and its obliquity has been recognised in the 7th century BC at the latest. There is also sparse evidence for knowledge of some of the characteristics of circumpolar stars, although the precession has not been discovered yet. Due to the wide spectrum of source material, however, this can only be a preliminary result. With regard to further studies on ancient Mesopotamian cosmology, the integration of further source material from the ancient Near East, and the application of modern computations tools offers a prospect for the future.

Acknowledgements

Special thanks go to the museum staff (account managers, archivists, photographers), who kindly allowed reprinting of images. Special thanks also go to individual researchers who provided feedback on various versions of this article, as did Jeanette C. Fincke (Leiden) and Gudrun Wolfschmidt (Hamburg).

9.6 Literatur

ASCALONE, ENRICO & LUCA PEYRONEL: Two Weights from Temple N at Tell Mardikh-Ebla, Syria: A Link between Metrology and Cultic Activities in the Second Millennium BC? In: *Journal of Cuneiform Studies* **J53** (2001), p. 1–12.

ARUZ, JOAN, BENZEL, KIM & JEAN M. EVANS: *Beyond Babylon: Art, Trade, and Diplomacy in the Second Millennium BC.* New York: The Metropolitan Museum of Art; New Haven, Conn., London: Yale University Press 2008.

BELMONTE, JUAN A. & JOSÉ LULL: *Astronomy of Ancient Egypt. A Cultural Perspective.* Cham: Springer 2023.

BEZOLD, CARL; KOPFF, AUGUST & FRANZ BÖLL: *Zenit- und Äquatorialgestirne am babylonischen Fixsternhimmel.* Sitzungsberichte der Heidelberger Akademie der Wissenschaften, Philosophisch-Historische Klasse. Heidelberg: Winter 1913.

BLACK, JEREMY A. & ANTHONY GREEN: *Gods, Demons and Symbols in Ancient Mesopotamia: An Illustrated Dictionary.* London: British Museum Press 1992.

ELSEN-NOVÁK, GABRIELE & MIRKO NOVÁK: Der "König der Gerechtigkeit." Zur Ikonologie und Teleologie des 'Codex' Ḫammurapi. In: *Baghdader Mitteilungen* **37** (2006), p. 131–155.

ESPENAK, FRED & JEAN MEEUS: *Five Millennium Catalog of Solar Eclipses: -1999 to +3000 (2000 BCE to 3000 CE) (Revised).* Greenbelt, Md.: National Aeronautics and Space Flight Administration. NASA technical papers TP-2006-2141141 (2006) und TP-2009-214172 (überarbeitete Ausgabe; 2009) `https://eclipse.gsfc.nasa.gov/SEpubs/5MCSE.html`.

FINKEL, IRVING & ALEXANDRA FLETCHER: Thinking outside the Box: The Case of the Sun-God Tablet and the Cruciform Monument. In: *Bulletin of the American School of Research* **375** (2016), p. 215–248.

FINCKE, JEANETTE C.: Additions to Already Edited Enūma Anu Enlil (EAE) Tablets, Part II: The Tablets Concerning the Appearance of the Sun Published in PIHANS 73, Part I. In: *KASKAL* **11** (2014), p. 103–139.

FOSTER, BENJAMIN R.: *Before the Muses: An Anthology of Akkadian Literature.* Potomac, Maryland, USA: CDL Press 1996.

GEORGE, ANDREWS: *The Epic of Gilgamesh. The Babylonian Epic Poem and Other texts in Akkadian and Sumerian.* London, New York: Penguin 2003.

GÖSSMANN, FELIX: Planetarium Babylonicum oder die Sumerisch-babylonischen Stern-Namen. In: DEIMEL, ANTON: *Sumerisches Lexikon 4/2.* Rome: Verlag des Päpstlichen Bibelinstituts 1950, p. 23–24.

HEIMPEL, WOLFGANG: The Sun at Night and the Doors of Heaven in Babylonian Texts. In: *Journal of Cuneiform Studies* **38** (1986), 2, p. 127–151.

HOROWITZ, WAYNE: *Mesopotamian Cosmic Geography.* Winona Lake, Indiana: Eisenbrauns (2nd edition) 2011.

HUBER, PETER J. & SALVO DE MEIS: *Babylonian Eclipse Observations.* Milan: IsIAO-Mimesis 2004.

HUNGER, HERMANN: *Astrological Reports to the Assyrian Kings. State Archives of Assyria III.* Helsinki: Helsinki University Press 1992.

HUNGER, HERMANN & DAVID PINGREE: *MUL.APIN. An Astronomical Compendium in Cuneiform.* Horn: Berger & Söhne (Archiv für Orientforschung, Beiheft 24) 1989.

HUNGER, HERMANN & DAVID PINGREE: *Astral Sciences in Mesopotamia.* Leiden: Brill (Handbuch der Orientalistik; Bd. 44) 1999.

HUNGER, HERMANN & JOHN STEELE: *The Babylonian Astronomical Compendium MUL.APIN.* London, New York: Routledge 2019.

KOCH, JOHANNES: *Neue Untersuchungen zur Topographie des babylonischen Fixsternhimmels.* Wiesbaden: Harrassowitz 1989.

KOCH-WESTENHOLZ, ULLA: *Mesopotamian Astrology. An Introduction to Babylonian and Assyrian Celestial Divination.* Copenhagen: Museum Tusculum Press and University of Copenhagen, Carsten Niebuhr Institute of Near Eastern Studies 1995.

KOCH-WESTENHOLZ, ULLA: Babylonian Views of Eclipses. In: *Res Orientales* XIII (2001), p. 71–84.

KREBERNIK, MANFRED: Die Götterlisten aus Fāra. In: *Zeitschrift für Assyriologie und vorderasiatische Archäologie* **76** (1986), p. 161–204.

KREBERNIK, MANFRED: Šer(i)da. In: *Reallexikon der Assyriologie und Vorderasiatischen Archäologie* (RlA), Vol. **12** (2009/10), p. 394–395. This lexikon entry is also available on the website of *Bayerische Akademie der Wissenschaften*: `https://publikationen.badw.de/en/rla/index#10696`.

LABAT, RENÉ & FLORENCE MALBRAN-LABAT: *Manuel d'épigraphie akkadienne: signes, syllabaires, idéogrammes.* Paris: Librairie orientaliste Paul Geuthner (fifth edition) 1976.

ŁAWECKA, DOROTA: The contest scene in EDIII Mesopotamian glyptic art. In: *Polish Archaeology in the Mediterranean* **31** (2022), 2, p. 83–121.

LEWY, HILDEGARD: Ištar-ṣad and the Bow Star. In: GÜTERBOCK, HANS.-G. & THORKILD JACOBSEN (eds.): *Studies in Honor of Benno Landsberger on His Seventy-Fifth Birthday.* Chicago: The University of Chicago Press (Assyriological Studies; 16) 1965, p. 273–281.

MEISSNER, BRUNO: *Babylon und Assyrien. Zweiter Band.* Heidelberg: Winter 1925.

NEUGEBAUER, OTTO: *Astronomical Cuneiform Texts. Babylonian Ephemerides of the Seleucid Period for the Motion of the Sun, the Moon, and the Planets.* New York: Springer 1955.

OELSNER, JOACHIM & WAYNE HOROWITZ: The 30-Star-Catalogue HS 1897 and the Late Parallel BM 55502. In: *Archiv für Orientforschung* **44/45** (1997/98), p. 176–185.

NEUMANN, HEINZ: Anmerkungen zu Johannes Koch, Neue Untersuchungen zur Topographie des babylonischen Fixsternhimmels. In: *Archiv für Orientforschung* **38/39** (1991/92), p. 110–124.

PAULE, ANNA: "Signs from Above? Research on the Observational Value of Solar Symbols and Solar Eclipse Records of the Ancient Near East and/or Cyprus." In: FRINCU, MARC (ed.): *Signs and Symbols: Above and Below.* Proceedings of the the 29th SEAC (European Society for Astronomy in Culture) Conference, held at Timisoara, Romania, from the 5th to 9th of September 2022. Submitted for publication.

REINER, ERIKA & DAVID PINGREE: *Babylonian Planetary Omens: Part Two. Enūma Anu Enlil, Tablets 50–51.* Malibu: Undena Publications 1981.

ROCHBERG, FRANCESCA: The Catalogues of Enūma Anu Enlil. In: STEINERT, ULRIKE (ed.): *Assyrian and Babylonian Scholarly Text Catalogues: Medicine, Magic and Divination.* Berlin, Boston: De Gruyter 2018, p. 121–136.

ROGERS, JOHN H.: Origins of the ancient constellations: I. The Mesopotamian traditions. In: *Journal of the British Astronomical Association* **108** (1998), 1, p. 9–28.

ROTH, MARTHA: *Law Collections from Mesopotamia and Asia Minor.* Atlanta, Georgia: Scholars Press (Writings from the Ancient World, Society of Biblical Literature, vol. 6) 1995.

RUGGLES, CLIVE L. N. & NICHOLAS J. SAUNDERS: The Study of Cultural Astronomy. In: RUGGLES, CLIVE L. N. & NICHOLAS J. SAUNDERS (eds.): *Astronomies and Cultures.* Niwot, CO: University Press of Colorado 1993, p. 1–31.

RUTZ, MATTHEW T.: Textual Transmission between Babylonia and Susa: A New Solar Omen Compendium. In: *Journal of Cuneiform Studies* **58** (2006), p. 63–96.

SCHAUMBERGER, JOHANN: *Sternkunde und Sterndienst in Babel. Assyriologische, astronomische und astralmythologische Untersuchungen von Franz Xaver Kugler. 3. Ergänzungsheft zum ersten und zweiten Buch.* Münster, Westfalen: Aschendorffsche Verlagsbuchhandlung 1935.

SCHRAMM, WOLFGANG: *Akkadische Logogramme.* Göttingen: Universitätsverlag Göttingen (Göttinger Beiträge zum Alten Orient; 5) 2010.

SHALTOUT, MOSALAM & JUAN A. BELMONTE: On the Orientation of Ancient Egyptian Temples: (1) Upper Egypt and Lower Nubia. In: *Journal for the History of Astronom* **XXXVI** (2005), p. 273–298.

SINCLAIR, ROLF M.: The Nature of Archaeoastronomy. In: BOSTWICK, TODD, W. & BRYAN BATES (eds.): *Viewing the Sky Through Past and Present Cultures; Selected Papers from the Oxford VII International Conference on Archaeoastronomy.* Phoenix, Arizona: City of Phoenix, Parks, and Recreation Department (Pueblo Grande Museum Anthropological Papers, vol. 15) 2006, p. 13–26.

SOŁTYSIAK, ARKADIUSZ: The Bull of Heaven in Mesopotamian Sources. In: *CULTURE and COSMOS. A Journal of the History of Astrology and Cultural Astronomy* **5** (2001), 2, p. 3–21.

STEELE, JOHN M.: *Observations and Predictions of Eclipse Times by Early Astronomers.* Dordrecht: Kluwer Academic Publishers (Archimedes: New Studies in the History and Philosophy of Science and Technology; vol 4) 2000.

VAN DER WAERDEN, BARTEL L.: Babylonian Astronomy. II. The Thirty-Six Stars. In: *Journal of Near Eastern Studies* **8** (1949), p. 6–26.

VAN DER WAERDEN, BARTEL L.: History of the Zodiac. In: *Archiv für Orientforschung* **16** (1952/53), p. 216–230.

VAN DER WAERDEN, BARTEL L.: *Erwachende Wissenschaften 2: Die Anfänge der Astronomie.* Stuttgart: Birkhäuser 1968.

VAN DIJK, JOHANNES J. A.: Inanna raubt den "großen Himmel". Ein Mythos. In: MAUL, STEFAN M. (ed.): *Festschrift für Rykle Borger zu seinem 65. Geburtstag am 24. Mai 1994. Tikip santakki mala bašmu.* Groningen: STYX Publications 1998, p. 9–38.
An English version of the mythological text is available on the website of *The Electronic Text Corpus of Sumerian Literature*: `https://etcsl.orinst.ox.ac.uk/section1/tr135.htm`.

VAN SOLDT, WILFRED: *Solar Omens of EAE: Tablets 23 (24)–29 (39).* Netherlands: Instituut voor het Nabije Osten 1995.

VERDERAME, LORENZO: The Halo of the Moon. In: FINCKE, JEANETTE C. (ed.): *Divination in the Ancient Near East. A Workshop on Divination Conducted during the 54th Rencontre Assyriologique Internationale, Würzburg 2008.* Winona Lake, Indiana: Eisenbrauns 2014, p. 91–104.

WEIDNER, ERNST F.: Der Tierkreis und die Wege am Himmel. In: *Archiv für Orientforschung* **7** (1931/32), p. 170–187.

WEIDNER, ERNST F.: Die astrologische Serie Enûma Anu Enlil. In: *Archiv für Orientforschung* **22** (1968/69), p. 65–75.

WIGGERMANN, FRANS A. M.: *Mesopotamian Protective Spirits: The Ritual Texts.* Groningen: STYX and PP Publications (Cuneiform Monographs 1) 1992.

WOODS, CHRISTOPHER E.: The Sun-God Tablet of Nabû-Apla-Iddina Revisited. In: *Journal of Cuneiform Studies* **56** (2004), p. 23–103.

ZOTTI, GEORG; HOFFMANN, SUSANNE M.; WOLF, ALEXANDER V.; CHÉREAU, FABIEN & GUILLAUME CHÉREAU: The Simulated Sky: Stellarium for Cultural Astronomy Research. In: *Journal of Skyscape Archaeology* **6** (2021), 2, p. 221–258.

Abbildung 10.1:
Tablet with the Epic of Gilgamesh

(Creative Commons CC BY 4.0, Osama Shukir Muhammed Amin)

Rätsel der alten Astronomie

Jörg Bäcker (Gummersbach)

Abstract: Enigmas of ancient astronomy

In the Epic of Gilgamesh it is said that Gilgamesh came to the Mashu (Twin) Mountains where the Sun rises and sets. In the same way, it is also told of a Sun that was always in the middle of the sky. Such descriptions are found especially in the oldest Sumerian texts, but hardly at all in the later Akkadian texts. The same peculiar static view of the world is also found in connection with the moon and Venus. The latter is described as standing in the centre of the sky and radiating down from there with extraordinary intensity. How does all this go together? The Gilgamesh epic is linguistically comprehensible to a large extent, but the cosmological relationships there, which incidentally also occur in the oldest Indian texts as well as in Indian traditions of North America, still remain puzzling and are to be presented here to the participants for discussion.

Zusammenfassung

Im Gilgamesch-Epos heisst es, dass Gilgamesch zu den Maschu (Zwillings)-Bergen kam, wo die Sonne auf- und untergeht. In gleicher Weise wird auch von einer Sonne erzählt, die immer in der Mitte des Himmels stand. Solche Beschreibungen finden sich besonders in den ältesten sumerischen, aber kaum noch in den späteren akkadischen Texten. Dasselbe eigenartige statische Weltbild findet man auch in Verbindung mit dem Mond und mit Venus. Letztere wird als in der Mitte des Himmels stehend und von dort als ausserordentlich intensiv herabstrahlend beschrieben. Wie geht das alles zusammen? Das Gilgamesch-Epos ist zwar sprachlich weitgehend verständlich, aber die dortigen kosmologischen Verhältnisse, die im übrigen auch in den ältesten indischen Texten wie auch in indianischen Überlieferungen Nordamerikas vorkommen, bleiben nach wie vor rätselhaft und sollen hier den Teilnehmern zur Diskussion gestellt werden.

Abbildung 10.2:
Enkidu, Freund von Gilgamesh, Ur, 2027–1763 v. Chr. (Iraq-Museum)

10.1 Literatur

BÄCKER, JÖRG: The Cosmic Pillar and the Cosmic Tree. Macrocosmos and Microcosmos – Types and Areas – Questions of Origin. In: WOLFSCHMIDT, GUDRUN (Hg.): *Orientierung, Navigation und Zeitbestimmung – Wie der Himmel den Lebensraum des Menschen prägt.* Hamburg: tredition (Nuncius Hamburgensis; Band 42) 2019, S. 210–218.

HOROWITZ, WAYNE: *Mesopotamian Cosmic Geography.* Winona Lake, Ind.: Penn State University Press, Eisenbrauns 1998, (2nd revised edition) 2012.

RÖLLIG, WOLFGANG: *Das Gilgamesch Epos.* Übersetzt und herausgegeben von Wolfgang Röllig. Stuttgart: Philipp Reclam jun. (Reclams Universal-Bibliothek) 2015.

Abbildung 10.3:
Gilgamesch mit einem Löwen
(assyrische Statue vom Palast von Sargon II., Khorsabad (713–706 v. Chr.)

(Louvre Paris, CC3, Urban)

Abbildung 11.1:
Diskos von Phaistos, oben Seite A, unten Seite B

(Bearbeitetes Foto des Autors vom Original im Archäologischen Museum Heraklion)

Der kretisch-minoische *Diskos von Phaistos*, ein herausragendes Artefakt der Archäoastronomie auf Basis einer dreiteiligen Kosmologie „Geist-Leben-Sache“

Hermann Wenzel (München)

Abstract: Enigmas of ancient astronomy

The subject of the treatise are the textures of the two sides of the so-called "Diskos of Phaistos", a drawing disk from the ancient Minoan palace of Phaistos (Crete). Prepared through analysis and hypotheses, the signs turn out to be daily numbers and a mathematical and astronomical compendium created by the authors. Through multiple overlapping patterns of textures (layer system), character positions of astronomical data (day numbers) are added and generated individually or synodically and sidereal in the network of the periods of all planets known at the time.

The process is divided into: permutations of the original characteristics in various matrices, organization of the distribution of characters, content differentiation of the character in the genders of the spiritual, physical and material. Deciphering the characters as daily numbers, proving the correct assignment of daily numbers to the character through the discovery of an algorithm. Planetary periods with various interactions e. g. period relations, eclipse periods and so-called "Great Years". Structural relationships between the sides of the discos.

Zusammenfassung

Gegenstand der Abhandlung sind die Texturen der beiden Seiten des so genannten „*Diskos von Phaistos*“, Zeichenscheibe aus dem alten minoischen Palast von Phaistos (Kreta). Vorbereitet durch Analyse und Hypothese erweisen sich die Zeichen als Tageszahlen und ein von den Urhebern angelegtes mathematisches und astronomisches Kompendium. Durch vielfach sich überlagernde Bemusterungen der Texturen (Layer-System) werden Zeichenpositionen astronomischer Daten (Tageszahlen) addiert und generieren einzeln oder im Geflecht die Perioden aller damals bekannter Planeten synodisch und siderisch.

Das Prozedere gliedert sich in: Permutationen der originären Zeichenfolge, in diverse Matrizen, Organisation der Zeichenverteilung, inhaltliche Unterscheidung der Zeichen in die Geschlechter des Geistigen, Leiblichen und Sächlichen. Entzifferung der Zeichen als Tageszahlen, Beweisführung der zutreffenden Zuordnung von Tageszahlen zu den Zeichen durch die Entdeckung eines Algorithmus. Planetenperioden mit diversen Interaktionen z. B. Periodenrelationen, Finsternis-Perioden und so genannte „Große Jahre“. Strukturelle Beziehungen zwischen den Seiten des Diskos.

11.1 Einführung

Seit über 100 Jahren gibt es Bemühungen die ‚Schrift‘ des geheimnisumwitterten *Diskos von Phaistos* zu deuten. Nun hat die akademische Wissenschaft es gänzlich als sprachlich unentzifferbar aufgegeben. Begründung: Der Zeichenvorrat sei zu gering, als dass man mit statistischen Methoden die Lautwerte der Zeichen ermitteln könne. Dagegen ist grundsätzlich einzuwenden, dass bisher als selbstverständlich angenommen wurde und wird, dass die Zeichen des Diskos Sprachzeichen seien für Silben oder Worte und es sich um einen Text mit verbaler Botschaft handele.

Ein methodischer Fehler liegt nun darin, dass eine Sprachlichkeit der Zeichen nur dann als erwiesen gelten kann, wenn sich ohne jeden Zweifel ein Text erschließen ließe, wie bei den Hieroglyphen auf dem Stein von Rosette durch die parallel vorhandene Übersetzung ins Griechische. Solange derartiges oder Texte in größerem Umfang beim *Diskos von Phaistos* aber nicht gegeben sind, muss die Bedeutung der Zeichen auch für andere Denkmodelle offen bleiben.

Die zu frühe Festlegung auf einen sprachlichen Charakter des *Diskos* verhinderte eine Strukturanalyse, die sprachlich nicht relevante Phänomene mit einbezogen hätte. Ein wissenschaftlich einwandfreier Ansatz beginnt daher mit der Bestandsaufnahme aller unterscheidbarer Gegebenheiten auf beiden Seiten des *Diskos*: Wie viele Zeichen finden sich auf jeder Seite, wie viele Zeichengruppen, wie viele von jeder Größe der Zeichengruppen, wie viel verschiedene Zeichen auf jeder Seite, wie viel Zeichenwiederholungen, usf., eine mühsame Arbeit, die viele Tabellen füllt, die wiederum auszuwerten sind. Mit dieser Methode wäre man frühzeitig auf Erstaunliches gestoßen, welches durch numerische Strukturen der Teilbarkeit, der Folge und vor allem der Symmetrie zunehmend die Möglichkeit einer Sprachlichkeit der Zeichen in Frage gestellt hätte.

Und so zeigt sich jetzt, dass der *Diskos von Phaistos* sehr wohl entzifferbar ist, allerdings in einer ganz anderen Richtung als der bisher verfolgten. Nicht die Sprachwissenschaft ist gefragt, sondern Mathematik und Astronomie, die in den beiden Seiten des Diskos hunderte von übereinander gelagerten Gleichungen und Daten entdecken, entwirren und thematisch zusammenstellen lassen.

11.2 Analyse und Entzifferung des *Diskos von Phaistos*

Abbildung 11.2:
Diskos von Phaistos,
Zeichnung der Ausgräber beider Seiten des Diskos, links Seite A, rechts Seite B

(Evans: Scripta Minoa I, 1909, p. 280–282)

Die Anordnung der Zeichen in zwei peripheren Kreiszeilen und zwei Spiralzeilen führen bald zur Vermutung, dass es sich auch um eine Kryptographie, eine Verschlüsselung von Vorläuferstrukturen handeln könne; etwa in Art diverser Matrizen, die z. B. die Zeichengruppen untereinander anordnen oder im Blocksatz Spalten und Zeilen.

Beide Seiten des Diskos verfügen über ein Inventar von 242 mittels Stempel geprägter Zeichen, die in 61 Gruppen gegliedert wurden. Die Begrenzung der Gruppen erfolgt mittels senkrecht gravierter Striche in den Zeilen.

Neben den geprägten Zeichen gibt es 17 so genannte Dorne, Strichmarkierungen unter Anfangszeichen der Gruppen ab dem Zentrum gesehen. Es sind 8 auf Seite B und 9 auf Seite A. In gleicher Sicht enden die Kreiszeilen jeder Seite mit 5 Punkten auf den letzten Begrenzungen, die möglicherweise jeweils als 5 Tage gewertet werden könnten.

242 Zeichen lassen sich in 22×11 Zeichen zerlegen, ein erster Hinweis auf die Bedeutung der „11“ als Gliederungsprinzip. Die Anzahl 61 der Zeichengruppen ist als Primzahl nur durch 1 und sich selbst teilbar; aber sie zerlegt sich in zwei Quadratzahlen, 25 und 36, was von Bedeutung ist. Die Zahlen 11 und 61 beherrschen die Oberflächen der beiden Seiten des Diskos: Die beiden peripheren Randzeilen weisen 99 ($= 9 \times 11$) Zeichen in 25 Gruppen auf. Die beiden Spiralzeilen der Binnenbereiche zählen 143 ($= 13 \times 11$) Zeichen in 36 Gruppen, 18 auf jeder Seite.

Die Anzahl der 242 geprägten Zeichen zerlegt sich in 44 Zeichen, die nur auf einer der beiden Seiten vorkommen und 2×99 ($= 2 \times 9 \times 11$) Zeichen die beidseitig auftreten, 99 auf jeder Seite.

Eines der 242 geprägten Zeichen wurde vor dem Ofenbrand ausgekratzt und nicht nachgestempelt, wie dies an anderen Positionen der Fall war. Das Zeichen ist aber wichtig und muss und kann auf verschiedene Weisen rekonstruiert werden. In den Spuren der ausgekratzten Stelle ist unter anderem ein kleiner Winkel zu erkennen.

Yves Duhoux, einem belgischer Forscher, dem es gelang den Diskos aus der Vitrine zu nehmen, spricht von einer geknickten Kontur im Bereich der ausgekratzten Stelle. Man könnte annehme, hier haben die Verfasser des Diskos einen Hinweis hinterlassen, welches Zeichen, nämlich ein Winkel, an dieser Stelle nach zu stempeln sei, was dann jedoch vergessen oder aus bestimmten Gründen absichtlich unterlassen wurde.

Eine andere Möglichkeit das ausgewischte Zeichen zu rekonstruieren besteht in einem Ausschlussverfahren mit Hilfe des Teilers „11“.

Eine dritte Methode, berücksichtigt etliche Daten, denen jeweils ein *Winkel zu 13 Tagen* fehlt, bliebe das ausgewischte Zeichen unberücksichtigt.

Abb. 11.3 zeigt den Ablauf der Zeichen in einer Kreismatrix zu der die untereinander geschichteten 61 Zeichengruppen ab Seite B (oben rechts) und Seite A (unten links) gebogen wurden. 10 Achsen – $x1 - x1$ bis $x10 - x10$ – halbieren jeweils das Zeicheninventar im Verhältnis 121 : 121 Zeichen ($11 \times 11 : 11 \times 11$) und die Anzahl der Zeichengruppen im Verhältnis 30 : 31. Der innere Zahlenkranz entspricht jeweils der Anzahl der Zeichen im daneben liegenden Abschnitt der Kreismatrix zwischen zwei Achsen. Gegenüber liegende Zahlen sind gleich groß. Mit einer Primzahl bedingten Ausnahme zwischen den Achsen x2 und x3 ist auch die Anzahl gegenüberliegender

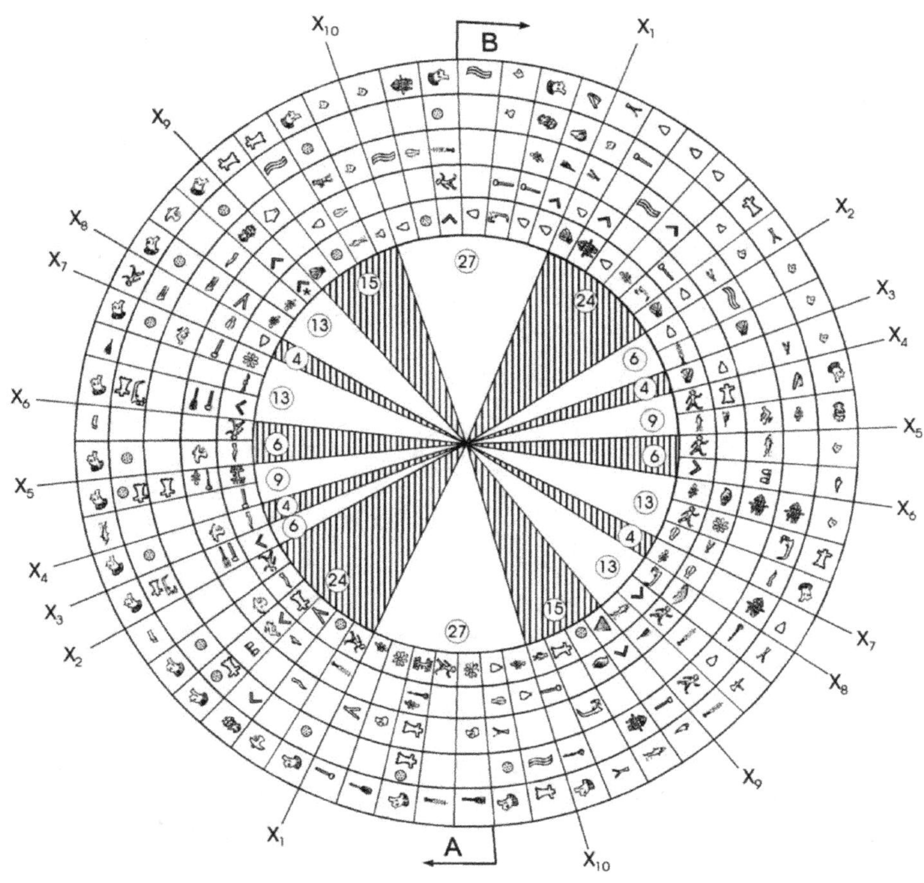

Abbildung 11.3:
Ordnung der Zeichenverteilung:
Kreismatrix der 61 Zeichengruppen des *Diskos von Phaistos*.
Rechte Hälfte Seite B, linke Hälfte Seite A.

(Grafik: Hermann Wenzel)

Zeichengruppen gleich groß. Abgesehen von der Anomalie handelt es sich um eine Drehsymmetrie.

Eine derartige Harmonie zwischen den Seiten wurde offenbar mit Bedacht eingerichtet.

Auf die Frage, ob die 22×11 Zeichen des Inventars der Zeichen graphisch als ein überzeugendes Muster darstellbar sind, begann ich die Quadrate zu 5×5 und 6×6 Zeichengruppen gesondert darzustellen. Nach etlichen Versuchen kam ich zu einer

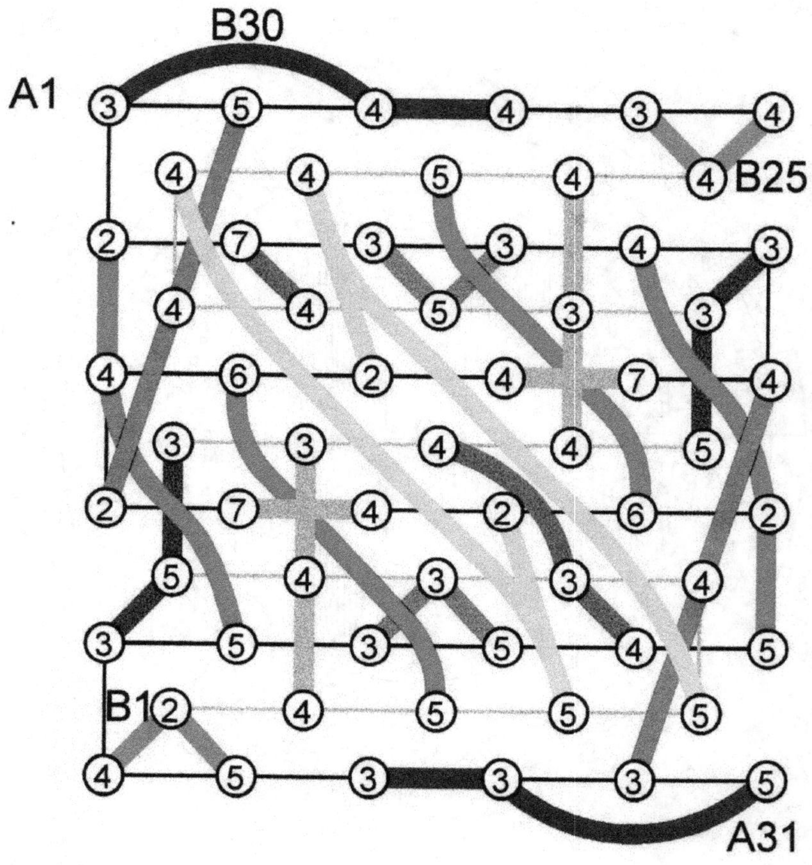

Abbildung 11.4:
22 × 11 Zeichen im verschachteltes Doppelquadrat, boustrophedon (pflugwendig)

(Grafik: Hermann Wenzel)

Darstellung, die ab Seite B/Zentrum erst das kleinere Quadrat boustrophedon (pflugwendig) zeigte und darunter das größere zu 6 × 6 Zeichengruppen ebenfalls boustrophedon. Dabei ergaben sich schon eine Reihe von Mustern mit einem Vielfachen von 11 Zeichen.

Im nächsten Schritt klappte ich das kleinere Quadrat (roter Linienzug) von oben nach unten, in die Zwischenräume von je vier Positionen des größeren Quadrats (schwarzer Linienzug). Es entstand ein verschachteltes Doppelquadrat, in welchem sich als frappierendes Ergebnis die Ausgangsfrage beantworten ließ.

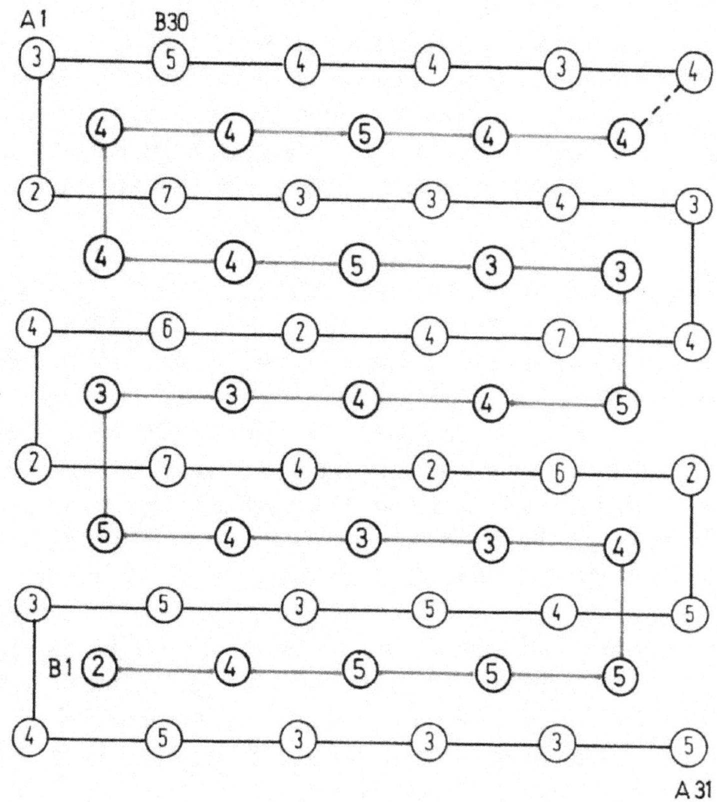

Abbildung 11.5:
Schema des Ablaufs der 61 Zeichengruppen im verschachtelten Doppelquadrat

(Grafik: Hermann Wenzel)

In einem optisch zunächst verwirrend erscheinenden Muster (Abb. 11.4) stellten sich durch beliebig anmutende Verbindungen von 2 bis 3 Positionen 22-mal 11 Zeichen in den ausgewiesenen 61 Zeichengruppen ein. Dabei wurden beliebig Positionen des größeren und des eingelagerten kleineren Quadrats miteinander verbunden. Das Gebilde erscheint wie ein Monitor, dessen Strukturen möglicherweise als Vorlage zu einer bildlichen Darstellung dienten.

Bei näherer Betrachtung der 22 Partitionen zeigt sich, dass sie aus 11 verstreuten, formgleichen Paaren bestehen, deren scheinbar beliebig verteilten Partner bei 180 Grad Drehung deckungsgleich werden. Es besteht also eine Drehsymmetrie um das

Zentrum der Anlage auf Achse A1 – A31, – eine „4“ – welche gleichzeitig das Problem der Primzahl 61 löst. Das betroffene rotbraune Paar hat als Anomalie zwei ungleich große Partner mit einmal 2 und einmal 3 Positionen, natürlich jeweils mit 11 Zeichen. Eine derart komplexe Drehsymmetrie kann sich nicht zufällig bilden. So etwas auszuklügeln bedarf ganz besonderer arithmetischer Fähigkeiten mit dem Vermögen Muster zu erkenne, über welche die Verfasser des Diskos offenbar verfügten.

11.3 *Diskos von Phaistos* – Arithmetik und Astronomie

Das Planetarium im arithmetischen Zahlenkreis und die Herkunft der Eckdaten des *Diskos von Phaistos*.

Die Minoer haben ihre Aufmerksamkeit offenbar intensiv einem vermuteten Zusammenhang zwischen Arithmetik und Astronomie gewidmet. Dabei sind sie auf das Phänomen der unendlichen Abläufe der additiven Zahlen-Kreise gestoßen. Im Zahlen-Kreis 1 bis 45 stießen sie durch fortlaufende Addition der Zahlen bei wechselndem Einsatz auf die Umlaufzeiten aller damals bekannter Planeten, und zwar synodisch und siderisch.

Einige Beispiele: So ergibt die additive Folge der Zahlen 44 bis 23 (über Ende und Anfang des Zahlen-Kreises hinaus) die 365 Tage des Sonnenjahres und gleich anschließend von 24 bis 35 die 354 Tage des synodischen Mondjahres.

Von 1 bis 27 stellen sich die Tage der synodischen Saturnperiode zu 378 Tagen ein. Von 27 bis 2 addieren sich die Tage der siderischen Periode des Mars zu 687 Tagen.

Die übergroßen Zeiträume der siderischen Perioden von Jupiter (4.332 d) und Saturn (10.759 d) können durch Bruchteile dargestellt werden. 12×361 für Jupiter von 10 bis 28 und 29×371 für Saturn von 20 bis 33.

Dieses Phänomen dürfte die minoischen Verfasser des Diskos derart fasziniert haben, dass es den Anstoß für die Entwicklung des Diskos gab. Die Notwendigkeit bei größeren Zahlen die Zahlen des Zahlenkreises mehrfach zu nutzen führte zur Überlegung dies durch Spiralbildungen zu ermöglichen, Einfacher war es aber zu einer Darstellung in Matrizen überzugehen.

Eine geeignete Matrix zeigt Abb. 11.7 (rechts) mit Bezug zu Eckdaten des Algorithmus (links).

Die Matrize besteht aus 7 Spalten und 37 Zeilen. Sie beginnt indes mit der Zahl 25 (Kopf einer Löwin), weil sich so die wesentlichen Eckdaten des Diskos einstellen. Nach der Zahl 45 geht es indes mehrfach mit „1 bis 45“ weiter.

Nach 37 Zeilen addieren sich die 6001 Tage der Gesamtzeit des Diskos ohne Dorn. Nach 19 Zeilen stellt sich der Zeitraum der Seite B mit 3.058 Tagen ein und nach weiteren 18 Zeilen der Zeitraum der Seite A mit 2.943 Tagen. Das ist schon ein überraschendes Ergebnis. Doch auch die Teilsummen der Geschlechter des Geistigen, Leib-

Kreis der Tageszahlen 1 bis 45
mit synodischen Planeten-Perioden

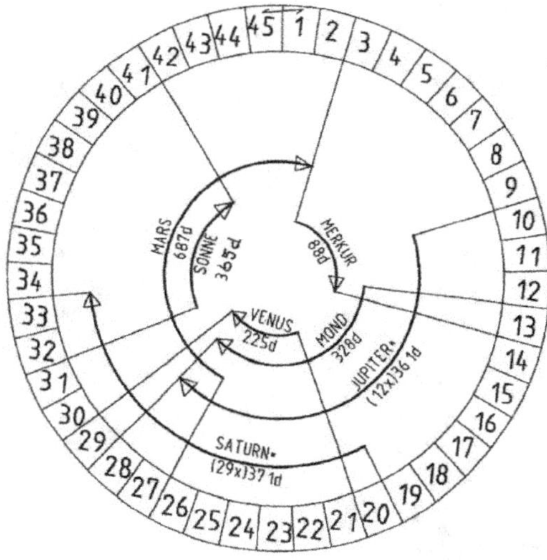

Kreis der Tageszahlen 1 bis 45
mit siderischen Planeten-Perioden

Abbildung 11.6:
Kreis mit Tageszahlen 1 bis 45 mit synodischen Planetenperioden
Kreis mit Tageszahlen 1 bis 45 mit siderischen Planetenperioden

(Grafik: Hermann Wenzel)

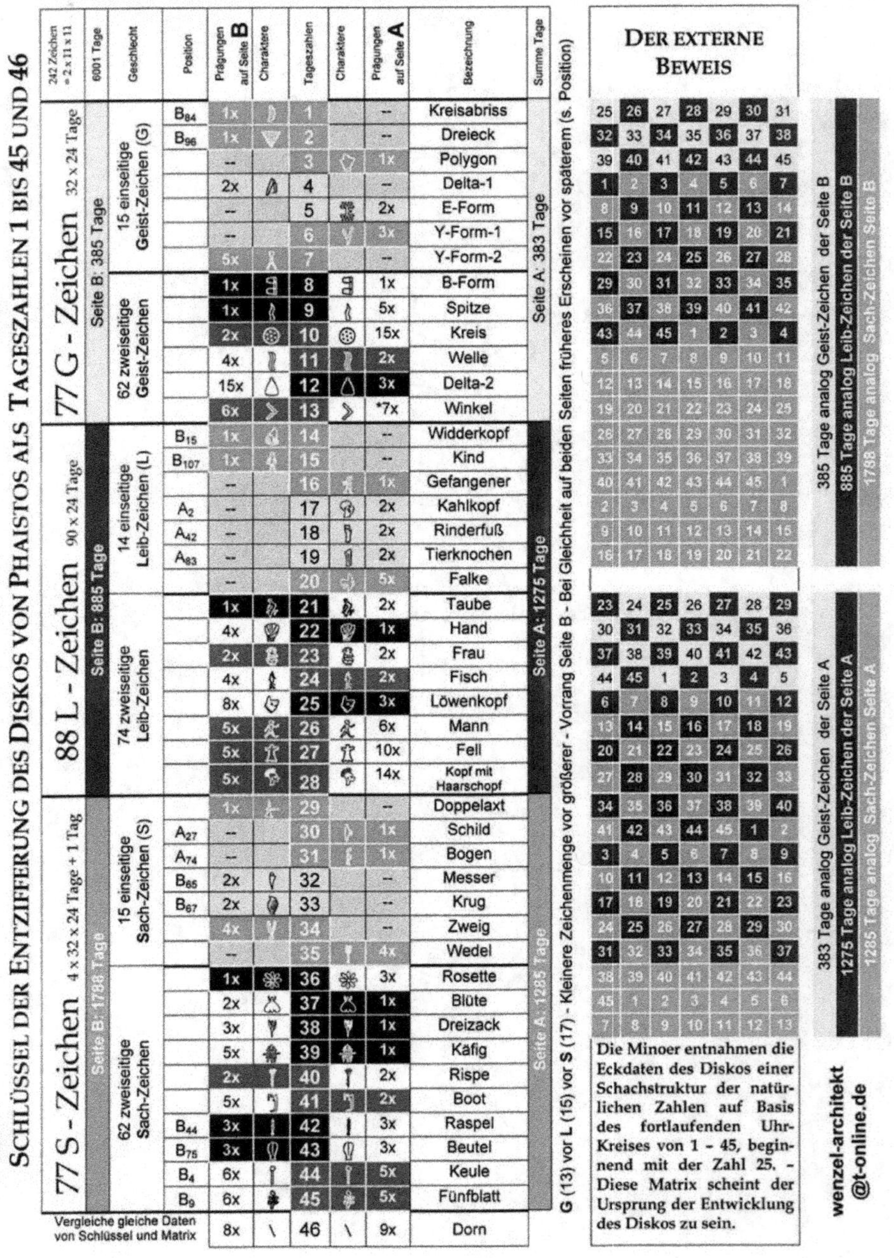

SCHLÜSSEL DER ENTZIFFERUNG DES DISKOS VON PHAISTOS ALS TAGESZAHLEN 1 BIS 45 UND 46

242 Zeichen = 2 x 11 x 11	6001 Tage	Geschlecht	Position	Prägungen auf Seite B	Charaktere	Tageszahlen	Charaktere	Prägungen auf Seite A	Bezeichnung	Summe Tage
77 G - Zeichen 32 x 24 Tage	Seite B: 385 Tage	15 einseitige Geist-Zeichen (G)	B_{84}	1x		1		--	Kreisabriss	Seite A: 383 Tage
			B_{96}	1x		2		--	Dreieck	
				--		3		1x	Polygon	
				2x		4		--	Delta-1	
				--		5		2x	E-Form	
				--		6		3x	Y-Form-1	
				5x		7		--	Y-Form-2	
		62 zweiseitige Geist-Zeichen		1x		8		1x	B-Form	
				1x		9		5x	Spitze	
				2x		10		15x	Kreis	
				4x		11		2x	Welle	
				15x		12		3x	Delta-2	
				6x		13		*7x	Winkel	
88 L - Zeichen 90 x 24 Tage	Seite B: 885 Tage	14 einseitige Leib-Zeichen (L)	B_{15}	1x		14		--	Widderkopf	Seite A: 1275 Tage
			B_{107}	1x		15		--	Kind	
				--		16		1x	Gefangener	
			A_2	--		17		2x	Kahlkopf	
			A_{42}	--		18		2x	Rinderfuß	
			A_{83}	--		19		2x	Tierknochen	
				--		20		5x	Falke	
		74 zweiseitige Leib-Zeichen		1x		21		2x	Taube	
				4x		22		1x	Hand	
				2x		23		2x	Frau	
				4x		24		2x	Fisch	
				8x		25		3x	Löwenkopf	
				5x		26		6x	Mann	
				5x		27		10x	Fell	
				5x		28		14x	Kopf mit Haarschopf	
77 S - Zeichen 4 x 32 x 24 Tage + 1 Tag	Seite B: 1788 Tage	15 einseitige Sach-Zeichen (S)		1x		29		--	Doppelaxt	Seite A: 1285 Tage
			A_{27}	--		30		1x	Schild	
			A_{74}	--		31		1x	Bogen	
			B_{65}	2x		32		--	Messer	
			B_{67}	2x		33		--	Krug	
				4x		34		--	Zweig	
				--		35		4x	Wedel	
		62 zweiseitige Sach-Zeichen		1x		36		3x	Rosette	
				2x		37		1x	Blüte	
				3x		38		1x	Dreizack	
				5x		39		1x	Käfig	
				2x		40		2x	Rispe	
				5x		41		2x	Boot	
			B_{44}	3x		42		3x	Raspel	
			B_{75}	3x		43		3x	Beutel	
			B_4	6x		44		5x	Keule	
			B_9	6x		45		5x	Fünfblatt	
Vergleiche gleiche Daten von Schlüssel und Matrix				8x	\	46	\	9x	Dorn	

G (13) vor L (15) vor S (17) - Kleinere Zeichenmenge vor größerer - Vorrang Seite B - Bei Gleichheit auf beiden Seiten früheres Erscheinen vor späterem (s. Position)

DER EXTERNE BEWEIS

25	26	27	28	29	30	31
32	33	34	35	36	37	38
39	40	41	42	43	44	45
1	2	3	4	5	6	7
8	9	10	11	12	13	14
15	16	17	18	19	20	21
22	23	24	25	26	27	28
29	30	31	32	33	34	35
36	37	38	39	40	41	42
43	44	45	1	2	3	4
5	6	7	8	9	10	11
12	13	14	15	16	17	18
19	20	21	22	23	24	25
26	27	28	29	30	31	32
33	34	35	36	37	38	39
40	41	42	43	44	45	1
2	3	4	5	6	7	8
9	10	11	12	13	14	15
16	17	18	19	20	21	22

385 Tage analog Geist-Zeichen der Seite B
885 Tage analog Leib-Zeichen der Seite B
1788 Tage analog Sach-Zeichen Seite B

23	24	25	26	27	28	29
30	31	32	33	34	35	36
37	38	39	40	41	42	43
44	45	1	2	3	4	5
6	7	8	9	10	11	12
13	14	15	16	17	18	19
20	21	22	23	24	25	26
27	28	29	30	31	32	33
34	35	36	37	38	39	40
41	42	43	44	45	1	2
3	4	5	6	7	8	9
10	11	12	13	14	15	16
17	18	19	20	21	22	23
24	25	26	27	28	29	30
31	32	33	34	35	36	37
38	39	40	41	42	43	44
45	1	2	3	4	5	6
7	8	9	10	11	12	13

383 Tage analog Geist-Zeichen der Seite A
1275 Tage analog Leib-Zeichen der Seite A
1285 Tage analog Sach-Zeichen Seite A

Die Minoer entnahmen die Eckdaten des Diskos einer Schachstruktur der natürlichen Zahlen auf Basis des fortlaufenden Uhr-Kreises von 1 - 45, beginnend mit der Zahl 25. - Diese Matrix scheint der Ursprung der Entwicklung des Diskos zu sein.

wenzel-architekt @t-online.de

Abbildung 11.7:
Algorithmus der Entzifferung des *Diskos von Phaistos*.
Korrektur: Zeile 1, Spalte 2 muss heißen: „Summe Tage".

(Grafik: Hermann Wenzel)

lichen und Sächlichen, das G-L-S-System, taucht nach etlichen Versuchen als zweimal gleiches Muster unter Beteiligung des Schachprinzips getrennt nach den Seiten auf: Ganz rechts in Abb. 11.7 geben drei farbige Spalten die Daten getrennt nach den Seiten, entsprechend gleichfarbigen Spalten 2 und 11 im Algorithmus (links) an.

Das zeigt nun eindeutig die Herkunft der Daten des Diskos aus dieser Matrix der natürlichen Zahlen.

Die Entzifferung der Zeichen als Tageszahlen durch die Rekonstruktion der Reihenfolge der 45 verschiedenen Zeichen zeigt ein Algorithmus (Abb. 11.7 links), den die Minoer selbst angelegt haben dürften.

Die 45 verschiedenen Zeichen gliedern sich entsprechend ihrer optischen Erscheinung in drei Geschlechter:

1. Abstrakta, dem Geistigen assoziierbar (13 verschiedene Zeichen, 77 insgesamt)
2. Dem Leiblichen von Mensch und Tier assoziierbar (15 verschiedene Zeichen, 88 insgesamt)
3. Sächliche Zeichen, Gerätschaften, Waffen, Pflanzenteile (17 verschiedene Zeichen, wiederum 77 Zeichen insgesamt).

Jedes dieser Zeichen hat Koeffizienten, die angeben, wie oft es, getrennt nach den Seiten wiederholt wird, und ob es überhaupt wiederholt wird. Zum Beispiel 36 Rosette 1/3, einmal auf Seite B dreimal auf Seite A.

Die Reihenfolge der 45 Zeichen folgt den steigenden Häufigkeiten der Koeffizienten, getrennt nach den drei Geschlechtern. Sind diese gleich groß gelten einige Regeln für den Vorrang eines Zeichen. Z. B. 23 Frau hat auf beiden Seiten die gleiche Häufigkeit – 2 – Hier hat Seite B, als die kleinere Seite Vorrang vor Seite A und gibt seinem Zeichen den Vortritt.

Seite B ist die kleinere Seite an diversen Eigenschaften. Sie hat insgesamt 119 Zeichen, Seite A 123. Seite B hat 34 verschiedene Zeichen, Seite A 35. Seite B hat 30 Zeichengruppen, Seite A 31. Seite B hat 8 Dorne, Seite A 9. Seite B hat 12 Zeichengruppen in der peripheren Kreiszeile, Seite A 13.

Einen weiteren Hinweis auf die Priorität der Seite B erhält man, wenn die 61 Zeichengruppen beider Seiten des Diskos als Kreismatrix mit 5 konzentrischen Kreisen dargestellt werden. In Gruppen mit 6 und 7 Zeichen teilen sich in den Zwischenzeichen jeweils zwei Zeichen eine Position, nach dem Prinzip: Randzeichen unter Randzeichen Mittelzeichen unter Mittelzeichen und Zeichen zwischen Rand und Mitte unter ebensolchen. Wird diese Kreismatrix nun in vier Quadranten gegliedert, die sich ab Seite B/Zentrum, B/Ende, A/Zentrum, A/Ende in die Quadranten I, II, III, IV, ausrichten, entfallen auf die Quadranten I/59, II/60, III/61, IV/ 62 Zeichen. In dieser Steigerung der Zeichenmengen erweist sich Seite B eindeutig als die kleinere und nach dem Prinzip der Priorität „kleiner vor größer“ als die primäre Seite.

Die Zeichenfolge – Mann, Fell, Kopf mit Haarschopf – hat auf Seite B jeweils 5 Wiederholungen. Hier regelt die steigende Folge auf Seite A mit – 6, 10, 14 – Zeichen die Rangfolge auf Seite B. Bei Blüte, Dreizack, Käfig, jeweils einmal auf Seite A ist es ähnlich. Hier regelt die steigende Folge auf Seite B mit 2, 3, 5 die Reihenfolge auf Seite A.

Zeichen, bei denen die Häufigkeiten auf beiden Seiten gleich groß sind, regeln ihr Nacheinander durch das frühere Erscheinen in den Texturen der Seiten B und A. In der Folge B/Zentrum, B/Rand – A/Zentrum, A/Rand erhalten die Zeichen fortlaufende Positionsnummern, die das frühere Erscheinen eines Zeichens angeben. In der 4. Spalte von links der Abb. 11.7 sind die Positionsnummern der betroffenen Zeichen angegeben.

Damit wurde den 45 verschiedenen Zeichen jeweils eine Tageszahl zugewiesen.

Der Dorn wurde zunächst als 46. Zeichen unten angehängt. Es wird sich indes zeigen, dass er dreiwertig ist: „0“ Null, dann wird er nicht mitgerechnet, „1“ oder 46. Für diese drei Werte gibt es eindeutige Belege.

Doch noch fehlen insgesamt die Beweise der Stimmigkeit. Diese sind entweder in den Strukturen der Nomenklatur (Abb. 11.7) oder in den Texturen der beiden Seiten des Diskos zu finden.

Werden nun die ermittelten Tageszahlen der Zeichen mit ihren Koeffizienten multipliziert stellen sich insgesamt 6.001 Tage ein. Sie gliedern sich nach den beiden Seiten des Diskos und den jeweiligen Geschlechtern des G-L-S-Systems (Geist, Leib, Sache) in 6 Teilsummen, deren Tage in den farbig angelegten Spalten 2 und 11 der Abb. 11.7 für die Seiten angegeben sind. Auf Seite A entfallen 2.943 Tage und auf Seite B 3.058 Tage.

Dennoch müssen sich die ermittelten Tageszahlen der 45 und mit Dorn 46 Zeichen des Diskos auch in den Texturen der beiden Seiten manifestieren.

Abb. 11.8 zeigt Seite A des Diskos als Matrix mit 31 Zeichengruppen und linksbündiger Anordnung der Zeichen in den Gruppen; in der linken Figur mit den bildlichen Zahlen des Diskos, rechte Figur mit den rekonstruierten arabischen Zahlen. Unterhalb der Matrizen wurden die Zahlen addiert.

Bei einer Auswertungen der Spaltensummen stellen sich diverse Planetendaten ein. Es handelt sich um eine einfache Verschlüsselungsmethode, die horizontal schreibt und transversal, senkrecht dazu, das Verschlüsselte angibt.

Die Spalten I und II addieren mit $4 \times 365\frac{1}{4}$ Tagen die solare Schaltperiode, die den Minoern offenbar bekannt war. Die Spalten III, IV, V addieren mit 1.312 Tagen 4 siderische Mondjahre zu je 328 Tagen. Die Spalten VI und VII addieren zusammen mit der vorgeschalteten Dornspalte 584 Tage. Es ist die synodische Periode der Venus.

Seite A ist damit der Zeitraum des Dreigestirns Sonne, Mond, Stern, das in bildlichen Darstellungen der Minoer mehrfach erscheint.

	D	I	II	III	IV	V	VI	VII
A1								
A2								
A3								
A4								
A5	D							
A6								
A7								
A8								
A9								
A10	D							
A11	D							
A12								
A13	D							
A14								
A15								
A16	D							
A17	D							
A18								
A19								
A20	D							
A21								
A22								
A23								
A24								
A25								
A26								
A27								
A28								
A29	D							
A30								
A31	D							
	9 x 46	**4 x 365¼ Sonne**		4 x 328 Mond			86	84

	D	I	II	III	IV	V	VI	VII
A1		36	17	35				
A2		26	42					
A3		5	40	45	27	27	10	28
A4		36	17	35				
A5	46	45	6	44				
A6		26	42	10	28			
A7		10	9	20				
A8		6	30	13	23			
A9		27	13	21	8	27	10	28
A10	46	9	20	10	28			
A11	46	26	18					
A12		13	44	35	41	27	28	
A13	46	9	20	10	28			
A14		44	24					
A15		5	40	45	27	27	10	28
A16	46	9	20	10	28			
A17	46	26	18					
A18		13	44	35	41	27	28	
A19		31	38					
A20	46	36	44	21	10	28		
A21		12	43	19	26			
A22		45	6	19	10	28		
A23		45	9	20				
A24		13	13	23	10	28		
A25		22	3	27				
A26		10	12	11	27			
A27		24	43	16	10	28		
A28		37	25	25				
A29	46	12	11	25				
A30		10	43	39				
A31	46	13	26	42	10	28		
	414	**681**	**780**	655	382	275	86	84

Abbildung 11.8:
Astronomische Daten in den Spalten der Matrizen
Diskos von Phaistos, Matrix der Seite A. Zeichengruppen linksbündig.
Linke Figur mit Originalzeichen, rechts die rekonstruierten Tageszahlen.

(Grafik: Hermann Wenzel)

Die II. Spalte der Matrix gibt mit 780 Tagen die synodische Periode des Mars an. Die erste und vierte Spalte bildet den Zeitraum von $681 + 382 = 1.063$ Tage. Es ist die synodische Schaltperiode des Mondjahres. Nach drei Mondjahren zu je 354 Tagen folgt ein Mondjahr zu 355 Tagen. Spalte I und die Dornspalte addieren $681 + 414$ Tage $= 1.095$ Tage. Es sind 3 Jahre zu je 365 Tagen.

Mit Beteiligung des Dorns zu 1 Tag – auf Seite A gibt es 9 Dorne – ergeben sich $2.943 + 9$ Tage = 2. 952 Tage. Es sind 100 synodische Monate zu je 29,52 Tage oder 108 siderische Monate zu je $27\frac{1}{3}$ Tage.

Wir sehen auf Seite A in dieser Anordnung bereits eine ganze Schar von Planetendaten von Sonne, Mond, Venus, Mars. Und es ist doch erst ein Anfang, wenngleich schon ein weiterer Beweis für die zutreffende Ermittlung der Tageszahlen zu den 45 verschiedenen Zeichen des Diskos.

Einseitige Zeichen beider Seiten = 826 Tage = 28 Lunationen zu je 29,5 Tage.

Zweiseitige Zeichen beider Seiten = 5.175 Tage =
23 siderische Venus-Zyklen zu je 225 Tagen.

Zweiseitige Zeichen der Seite A = 2.487 Tage =
91 siderische Monate, zu je 27,329...Tage.

Zweiseitige Zeichen der Seite B = 2.688 Tage =
91 synodische Monate zu je 29,538...Tage

15 synodische Perioden des Merkurs + 15 siderische Perioden =
Zeitraum der Seite B ohne Dorn:
$15 \times 115,933...$ Tage $+15 \times 87,933...$ Tage = 3.058 Tage
oder $15 \times 116 - 1 + 15 \times 88 - 1 = 3.058$ Tage.

Die Abb. 11.9 zeigt in einem steganographischen Muster-Daten von Jupiter und Saturn, links mit den Zeichen des Diskos und rechts mit den rekonstruierten arabischen Zahlen. Steganographisch heißt versteckt verschlüsselt. Die blauen Positionen der linken Abbildung addieren 399 Tage einer synodischen Periode des Jupiters und die grünen, dazu vertikal gespiegelt, 378 Tage einer synodischen Periode des Saturns. Zusammen ergibt sich die magische Zahl von 777 Tagen.

Der hellgraue Hintergrund der linken Abbildung (Abb. 11.9) zeigt rechts, dunkelblau in einem Flechtmuster die Hälfte des siderischen Zyklus des Jupiters zu 2.166 Tagen (oder $\frac{1}{2} \times 4.332$ Tage). Damit sind alle Positionen der Seite A mit den Leerfeldern in dieser Anordnung abgedeckt und drei weitere Planetendaten erschlossen. Eine solche Konstellation in dieser Anordnung zu finden war sicher nicht einfach und bedurfte einiges an kombinatorischer Rechenarbeit mit etlichen Versuchen.

	D	I	II	III	IV	V	VI	VII
A1								
A2								
A3								
A4								
A5	D							
A6								
A7								
A8								
A9								
A10	D							
A11	D							
A12								
A13	D							
A14								
A15								
A16	D							
A17	D							
A18								
A19								
A20	D							
A21								
A22								
A23								
A24								
A25								
A26								
A27								
A28								
A29	D							
A30								
A31	D							

	D	I	II	III	IV	V	VI	VII
A1		36	17	35				
A2		26	42					
A3		5	40	45	27	27	10	28
A4		36	17	35				
A5	46	45	6	44				
A6		26	42	10	28			
A7		10	9	20				
A8		6	30	13	23			
A9		27	13	21	8	27	10	28
A10	46	9	20	10	28			
A11	46	26	18					
A12		13	44	35	41	27	28	
A13	46	9	20	10	28			
A14		44	24					
A15		5	40	45	27	27	10	28
A16	46	9	20	10	28			
A17	46	26	18					
A18		13	44	35	41	27	28	
A19		31	38					
A20	46	36	44	21	10	28		
A21		12	43	19	26			
A22		45	6	19	10	28		
A23		45	9	20				
A24		13	13	23	10	28		
A25		22	3	27				
A26		10	12	11	27			
A27		24	43	16	10	28		
A28		37	25	25				
A29	46	12	11	25				
A30		10	43	39				
A31	46	13	26	42	10	28		

Abbildung 11.9:
Steganographische Muster:
Links synodische Perioden von Jupiter und Saturn,
rechts eine halbe Periode des siderischen Jupiterzyklus.

(Grafik: Hermann Wenzel)

11.3.1 Das *Große Jahr der siderischen Planetenperioden*

Das *Große Jahr der siderischen Planetenperioden* auf Seite A des Diskos:
Seite A mit 9 Dornen zu je 46 Tagen zeigt als Bruchteil zu einem Fünftel das (hier von mir so genannte) *Große Jahr der siderischen Planetenperioden.*

Ich verstehe darunter die Summe der 7 siderischen Planetenperioden:
Merkur 88 Tage + Venus 225 Tage + Mond 328 Tage + Sonne 365 Tage + Mars 687 Tage + Jupiter 4.332 Tage + Saturn 10.759 Tage = 16.785 Tage = 5 × 3.357 Tage. 3.357 Tage entsprechen Seite A einschließlich der dort vorhandenen 9 Dorne zu je 46 Tagen (2.943 d $+9 \times 46$ d = 3.357 d).

Mit einem Fünftel bezieht sich also Seite A auf die siderischen Perioden aller 7 Planeten, des ganzen damals bekannten Planetariums.

In den Matrizen, wie zuvor sind ebenfalls steganographisch, links 25 synodische Monate, in einem fünfteiligen stilisierten Schmetterlingsmotiv zu $25 \times 29,52$ Tagen und rechts 50 synodische Monate in einem vertikalen, teils unterbrochenem, Streifenmuster zu ebenfalls je 29,52 Tagen verschlüsselt. Das linke Muster kann in das rechte eingelagert werden (hellgrau). Beide Muster addieren sich zu 2.212,5 Tagen = 75 synodische Monate. Werden diese von der ganzen Seite A (2.943 d) abgezogen, verbleiben 730,5 Tage in den nicht besetzten Feldern. Es sind 2 Jahre zu je $365\frac{1}{4}$ Tage. Auf die linke Seite entfallen ferner 27 siderische Monate zu je $27\frac{1}{3}$ Tage und auf die rechte 54 siderische Monate ebenfalls zu je $27\frac{1}{3}$ Tage.

Auf die rechte Seite entfallen einschließlich der 9 Dornmarkierungen der Seite A $1.476 + 9 \times 46 = 1.890$ Tage. Sie entsprechen 5 synodischen Saturnperioden zu je 378 Tagen.

Insgesamt zeigt Seite A somit eine Anzahl von Planetenperioden. Das Besondere ist jedoch die Periodenrelation zwischen synodischen und siderische Monaten, das weiterhin auf eine ausgereift intensive Himmelsbeobachtung der Minoer schließen lässt.

Die Matrizen beider Seiten des Diskos wandeln sich hier in eine zentrierte Anordnung der Zeichen in den Zeichengruppen (Abb. 11.11) : Randzeichen stehen unter Randzeichen, mittlere Zeichen unter mittleren Zeichen, ich nenne sie die „Seele“, und Zeichen zwischen Rand und Mitte, die Zwischenzeichen unter ebensolchen. Auf Seite B wurden aus optischen Gründen analog zu Seite A zwei zusätzliche leere Spalten eingefügt, so dass beide Seiten 7 Spalten zeigen.

Teilbereiche der Spalten wurden jeweils folgendermaßen benannt: Seele (= mittlere Zeichen), rechter Flügel, linker Flügel und links vorgesetzte Dornspalte. Die Zeichen sind mit ihren zahlenmäßigen Entsprechungen dargestellt und unter den Matrizen summiert.

Dabei gibt es nun ein großes Erstaunen über die größenmäßigen Verwandtschaften der beiden Seiten mit einer mysteriösen Arithmetik.

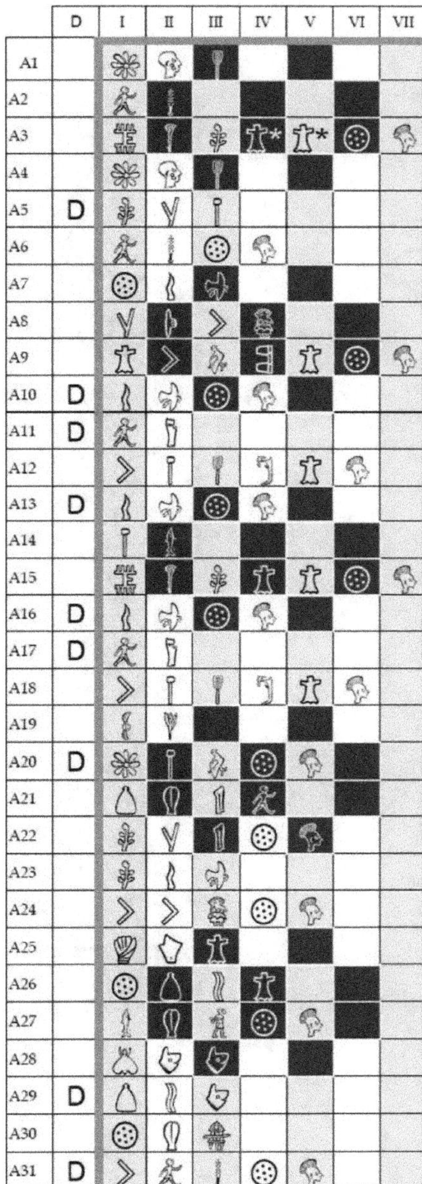

	D	I	II	III	IV	V	VI	
A1		36	17	35				
A2		26	42					
A3		5	40	45	27	27	10	2
A4		36	17	35				
A5	46	45	6	44				
A6		26	42	10	28			
A7		10	9	20				
A8		6	30	13	23			
A9		27	13	21	8	27	10	2
A10	46	9	20	10	28			
A11	46	26	18					
A12		13	44	35	41	27	28	
A13	46	9	20	10	28			
A14		44	24					
A15		5	40	45	27	27	10	2
A16	46	9	20	10	28			
A17	46	26	18					
A18		13	44	35	41	27	28	
A19		31	38					
A20	46	36	44	21	10	28		
A21		12	43	19	26			
A22		45	6	19	10	28		
A23		45	9	20				
A24		13	13	23	10	28		
A25		22	3	27				
A26		10	12	11	27			
A27		24	43	16	10	28		
A28		37	25	25				
A29	46	12	11	25				
A30		10	43	39				
A31	46	13	26	42	10	28		

Abbildung 11.10:
Steganographische Muster für synodische Monate
Links: 25 synodische Monate. Rechts: 50 synodische Monate.

(Grafik: Hermann Wenzel)

	Dorn	Linker Flügel			*Seele*	Rechter Flügel		
A31	46	13	26		42		10	28
A30		10			43			39
A29	46	12			11			25
A28		37			25			25
A27		24	43		16		10	28
A26		10	12				11	27
A25		22			3			27
A24		13	13		23		10	28
A23		45			9			20
A22		45	6		19		10	28
A21		12	43				19	26
A20	46	36	44		21		10	28
A19		31						38
A18		13	44	35		41	27	28
A17	46	26						18
A16	46	9	20				10	28
A15		5	40	45	27	27	10	28
A14		44						24
A13	46	9	20				10	28
A12		13	44	35		41	27	28
A11	46	26						18
A10	46	9	20				10	28
A9		27	13	21	8	27	10	28
A8		6	30				13	23
A7		10			9			20
A5		26	42				10	28
A5	46	45			6			44
A4		36			17			35
A3		5	40	45	27	27	10	28
A2		26						42
A1		36			17			35
Σ	414	1362			323	1258		
		1685				1672 (mit Dorn)		
	1776 = 2 x 888							

SEITE A

SEITE B

Dorn	Linker Flügel			*Seele*	Rechter Flügel			
46	12						11	B1
	41	44				37	25	B2
	12	44		45		23	28	B3
	12	13		38		14	4	B4
46	22	12		34		25	7	B5
	39	13				44	12	B6
46	12			11			12	B7
	45			13			12	B8
	41	44				37	27	B9
46	22	12		34		25	7	B10
46	12			11			25	B11
	42			22			25	B12
46	22	12				34	25	B13
	26	27				4	28	B14
	24	38		21		45	23	B15
	26			24			25	B16
	13			8			32	B17
	45	33		39		39	25	B18
	26	36				41	27	B19
	43	34				9	28	B20
	45	43				39	12	B21
	41	1				40	7	B22
	13	26		42		12	29	B23
	24	38				26	42	B24
46	2	13				44	32	B25
	10	33				39	24	B26
	27			41			7	B27
46	15	44				40	28	B28
	45	12				11	27	B29
	12	43		7		10	28	B30
368	1386			390	1282			Σ
				1672				
	1776 = 2 x 888							

Abbildung 11.11:
Matrizen der Seiten des Diskos mit zentrierter Anordnung der Zahlentsprechungen der Zeichen und Seitenvergleich der Zeiträume in den Spaltengruppen.

(Grafik: Hermann Wenzel)

Teilbereiche der Seiten sind untereinander gleich:
Rechter Flügel und Seele der Seite B ist gleich rechtem Flügel und Dornspalte der Seite A (1.282 d + 390 d = 1.258 d + 414 d). Linker Flügel und Seele der Seite B ist gleich linkem Flügel und Dornpalte der Seite A (1.386 d + 390 d = 1.362 d + 414 d). Das heißt, beide Seite haben, vertauscht vermittelt durch die Seelen und die Dornspalten, innigste zeitgleiche Bezüge zu einander.

Ein Kuriosum entsteht, wenn der rechte Flügel der Seite B + Seele B (1.282 d + 390 d) mit dem linken Flügel der Seite A + Seele A (1.362 d + 323 d) vereinigt werden. Es stellt sich der gesamte Zeitraum der Seite A mit Dornspalte ein (3.357 d). obwohl in der Rechnung der Dorn nicht einbezogen wurde.

Das gleiche Ergebnis stellt sich im umgekehrten Fall ein: Linker Flügel B + Seele B (1.386 + 390 d) + rechter Flügel A + Seele A (1.258 d + 323 d) = vollständige Seite A einschließlich Dorn (3.357 d), obwohl auch hier der Dorn unberücksichtigt blieb. Wo da jeweils der Dorn herkommt ist noch nicht geklärt. Jedenfalls müssen die 9 Dorne der Seite A in den jeweils erfassten Flügeln der Spalten bereits enthalten sein.

Es stellt sich die Frage nach dem Verbleib von Seite B in diesem Zusammenhang. Die Antwort ist einfach. Seite B, ebenfalls mit den in B vorhandenen 8 Dornen entspricht den oben unberücksichtigten Spalten, ohne Seelen, aber mit Dornspalten. Hier fehlen in der Berechnung die jeweiligen Seelen, sind aber im Ergebnis der vollständigen Seite B wieder enthalten. Doch woher kommen nun die jeweiligen Seelen, obwohl Seelen in der Berechnung nicht berücksichtigt wurden. Auch das ist vorerst ungeklärt.

Die Abb. 11.12 zeigt beide Seiten des Diskos mit zentrierten Zeichengruppen und Anordnung der Zeichen in 5 Spalten. Das bedeutet, dass bei Zeichengruppen mit 6 und 7 Zeichen auf Seite A insgesamt 10 Positionen mit zwei Zeichen besetzt werden (nach dem Prinzip *„zwischen Rand und Seele"*), aber in Bezug auf die Schachfelder als eine Position gelten. Die Matrizen der Seiten verlaufen rechts von oben nach unten und links von unten nach oben.

Werden beide Seiten nun dem Schachmuster unterworfen ergibt sich für dunkle und helle Felder eine genaue Aufspaltung in die Seiten-Zeiträume ohne Dorn: A (ocker) 2.943 Tage und B (dunkelblau) 3.058 Tage.

Wird der Dorn einbezogen und den Zeichen zugeordnet an denen er im Original haftet, ohne in das Schachmuster eingeordnet zu werden, stellen sich auch die seitenbezogenen Zeiträume mit den 17 Dornen ein (3.357 Tage für Seite A und 3.426 Tage für Seite B). Dieses Schachmuster stellt die innigste Verknüpfung der beiden Seiten des Diskos dar.

Paradigma der 45 verschiedenen Zeichen mit Häufigkeiten auf den beiden Seiten und rekonstruierten Tageszahlen. Zahlen in Klammern unter den Zeichen zeigen die Wiederholungen an: links auf Seite B, rechts auf Seite A.

Seite A

	D	Rand	Z	Seele	Z	Rand
A31		13	26	42	10	28
A30		10		43		39
A29	46	12		11		25
A28		37		25		25
A27		24	43	16	10	28
A26		10	12		11	27
A25		22		3		27
A24		13	13	23	10	28
A23		45		9		20
A22		45	6	19	10	28
A21		12	43		19	26
A20	46	36	44	21	10	28
A19		31				38
A18		13	44/3		41/2	28
A17	46	26				18
A16	46	9	20		10	28
A15		5	40/4	27	27/1	28
A14		44				24
A13	46	9	20		10	28
A12		13	44/3		41/2	28
A11	46	26				18
A10	46	9	20		10	28
A9		27	13/2	8	27/1	28
A8		6	30		13	23
A7		10		9		20
A6		26	42		10	28
A5	46	45		6		44
A4		36		17		35
A3		5	40/4	27	27/1	28
A2		26				42
A1		36		17		35

Seite B

D	Rand	Z	Seele	Z	Rand
46	12				11
	41	44		37	25
	12	44	45	23	28
	12	13	38	14	4
46	22	12	34	25	7
	39	13		44	12
46	12		11		12
	45		13		12
	41	44		37	27
46	22	12	34	25	7
46	12		11		25
	42		22		25
46	22	12		34	25
	26	27		4	28
	24	38	21	45	23
	26		24		25
	13		8		32
	45	33	39	39	25
	26	36		41	27
	43	34		9	28
	45	43		39	12
	41	1		40	7
	13	26	42	12	29
	24	38		26	42
46	2	13		44	32
	10	33		39	24
	27		41		7
46	15	44		40	28
	45	12		11	27
	12	43	7	10	28

Abbildung 11.12:
Schachmuster bei zentrierter Anordnung der Zahlentsprechungen der Zeichen beider Seiten. Helle und dunkle Positionen unterscheiden die Zeiträume der Seiten des Diskos.

(Grafik: Hermann Wenzel)

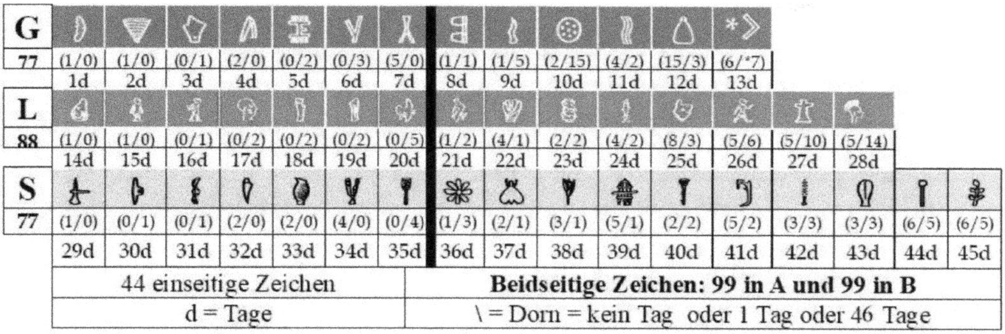

G																
77	(1/0)	(1/0)	(0/1)	(2/0)	(0/2)	(0/3)	(5/0)	(1/1)	(1/5)	(2/15)	(4/2)	(15/3)	(6/*7)			
	1d	2d	3d	4d	5d	6d	7d	8d	9d	10d	11d	12d	13d			
L																
88	(1/0)	(1/0)	(0/1)	(0/2)	(0/2)	(0/2)	(0/5)	(1/2)	(4/1)	(2/2)	(4/2)	(8/3)	(5/6)	(5/10)	(5/14)	
	14d	15d	16d	17d	18d	19d	20d	21d	22d	23d	24d	25d	26d	27d	28d	
S																
77	(1/0)	(0/1)	(0/1)	(2/0)	(2/0)	(4/0)	(0/4)	(1/3)	(2/1)	(3/1)	(5/1)	(2/2)	(5/2)	(3/3)	(3/3)	(6/5)	(6/5)
	29d	30d	31d	32d	33d	34d	35d	36d	37d	38d	39d	40d	41d	42d	43d	44d	45d
	44 einseitige Zeichen							**Beidseitige Zeichen: 99 in A und 99 in B**								
	d = Tage							\ = Dorn = kein Tag oder 1 Tag oder 46 Tage								

Abbildung 11.13:
Das G-L-S-System der 45 verschiedenen Charaktere
mit den rekonstruierten Tageszahlen oder 46 Tage

(Grafik: Hermann Wenzel)

Gliederung untereinander in die drei Geschlechter des Geistigen, Leiblichen und Sächlichen (G, L, S – System). Links des schwarzen Balkens stehen einseitige Zeichen, rechts beidseitige.

11.3.2 Drei Geschlechter – G-L-S

Ein weiteres Kuriosum handelt von den drei Geschlechtern – G-L-S – der Seiten. Die Addition der Tage, einschließlich der zugehörigen Dorne ergibt folgende Schemata. Werden darin die Leibzeichen (L) der Seiten miteinander vertauscht, ergeben sich Merkwürdigkeiten bezüglich des Dorns:

Seite A mit 9 Dornen			Seite B mit 8 Dornen	
G	**613**		569	G
L	1.367		**1.069**	L
S	**1.377**		1.788	S

Abbildung 11.14:
Verschränkte Zeiträume mit Dorn führen zu einem irrealen Ergebnis ohne Dorn

(Grafik: Hermann Wenzel)

Der verschränkte Zeitraum der grauen Felder (Abb. 11.14) entspricht Seite B des Diskos ohne Dorn + 1 Tag, obwohl in der Rechnung 11 Dorne enthalten sind.

Seite A mit 9 Dornen			Seite B mit 8 Dornen	
G	613		**569**	G
L	**1.367**		1.069	L
S	1.377		**1.788**	S

Abbildung 11.15:
Verschränkte Zeiträume mit Dorn
führen zu einem irrealen Ergebnis mit allen 17 Dornen

(Grafik: Hermann Wenzel)

Der verschränkte Zeitraum der grauen Felder (Abb. 11.15) beträgt hier 3.724 Tage = 2.943 Tage – 1 Tag + 782 Tage (782 = 17 × 46 d). Das entspricht dem Zeitraum der Seite *A* – 1 Tag, nun aber mit allen 17 Dornen beider Seiten zu je 46 Tagen, obwohl nur 6 Dorne in der Rechnung enthalten sind.

Wie diese Merkwürdigkeiten zustande kommen, bleibt vorerst ungeklärt. Auch warum so etwas eingerichtet wurde. Denkbar, dass die Autoren Schüler damit verblüffen wollten oder die Frage nach den Gründen stellten.

11.3.3 Finsternis-Perioden

Der Gesamtzeitraum des Diskos ohne Dorn, vergrößert um eine synodische Periode der Venus ergibt die Saros-Periode:

6.001 Tage + 584 Tage = 6.585 Tage

Die Venusperiode lässt sich auf verschiedene Weisen darstellen und mit dem Gesamtzeitraum verbinden.

In der Ebene ohne Dorn bildet die Hälfte der Seite A, die 15 letzten Abteilungen, mit 1.388 Tagen eine 47-monatige Finsternisperiode, die auch den Babyloniern wohl vertraut war. Dazu van der Waerden:

> „... *Alles in allem ist die Periode von 47 Monaten die einzige, die alle bekannten Tatsachen befriedigend erklärt. Sie zeichnet sich auch unter*

allen Finsternisperioden, die kürzer als Saros sind, dadurch aus, dass sie allein eine ganze Zahl von Drachenmonaten enthält.“[1]

11.3.4 Der Dorn und die Endpunkte

Eine wesentliche Aussage dürfte im Zeitraum beider Seiten des Diskos mit Dorn und Endpunkten liegen:

Seite B + Seite A + 17 *Dorne* (à 46 d) + 10 Endpunkte (jeweils à 1 d) = 6.001 Tage + 782 Tage + 10 Tage = 6.793 Tage = $230 \times 29{,}535$ d = 230 synodische Monate.

6.793 Tage entsprechen dem Zeitraum des *Mondbahnkreisels* oder der sog. *Mondwende* von 18,6 Jahren bezüglich der Fixsterne.

Die umfassendste Aussage des Diskos könnte somit die Darstellung dieses *Mondbahnkreisels* sein, dessen Beobachtung durch die sich verschiebenden Mondaufgänge man den Minoern durchaus zutrauen darf.

11.4 Frühgeschichtliche und Antike Quellen mit Bezügen zum *Diskos von Phaistos*

- Die sog. Sternenscheibe von Nebra,[2] eine Bronzescheibe mit Gold-Applikationen, ist etwa zeitgleich mit dem *Diskos von Phaistos*. Doch während die Applikationen der Sternenscheibe, mit Unsicherheiten astronomisch gedeutet werden müssen, „spricht“ der *Diskos von Phaistos* mit exakten Zahlen Klartext.
- Der Mechanismus von Antikythera,[3] ein fantastisches astronomisches Räderwerk, datiert auf 70 bis 60 v. u. Z., ist wesentlich jünger als der *Diskos von Phaistos*, aber er enthält viele ähnliche Daten. So könnte vermutet werden, dass die Kenntnis vom *Diskos von Phaistos* bis weit in die damalige Zukunft ausstrahlte.
- Der interessanteste Bezug zum *Diskos von Phaistos* findet sich allerdings im Schildgleichnis von Homers Ilias.[4] Die Schichten dieses metaphorischen Schildes beschreiben unter anderem Sternbilder und ganz wesentlich einen Tanzplatz oder eine Choreographie, die Daidalos der Ariadne auf Kreta machte. Könnte diese Choreographie der *Diskos von Phaistos* selbst sein oder gar der Faden der Ariadne, der Theseus verhalf aus dem Labyrinth zu entkommen, und der somit ein Plan des Labyrinths sein müsste. Da liegt der Gedanke nicht fern, dass der

1 Waerden: Die Anfänge der Astronomie, 1968, S. 123.
2 Meller & Michel 2018.
3 Freeth (2010). Jones 2017. Bitsakis (2015).
4 Homer: Ilias, 1970, 18. Gesang, S. 590–605.

Diskos von Phaistos ein Plan des Labyrinths oder das Labyrinth in spiritueller Form selbst sei und Daidalos als historische und nicht mehr nur mythische Gestalt sein Urheber.

11.5 Fazit

Mit den vorgestellten Beobachtungen und Beweisen ist der Diskos von Phaistos im Wesentlichen stringent entziffert. Hauptbestandteile, die zu diesem Ergebnis führten waren: Die Isolierung der 45 (mit Dorn 46) verschiedenen Charaktere, deren Dreiteilung in das von mir so genannte G-L-S-System, der Zuordnung in die Geschlechter des Geistigen, Leiblichen und Sächlichen, sowie die Häufigkeiten ihrer Wiederholungen (den Koeffizienten) als wesentliches Kriterium zur Ermittlung der Reihenfolge der 45 verschiedenen Zeichen und damit einer Zuordnung von Tageszahlen zu den Zeichen, mit denen sich eine Vielzahl astronomischer Daten in den beiden Texturen erschließen lassen.

Einzig für die genaue Abfolge der 242 geprägten Zeichen lässt sich bisher kein arithmetisches System erschließen. Hier kommt die Möglichkeit eines sprachlichen Phänomens wieder ins Spiel; denn werden den 46 verschiedenen Zeichen Silben zugeordnet, werden die Texturen sprechbar, was der mündlichen Überlieferung der sehr komplexen Inhalte des Diskos zugutekommt. Ob dabei ein sinnvoller Text generiert werden könnte und in welcher Sprache bleibt fraglich.

11.6 Literatur

BARTHEL, THOMAS S.: *Forschungsperspektiven für den Diskos von Phaistos.* München: Hirmer (Münchner Beiträge zur Völkerkunde; Band I) 1988, S. 9–24.

BITSAKIS, YANIS: Ein antiker mechanischer Kosmos – der Mechanismus von Antikythera. In: *Antike Welt – Zeitschrift für Archäologie und Kulturgeschichte* **46** (2015), 5, S. 27–32.

DUHOUX, YVES: *Le disque de Phaestos. Archéologie, Epigraphie, Edition critique.* Louvain: Éditions Peeters 1977.

EVANS, ARTHUR J.: *Scripta minoa: the written documents of minoan Crete with special reference to the archives of Knossos (Band 1): The hieroglyphic and primitive linear classes.* Oxford 1909.

FAURE, PAUL: *Kreta: Das Leben im Reich des Minos.* Stuttgart: Reclam (2. Auflage) 1978.

FREETH, TONY: Die Entschlüsselung eines antiken Computers. In: *Spektrum der Wissenschaft* (2010), Nr. 05, S. 62–70.

HOMER: *Ilias.* Übersetzung von ROLAND HAMPE. Stuttgart: Reclam Philipp Jun. 1970.

JONES, ALEXANDER: *A Portable Cosmos. Revealing the Antikythera Mechanism. Scientific Wonder of the Ancient World.* Oxford: Oxford University Press 2017.

MELLER, HARALD & KAI MICHEL: *Die Himmelsscheibe von Nebra. Der Schlüssel zu einer untergegangenen Kultur im Herzen Europas.* Berlin: Propyläen 2018.

NEUMANN, GÜNTER: Zum Forschungsstand beim *Diskos von Phaistos.* In: *KADMOS* Bd. VII (1968), S. 27–44.

OLIVIER, JEAN-PIERRE: *Le disque de Phaistos. Édition photographique.* Paris: École Française d'Athènes 1975.

WAERDEN, BARTEL LEENDERT VAN DER: *Erwachende Wissenschaften 2: Die Anfänge der Astronomie.* Stuttgart: Birkhäuser 1968.

WENZEL, HERMANN: Trilogie über Strukturanalyse, Entzifferung und Astronomie des *Diskos von Phaistos* – in Stabi München, Uni Bibliothek Konstanz und Nationalbibliothek Stuttgart / Dresden.

WENZEL, HERMANN: 11 Beiträge über die Entzifferung des *Diskos von Phaistos* im Wissenschaftsportal der Gerda-Henkel-Stiftung Düsseldorf, 27.10.2010, 31.12.2010, . . . , `https://lisa.gerda-henkel-stiftung.de/beitraege?user_id=371`

Abbildung 12.1:
- Artistic Illustration of an Aboriginal Man and a Red Sprite

(© A. Ploum / P. Heemels (with permission of Kim Phillipsen and NCKU/NSPO, Taiwan)

Iconography, Science, and Lightning Figures

Albrecht Ploum (Nijmegen, Netherlands)

Abstract

Some Australian Aboriginal figurative paintings in the Kimberley and the Northern Territory, known as 'lightning figures', show remarkable resemblance in morphology with strange atmospheric phenomena such as *Red Sprites* (upper-atmospheric optical phenomena associated with thunderstorms) that sometimes can be perceived with the naked eye in those parts of the continent. I argue that some ancient markings can be related in a consistent way to real perceived atmospheric phenomena.

Zusammenfassung: Ikonographie, Wissenschaft und Blitzfiguren

Einige figurative Gemälde der australischen Aborigines in der Kimberley-Region und im Northern Territory, die als „Blitzfiguren" bekannt sind, weisen in ihrer Morphologie bemerkenswerte Ähnlichkeiten mit seltsamen atmosphärischen Phänomenen wie *Red Sprites* (optische Phänomene in der oberen Atmosphäre im Zusammenhang mit Gewittern) auf, die manchmal mit bloßem Auge in diesen Teilen des Kontinents wahrgenommen werden können. Ich argumetiere, dass einige alte Markierungen in konsistenter Weise mit real wahrgenommenen atmosphärischen Phänomenen in Zusammenhang gebracht werden können.

Preface

The article is a revised version of research report of 'Iconography Science and Lightning figures' (published 2012 by the Journal of the Australian Institute of Aboriginal

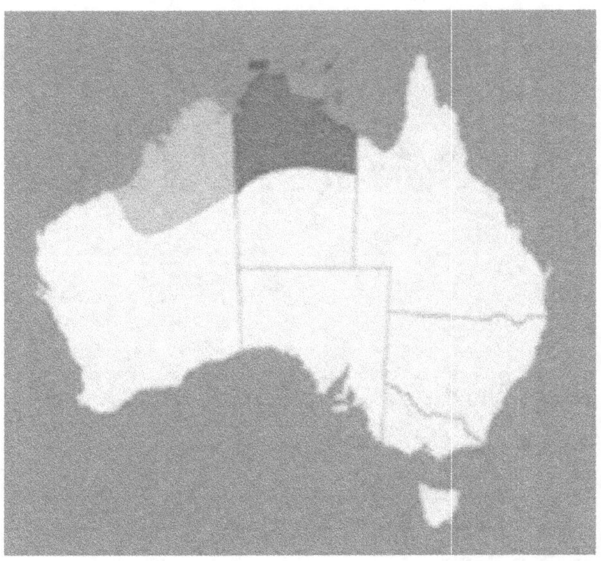

Abbildung 12.2:
Artistic Illustration Australia
(© Albrecht Ploum)

and Torres Strait Islander Studies 2012/2). The article was part of a fundamental study on the role of form and color similarities at the development of new theories. Because of the black/white tone printing restrictions and relative few color images of transient luminous events at the time (2012), it took several years to upgrade and revise my report. This revised version is for educational purposes.

12.1 Introduction: Iconography in Science

This paper is based on fundamental psychological, meteorological and anthropological research of specific global atmospheric phenomena such as Red Sprites that might have influenced indigenous rock art. There are remarkable resemblances between atmospheric phenomena and the morphology of some obsolete Australian Aboriginal figurative rock paintings in the Kimberley and the Northern Territory, known as 'Lightning Figures'. Red Sprites appear always together with thunderstorms but they totally differ from the common cloud to earth lightning. Although the hypothesis contains interpretational aspects, it should be clear that as long as interpretations are subject for falsification (open for conflicting arguments)[1] they still are part of an scientific approach.[2] In this paper an effort is made to test the null hypothesis (looking for conflicting arguments) that there is no relationship between Red Sprites and specific Australian rock art, known as 'Lightning Figures'. The statements to be tested are:

- The ability for rock-painters to observe the Red Sprite phenomena did not exist because such phenomena are too short-lived to be within the ability of the human eye.
- There is no climatological or geophysical opportunity for potential observation of the phenomena in question.
- There is no morphological resemblance between the lightning figures and the Red Sprite phenomena.
- There is no mythological relation with Red Sprites Since there has been found empirical evidence (Dowden 1997), that under the right natural conditions Red Sprites indeed can be observed by the naked eye it became possible to test the other propositions through falsification.

12.2 Australian Lightning Figures and other figures

There are many paintings in the Kimberley that Aborigines interpret as Lightning Figures or Lightnings (Figures 12.3a, b). According to Crawford (1977) they are associated with dry lightning and are usual smaller than Wanjina figures and are depicted in monochrome red, although in some cases they have small quantities of white or black pigment adhering to them. The paintings usually lack 'eyes', the 'head' is represented by a heavy encircling line, and the 'face' may be bisected. The 'headdress' is usually large and consists of radiating lines. The 'bodies' are coloured red, and little or no body decoration is now visible.

What kinds of realistic observation can be made in relation to the above lightning figures? Only in the last two decades (1990–2010) it has become possible to develop an idea of the possible natural phenomena by which the Lightning Figures could have been inspired. The phenomenon in question is called 'Red Sprite'.

1 According Karl Popper, (e. g. 1959, 1963) and popularised in archaeology by Louis Binford (e. g. 1967) a theory is scientific if the population of potential falsificators is not empty, in other words if basic statements can be formulated that make the theory testable.

2 For example the theory that Australia was a part of Antarctica, once started with the remarkable speculation of form resemblance (that is similarity) of the coastlines of South American and western Africa. As long as no conflicting geological and paleontological data were found the development of a scientific explanation could take place.

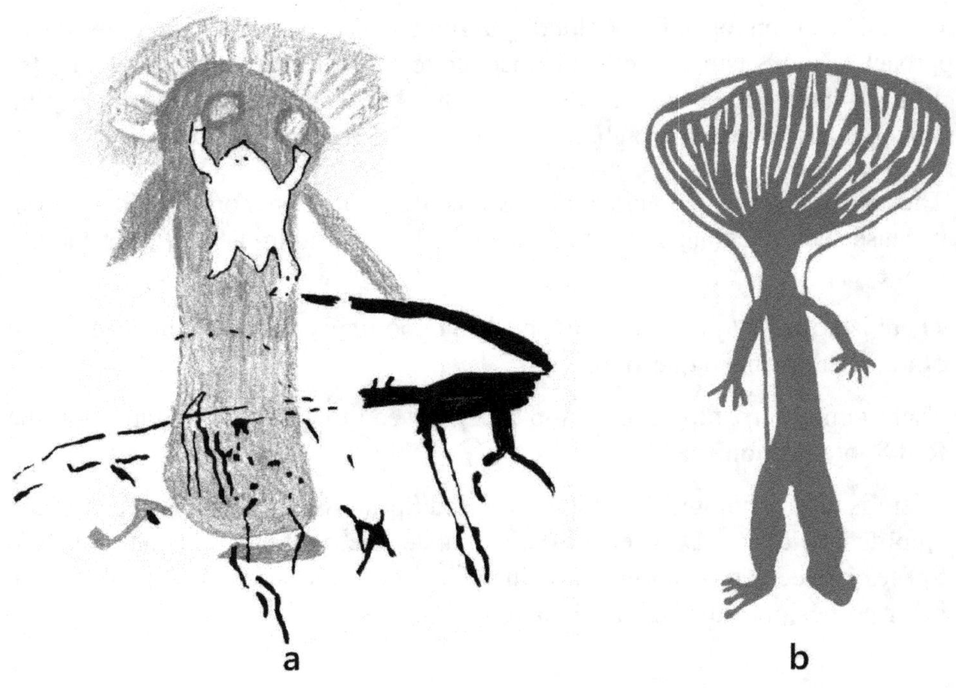

Abbildung 12.3:
(a) Lightning Figure near Kununurra; figure is eroded red; black lines are more recent paintings in white ochre. The bottom group probably depicts a string of 'yams';
(b) Lightning Figure in red, Admiralty Gulf

(a) Crawford 1977, 368; (b) Crawford 1977, 369).
(Images reproduced with permission of AIATSIS)

12.3 Atmospheric phenomena: Red Sprites, Elves, Blue Jets and Gigantic Jets

Red Sprites (see figure 4 and 5) together with Elves, Blue Jets and Gigantic Jets belong to the category of upper atmospheric transient luminous events (TLE). Red Sprites have been documented recently using low-light-level television technology. The first images of a Sprite were accidentally obtained in 1989 (Franz et al. 1990). By 1990, about twenty images had been obtained from the space shuttle (Vaughan et al. 1992; Boeck et al. 1995). Since then, video sequences of many thousands of Sprites have been captured. These include measurements from the ground (Lyons 1994; Winckler 1995), and from aircraft (Sentman and Wescott 1993; Sentman et al. 1995). Sprites only happen above large, active thunderstorm systems, mostly in small groups. To see them requires visual access to the region above the storm, unobstructed by intervening clouds, and viewing against a dark stellar background. In most locations these conditions happen only rarely.

Elves:[3] Sometimes Sprites appear together with 'Elves'. Elves often appear as a dim, flattened, expanding glow around 200 to 400 km in diameter that lasts, typically, for just one millisecond. (not visible to the naked eye)

Blue Jets: Numerous images have also been obtained from aircraft of so-called 'Blue Jets' (Wescott et al. 1995), also a previously unrecorded form of optical activity above thunderstorms. Blue Jets appear to emerge directly from the tops of clouds and shoot upward in narrow cones through the stratosphere (circa 300 milliseconds, [ms]) and theoretically might be visible to the naked eye. Their upward speed has been measured to be about 100 kilometres per second.

Gigantic Jets: are Blue Jet related, but less frequent, more extended and longer lasting (>500 ms), (visible to the naked eye).

It was a colour image of a Red Sprite that appeared as a 'burning tree', captured with a low-light-level camera in Taiwan (Figure 12.4), with features reminiscent of the two Lightning Figures reported from Admiralty Gulf and Near Kununnurra (Figure 12.3), which caught my attention.

Red Sprites characteristically take shapes of jellyfish, columns, fingers, paws, arrows , trees, insects and carrots. After gathering, over several years, many impressive (video) images of Red Sprites and Blue Jets, it became interesting to know whether not only the forms of Lightning Figures but also many stick-figures iconographically can be related to the Red Sprite phenomenon.

12.4 Pictorial views of Transient Luminous Events

Red Sprites are only barely detectable by the unaided human eye because of their short duration (10 to 200 ms) and low brightness, but in intensified television images obtained from the ground and from aircraft they appear as dazzlingly complex structures. Since there is discrepancy between images obtained from low light cameras and naked eye observations, it seems useful to show a scientific reconstruction of transient luminous events (TLE) related to lightning (Figure 12.5).

3 'Elves' is an acronym for Emissions of Light and Very Low Frequency Perturbations from Electromagnetic Pulse Sources; this term refers to the process by which the light is generated (the excitation of nitrogen molecules by electrons).

Abbildung 12.4:
Burning-tree Sprite (7 June 2001)

(credit ISUAL Project, NCKU/NSPO, Taiwan. Image is pseudo-colour rendered)

12.5 Climatological or geophysical opportunity for potential observation of the Red Sprite phenomena

Could Red Sprites be seen in the Kimberley and in Arnhem land? Climatologically, the Northern Territory is known for multiple Red Sprite observations. As early as 1997, Walter Lions and Russel A. Armstrong produced a world map of recorded Red Sprites distribution on which the Northern Territory was indicated as multiple Sprite observation area (see UW Sprite Ballon Experiment 2002).

There are no reasons to assume that there were long-lasting climate changes in the latitude of the Kimberley and Arnhem Land regions during the late Holocene. They are regions of monsoons, of dry and wet periods, where, especially between seasons and during the wet, lightning and thunderstorms can be dramatic. The landscape contains high vantage points from which – and over great distances – the atmosphere above thunderstorms can be observed.

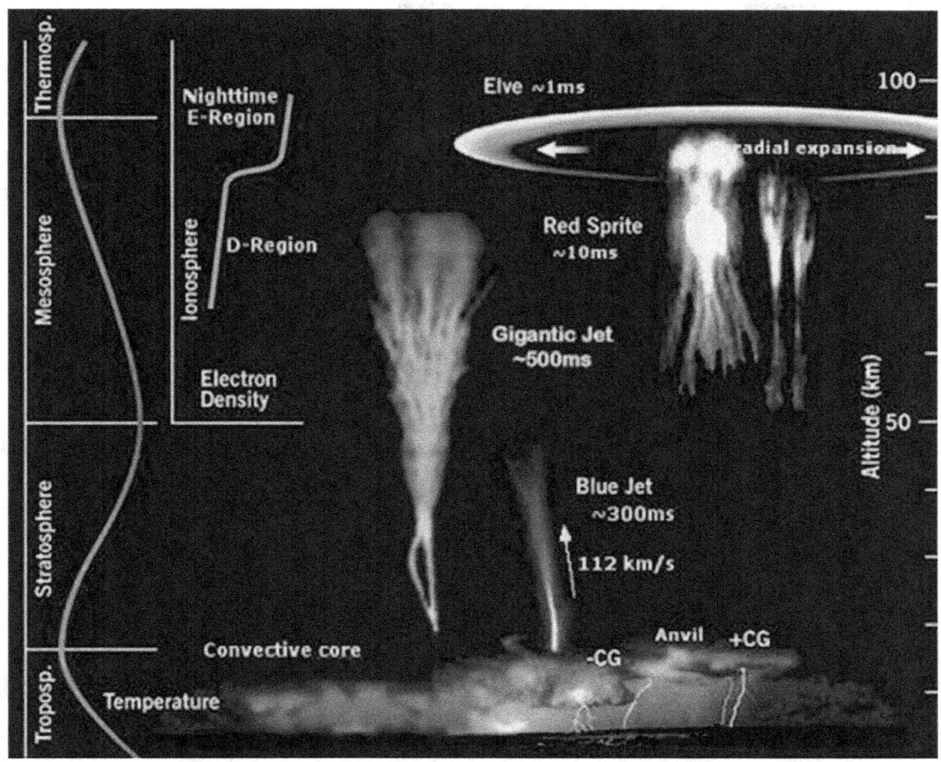

Abbildung 12.5:
Pictorial view of Elves, Sprites and Blue Jets
(from Neubert 2003; image reprinted with permission of Neubert)

12.6 The ability of rock-painters to observe the Red Sprite phenomena

It had been doubted that Red Sprites could be seen with the naked eye, but in 1997 unaided observation was reported by members of a team from the University of Otago researching near Darwin. The team leader, Professor Richard Dowden (1997), wrote that Red Sprites:

> *"must have occurred above [the Northern Territory] every Wet season for millions of years. They rarely last more than a tenth of a second and are generally faint but some – certainly one and maybe up to nine of the 72 caught on video on Wednesday night – can be seen by the naked eye* [...].

> *Such red sprites must surely have been seen by the Aboriginal peoples in the past, particularly before modern man polluted the night sky with artificial lighting. Is there an oral tradition of such sightings?"*

It was interesting to know if such a ghost-like fiery phenomenon with bizarre forms high in the sky, even on the threshold of visibility, was represented in Aboriginal cultures. However, the Otago researchers' attention was focused upon oral traditions and not on iconography. There was still the question how a naked eye observation of the Burning-tree Sprite might look, and how to verify the true colours of the Lightning Figures of Admiralty Gulf and Kununurra. However, at that time it was more important to me to find out more about the naked eye observations of sprites.

We knew from traditional psychology that the time needed to build up a complete image for recognition of patterns is approximately 200 ms. Recent neurological experiments of responses of single neurons in the human temporal lobe showed that the threshold of conscious recognition lies around 33 ms (Quiaroga et al. 2008). This means that at this duration the subjects consciously perceived the images in only about half of the trials (threshold). Responses under 33 ms are possible but are beyond the criteria for the absolute threshold. Positive responses under 33 ms not only are conditioned by intensity and duration of the stimulus but also influenced by variables such as the subject's experience, background, motivation, expectation, cognitive processes and level of adaptation to the stimulus. The colour of red sprites can be neon-light red, blue, green and white. The majority of light emitted by Sprites ranges beyond 650 nanometres wavelength, for which the eye is not as sensitive. Visibility studies (Taylor 1980) showed that some Australian Aborigines possess significant heightened visual acuity, probably enhanced by needs of hunting and their great patience and the long period of their observations. What might be visible of red sprites by the naked eye can be derived from scientific measurements. According to Stenbaek-Nielsen and McHargh (2008, 23):

> *"The active phase of red sprites is very short, a few milliseconds, while the decay phase is much longer typically lasting 10 to 100 ms. Observations, made at 1000 fps [frames per second], show that small structures, which we called beads (dotted lines), would remain visible in some events for almost a full second."*

Oscar van der Velde, Red Sprite researcher at the Technical University of Catalonia (Spain) (personal communication 23 December 2010) confirms that the 'hairy' parts above and below the more intense middle part last only a few milliseconds (say 3 to 6 ms). The middle part lasts much longer, approximately 30 to 100 ms depending on the duration of extinction. The middle part of a Sprite as a whole, can last sometimes 200 ms or more in the case of a dancing Sprites where elements or several groups pro-

Abbildung 12.6:
Colour photograph and simultaneous naked-eye observation of Red Sprites, 11 September 2006

(Camera settings: Canon EOS 5D, 50 mm lens, f/1.8, 4 seconds, ISO 1600; photograph reproduced with permission of Oscar van der Velde)

pagate horizontally. In tropical latitudes, large Jets may last 100 to 500 ms, sufficient for perception by the naked eye.[4]

4 Examples are at the EuroSprite blog at <eurosprite.blogspot.com> and at the 'Sprites Elves Blue Jets Lightnings' page of the Sonotaco.com website at `sonotaco.com/sample/sprite/e_index.html`.

From this it can be concluded that the naked eyed typically cannot perceive the complete Red Sprite emission. Only parts or groups lasting more than approximately 30 ms can be recognised by the naked eye under good visibility conditions. The middle parts of the schematic image (figure 5) might be best observed by the unaided eye. Because of significant differences of acuity among different observer groups, at optimum conditions, even beyond conscious recognition threshold (<30 ms), a few more red sprite structures might be detected. At a subconscious level, durations of 3 to 10 ms might be registered; however the precise effect on image development is still debatable.

The above duration analyses might be illustrated by some (impressions) of sprite observations (video) (`http://sonotaco.com/sample/sprite/e_index.html`). In September 2006, Van der Velde made a photograph of Red Sprites that also could be observed by the naked eye by a normal observer (Figure 12.6).

Reasons why naked eye observations of Red Sprite are rarely reported has been discussed by members of a research group of the University of Alaska (Heavner 2004). The observation rules from the research group are:

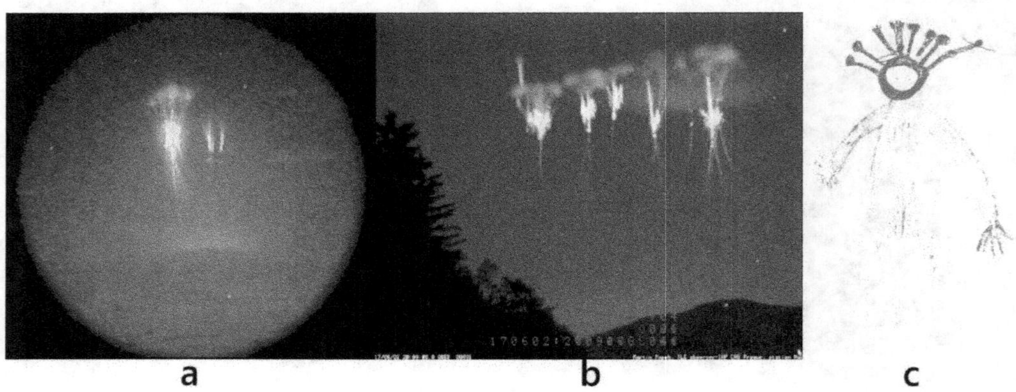

Abbildung 12.7:
(a) Sprites over thunderstorms in Kansas on 10 August 2000, observed in the mesosphere, with an altitude of 50 to 90 kilometres as a response to powerful lightning discharges from tropospheric thunderstorms. The true colour of sprites is 'pink-red´.
The middle part, durations from 10 to 200 ms
(b) Red Sprites over Europe, Slovenia, June 4, 2017
(c) Red figure from Mitchell River Falls. The face is blank suggesting that it was painted originally in another colour; radiating from the 'head' are lines that terminate in red blobs, a feature reminiscent of the King Edward River 'bird' site Wanjina

(a) Credit Walter A. Lyons; (b) Photo: Martin Opek;
(c) Crawford 1977, 367, reproduced with permission of AIATSIS)

- A clear view above a thunderstorm is required. This generally means the thunderstorm activity must be on the horizon. Additionally, there must be very little intervening cloud cover.
- Best viewing distance from the storm is 100–200 miles (200–300 km). At these distances sprites will subtend a vertical angular distance of 10–20 degrees. This is 2–4 times the separation of the pointer stars in the Big Dipper.
- For observing sprites, it must be completely dark. (i.e. no longer twilight)
- The eyes must be completely dark adapted. Use same criteria for this as for astronomical observing. If you can see the Milky Way, then it is probably dark enough and the eyes have adapted enough to see sprites.
- Fix your gaze on the space above an active thunderstorm. Do not be distracted by underlying lightning activity in the storm. Block out the lightning if necessary using a piece of dark paper in such a way as to still being able to view what is going on above the cloud.
- Sprites will be very brief flashes just on the edge of perceptibility. They occur too quickly to follow with the eyes, but their strange vertically striated structure and dull red colour may be perceived.
- Patience will be rewarded. If the right kind of storm is present and one's viewing geometry is favourable, then there is a greater likelihood of seeing a sprite than of seeing a shooting star or comet.

12.7 Morphological resemblance between Lightning Figures and sprite phenomena

The Red Sprite structures below are compared with depictions of figures found in north-western Australian regions which are known for frequent Red Sprite lightning. The first examples (Figure 12.7a, b) are typical universal sprite structures (here observed in Kansas and Slovenia), compared with one of the known red blob figures (Figure 12.7c) at Mitchell River Falls and the King Edward River 'bird' site (Crawford 1977, 365, 367).

Figure 12.8a is an often observed image of Red Sprites propagating horizontally over large thunderstorm systems. In one of his early papers (1996) Walter Lyons termed them 'dancing sprites', which might also be the impression of the actions of the typical stick figure scenes of the Cadell River (Figure 8b) recorded by Brandl (1977).

The suggested fishbone hair dress of the figures are even much better present in the hairy parts ending with a dot of another typical red sprite display (Figure 12.8c). The above examples of Red Sprites should be totally or partly visible within a flash by

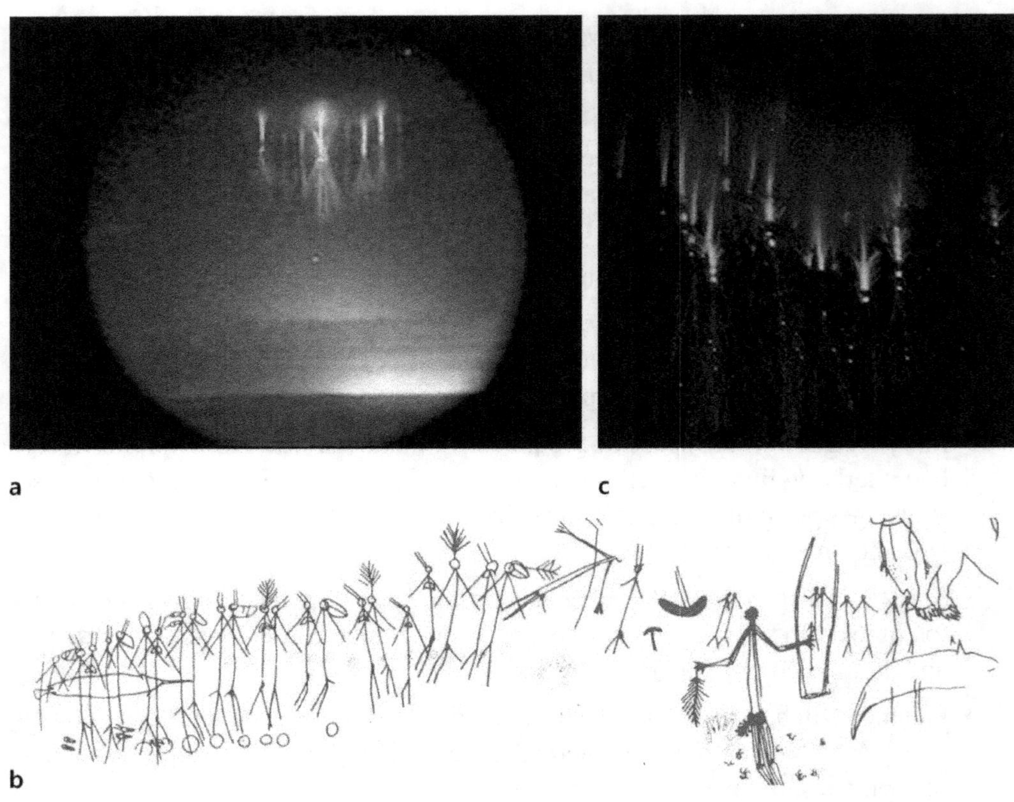

Abbildung 12.8:
(a) Sprites intensified by high-speed camera capable of recording more than 5000 frames per second at the Yucca Ridge Field Station in Fort Collins from July through August 2005
(b) Cadell River anthropomorphs and other designs (average height of figures at left 500 mm)
(c) Colour photo of a Red Sprite, 8 October 2009, The sprite appeared instantaneously and had a duration of less than 40 ms (threshold naked eye recognition 30 ms). The camera had its internal filter removed and was sensitive to near-infrared light emitted by sprites.

(a) credit Walter A. Lyons); (b) after Brandl 1977, 228. Reproduction permitted by AIATSIS; (c) Image reproduced with permission of Oscar van der Velde)

the naked eye, because they fall within the conscious recognition threshold duration between 10 to 200 ms.

The next black-and-white photograph is a Sprite captured by SonotaCo in 2005 (Figure 12.8a). (see real time video `http://sonotaco.com/sample/sprite/e_index.html`), supplemented with a photo of Red Sprites over Hiratsuka (Japan) 2021, McPott (Figure 12.9b). They are compared to painted images of figures, recorded by

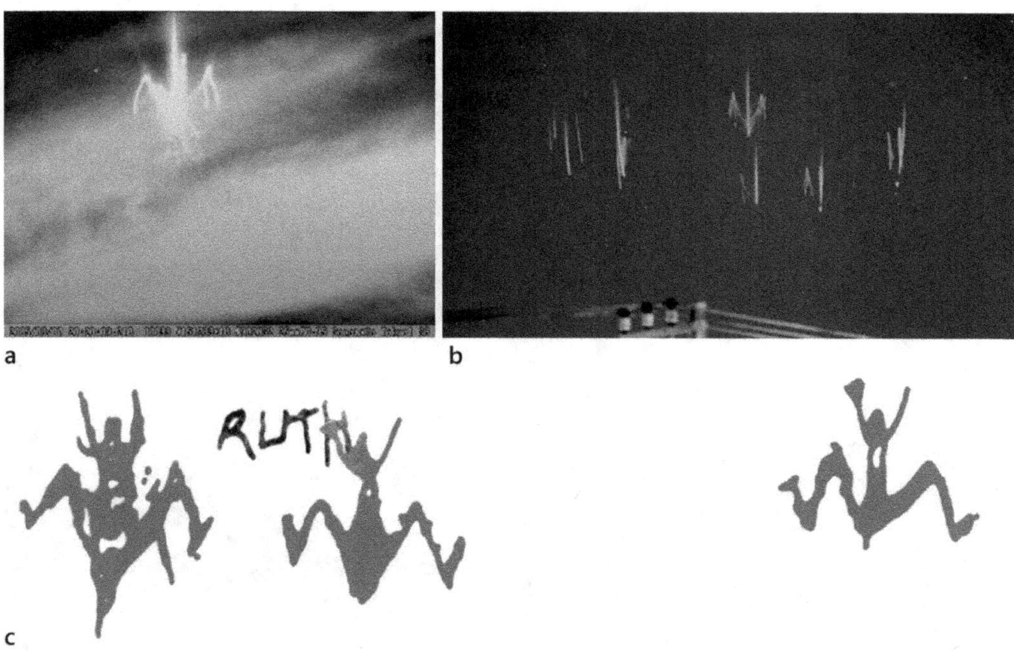

Abbildung 12.9:
(a) Red Sprite. captured by SonotaCo 2005 (duration 10 to 100 ms)
(b) (supplement): Red Sprites over Hiratsuka, Kanagawa Prefecture.
Japan 12 November 2021, 3:55 a.m.
(c) Upper Colo images (image size approximately two metres)

(a) reproduced with permission of SonotaCo; (b) photo: Smokey McPott; (c) after J. Clegg 1977, 273 (the middle figure is accompanied by a later graffito). Image reproduced with permission of AIATSIS)

John Clegg at Upper Colo (Figure 12.9c) and with the typical Grasshopper structure of the lightning figures Namarrgon and his wife Barrginj depicted at the Nourlangie Rock site (Figure 12.10).

12.8 Mythological relation with Sprites

Although these iconographical comparisons might at first glance appear speculative, there exist strong arguments that specific sprite structures (such as Figure 12.9a and Figure 12.9b), are central to the iconography of Aboriginal Lightning mythology. This statement can be elaborated by reference to the well-known text about the Nourlangie painting [Parks Australia n.d.]): Namarrgon and his wife Barrginj

Abbildung 12.10:
Creation Ancestors: middle: Namondjok;
right: Lightning Man Namarrgon; bottom: his wife Barrginj

(credit Department of the Environment, Water, Heritage and the Arts, Canberra)

> *"The Lightning Man rock art site (lower part of the Nourlangie Rock) shows several paintings of Creation Ancestors. An important one is Namondjok, of whom different clan groups tell different stories and who apparently now lives in the sky and can only be seen at night. Another important Creation ancestor is the Lightning Man Namarrgon who is still active today. He is responsible for the violent lightning storms that occur every wet season. He uses the axes on his head, elbows and feet to split the dark clouds and make lightning and thunder. The children of Namarrgon and his wife Barrginj are the Alyurr. They are associated with rare grasshoppers, with striking blue and orange colours, which are seen just before the wet season when they come out and call to their father to bring the wet-season storms."*

In this case the environmental factor is not at issue. The Lightning Man site is situated in a region that is known for its dazzling lightning of the Gudjew monsoon season (between January and March) where steep escarpments form ideal conditions for observation of transient luminous events such as Red Sprites, Elves, Blue Jets and Gigantic Jets.

The forms of the lightning man Namarrgon and his family show typical Red Sprite structure, as well, the description of activity above the thunderstorms is suggestive. Even the sprite structure seems to be related to the grasshopper anatomy and its colours. Is this not what we are looking for? Aboriginals artists, with remarkable good vision and ability of natural observation observed on the edge of visibility the ghostlike phenomena above the thunderstorms that bring rain and fertility. Their mythology might be seen to be based on these observations.

12.9 Conclusion

This study has no pretension to be a new speculative interpretation of prehistoric art. It only shows that similarities in forms of some depictions might be considered as a form of pareidolia (the illusive perception of a pattern or meaning were it does not actually exist) and that such morphological similarities might be used to test a scientific hypothesis. One of the tasks of a scientist should be to find out under which conditions a phenomenon must be classified as pareidolia. Sometimes it might be extremely difficult and on rare occasions it might happen that subject and object (for aboriginals) are direct related. In this paper criteria for falsification are given.

Note

I have not made references to earlier studies between prehistoric rock art and plasma physical phenomena. The first paper on this topic read was by the astrophysicist George L. Siscoe (Technology Review 1976) in which he related some prehistoric cave art to real observed meteorological phenomenon such as the aurora borealis. Martinus Van Der Sluijs and Anthony Peratt (2010) related some characteristic rock art figures to the observation of a 'super aurora' that would have been observable by prehistoric peoples.

12.10 Follow up research of Albrecht Ploum

"The role of lightning phenomena and TLE research at the interpretation of American prehistoric rock art".

Abstract: A scientific approach to the investigation of a class of ancient Australian Aboriginal rock art paintings suggests that their iconography can be interpreted as being stimulated by naked eye observation of lightning phenomena known as *Transient Luminous Events*. It is argued that, under comparable environmental – geophysical, meteorological and climatological conditions, this approach could be used to understand some style aspects of American Barrier Canyon rock art pictographs.

Acknowledgments

I am grateful to Dr Graeme K. Ward, Research Fellow, Australian Institute of Aboriginal and Torres Strait Islander Studies, for guidance and editorial advice, and for various advice and information to Dr Jean Clottes, Conservateur général du Patrimoine de France; Ing. Patrick Heemels, I.T. specialist, Netherlands; Dr Oscar van der Velde, meteorologist, Electrical Engineering Department of the Technical University of Catalonia, Spain; Professor Dr Frank Verbunt, Astronomer, Rijks Universiteit Utrecht, Netherlands.

Appendix A: Illustrations

Colour: Apart from the cover page, the illustrations in the research paper 'Iconography Science and Lightning Figures' were published in black/white, mainly because of the journal's publishing tradition. Unfortunately the lacking of (especially) the red colour, means missing an important element for imagining the colour relation between the Red Sprite phenomena and red rock paintings. In this revised supplement the colours used in models and rock paintings are more or less representative.

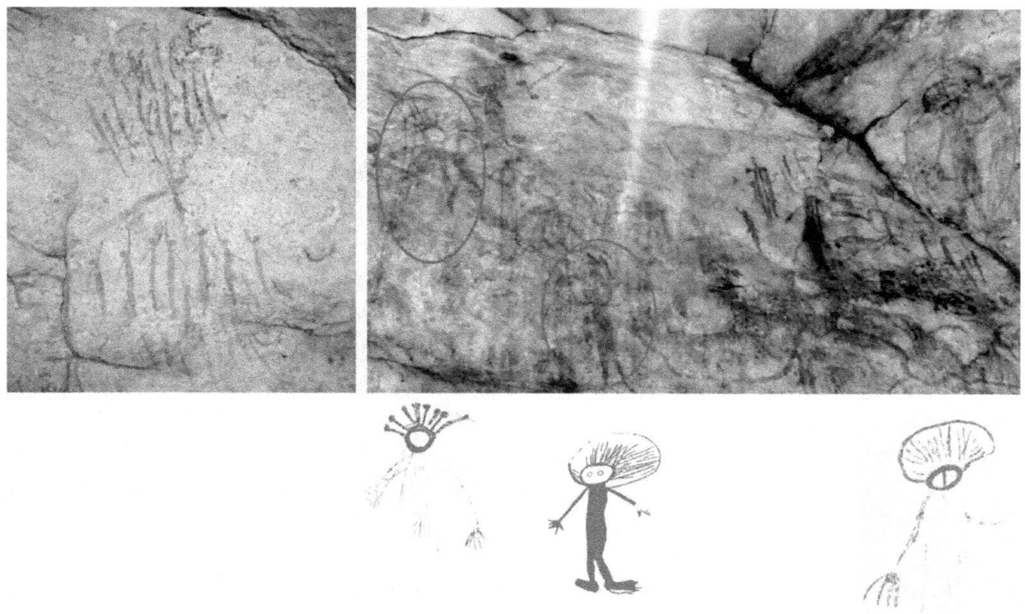

Abbildung 12.11:
Panel Mitchell River Falls. Spearing scene with elegant figures and three encircled lightning associated anthropomorphic red figures.

(Credit Department of the Environment, Water, Heritage and the Arts, Canberra)

Style: Some important figures (Fig. 12.7c) were extracted from a large panel (ca. 7 m × 3 m) at Mitchell River Falls (Fig. 12.11), Australia (Crawford 1977). The panel is very complicated because of aging and much overpainting. It shows some small anthropomorphic Wanjina figures associated with lightning figures, embedded and surrounded by many elegant streamlined beings which Crawford classified as *Gwion Style* figures.

12.11 Literature

BINFORD, LOUIS: Smudge Pits and Hide Smoking: The use of analogy in archaeological reasoning. In: *American Antiquity* **32** (1967), p. 1–12.

BOECK, WILIAM L.; VAUGHAN, O.H.; BALKESLEE, R.J.; VONNEGUT, B.; BROOKS, M. & J. MCKUNE: Observations of lightning in the stratosphere. In: *Journal of Geophysical Research* **100** (1995), p. 1465–1475.

BRANDL, ERIC: Human stick figures in rock art. In: UCKO, P.J. (ed.): *Form in Indigenous Art: Schematization in the Art of Aboriginal Australia and Prehistoric Europe.* Canberra: Australian Institute of Aboriginal Studies 1977, p. 220–242.

CLEGG, JOHN K.: A method of resolving problems which arise from style in art. In: UCKO, P.J. (ed.): *Form in Indigenous Art: Schematization in the Art of Aboriginal Australia and Prehistoric Europe.* Canberra: Australian Institute of Aboriginal Studies 1977, p. 260–276.

CRAWFORD, IAN M.: The relationship of Bradshaw and Wandjina art in north-west Kimberley. In: UCKO, P.J. (ed.): *Form in Indigenous Art: Schematization in the art of Aboriginal Australia and prehistoric Europe.* Canberra: Australian Institute of Aboriginal Studies 1977, p. 368–369.

DOWDEN, RICHARD L.: *Huge jellyfish-shaped lights in the sky above Australia's Northern Territory.* Otago, New Zealand: Otago Space Physics Group, University of Otago 1997, https://www.physics.otago.ac.nz/space/darwin97/press_release.html (accessed 9 February 2024).

FRANZ, ROBERT; NEMZEK, C.R.L. & L.R. WINCKLER: Television images of a large upward electrical discharge above a thunderstorm system. In: *Science* **249** (1990), p. 48–51.

HEAVNER, MATT JAMES: *Red Sprites and Blue Jets.* Geophysical Institute of the University of Alaska Fairbanks 2004, https://web.archive.org/web/20120207055606/http://elf.gi.alaska.edu/ (accessed 9 February 2024).

LYONS, WALTER A.: Characteristics of luminous structures in the stratosphere above thunderstorms as imaged by low-light video. In: *Geophysical Research Letters* **21** (1994), p. 875.

LYONS, WALTER A.: Sprite observations above the U.S. High Plains in relation to their parent thunderstorm systems. In: *Geophysical Research Letters* **101** (1996), p. 29 641–29 652.

LYONS, WALTER A.: 'Classic Carrot Sprite', personal communication, 14 February 2024.

LYONS, WALTER A.: 'Classic C-Sprites', personal communication', 14 February 2024.

NEUBERT, TORSTEN: On Sprites and Their Exotic Kin. In: *Science* **300** (2003), p. 747–749.

NGARJNO, UNGUDMAN & NYAWARRA BANNGAL [Ngarinyin munnumburra]: Gwion Gwion: Dulwan Mamaa. Secret and sacred pathways of the Ngarinyin Aboriginal people of Australia. Edited by JEFF DORING. Cologne: Könemann 2000.

PARKS AUSTRALIA N.D.: Nourlangie and Nanguluwur Art Sites, Kakadu National Park, Department oft he Environment, Water, Heritage and the Arts. Canberra, https://www.environment.gov.au/parks/kakadu/visitor-activities/rock-art-nourlangie.html (accessed 6 February 2024).

PLOUM, ALBRECHT: *Spiegel des Universums. Das kosmische Antlitz der Erde.* Landgraaf: Hoppers 1996.

POPPER, KARL: *The logic of Scientific Discovery.* London: Hutchinson 1959.

POPPER, KARL: *Conjectures and Refutations. The Growth of Scientific Knowledge.* London: Routledge and Kegan Paul 1963.

QUIAROGA, R. QUIAN; MUKAMEL, ROY; ISHAM, EVE A.; MALACH, RAFAEL & ITZHAK FRIED: Human single-neuron responses at the threshold of conscious recognition. In: *Proceedings of the National Academy of Sciences of the United States of America* **105** (2008), 9, p. 3599–3604.

SENTMAN, DAVID D. & EUGENE M. WESCOTT: Video observations of upper atmospheric optical flashes recorded from an aircraft. In: *Geophysical Research Letters* **20** (1993), p. 2857.

SENTMAN, DAVID D.; WESCOTT, EUGENE M.; OSBORNE, DAN L.; HAMPTON, DON L. & MATT JEFF HEAVNER: Preliminary results from the Sprites94 aircraft campaign: 1. Red sprites. In: *Geophysical Research Letters* **22** (1995), 10, p. 1205–1208, https://www.researchgate.net/publication/23899837_Preliminary_results_from_the_Sprites94_Aircraft_Campaign_1_Red_sprites (accessed 6 February 2024).

SISCOE, GEORGE I.: Solar-terrestrial relations – Stone age to space age. In: *Technology Review* 78 (1976), p. 26–37.

SONOTACO: A Sprite Captured By Ufocapturev2 at 7 August 2005, SonotaCo.com, https://sonotaco.com/e_index.html (accessed 9 February 2024).

STENBAECK-NIELSEN, HANS & MATTHEW G. MCHARG: High time-resolution sprite imaging: observations and implications. In: *Journal of Physics D: Appied Physics* **41** (2008), 234009.

TAYLOR, HUGH R.: Prevalence and causes of blindness in Australian Aborigines. In: *Medical Journal of Australia* **1** (1980), 2, p. 71–76.

UW SPRITE BALLOON EXPERIMENT: Where are Sprites observed? 2002, https://earthweb.ess.washington.edu/space/AtmosElec/spriteinfo.html (accessed 6 February 2024).

VAN DER SLUIJS, MARINUS A. & ANTHONY L. PERATT:Astronomical petroglyphs: Searching for rockart evidence for an ancient super aurora. In: *Expedition* **52** (2010), 2, https://www.penn.museum/documents/publications/expedition/52-2/van%20der%20sluijs%20peratt.pdf (accessed 6 February 2024).

VAN DER VELDE, OSCAR N.D.: http://www.lightningwizard.com (accessed 6 February 2024).

VAUGHAN JR., OTHA H.; BLAKESLEE, RICHARD; BOECK, WILLIAM L.; VONNEGUT, BERNARD; BROOK, MARX & JOHN MCKUNE JR.: A cloud-to-space lightning as recorded by the Space Shuttle payload-bay T-V camera. In: *Monthly Weather Review* **120** (1992), p. 1459–1461.

WESCOTT, EUGENE M.; SENTMAN, DAVID D.; OSBORNE, DAN L.; HAMPTON, DON L. & MATT J. HEAVNER: Preliminary results from the sprites94 aircraft campaign: 2. Blue Jets. In: *Geophysical Research Letters* **22** (1995), 10, p. 1209–1212. https://www.researchgate.net/publication/23900007_Preliminary_results_from_the_Sprites94_Aircraft_Campaign_2_Blue_jets(accessed 6 February 2024).

WINCKLER, JOHN R.: Further observations of cloud-ionosphere electrical discharges above thunderstorms. In: *Journal of Geophysical Research* **100** (1995), D7, p. 14335–14345.

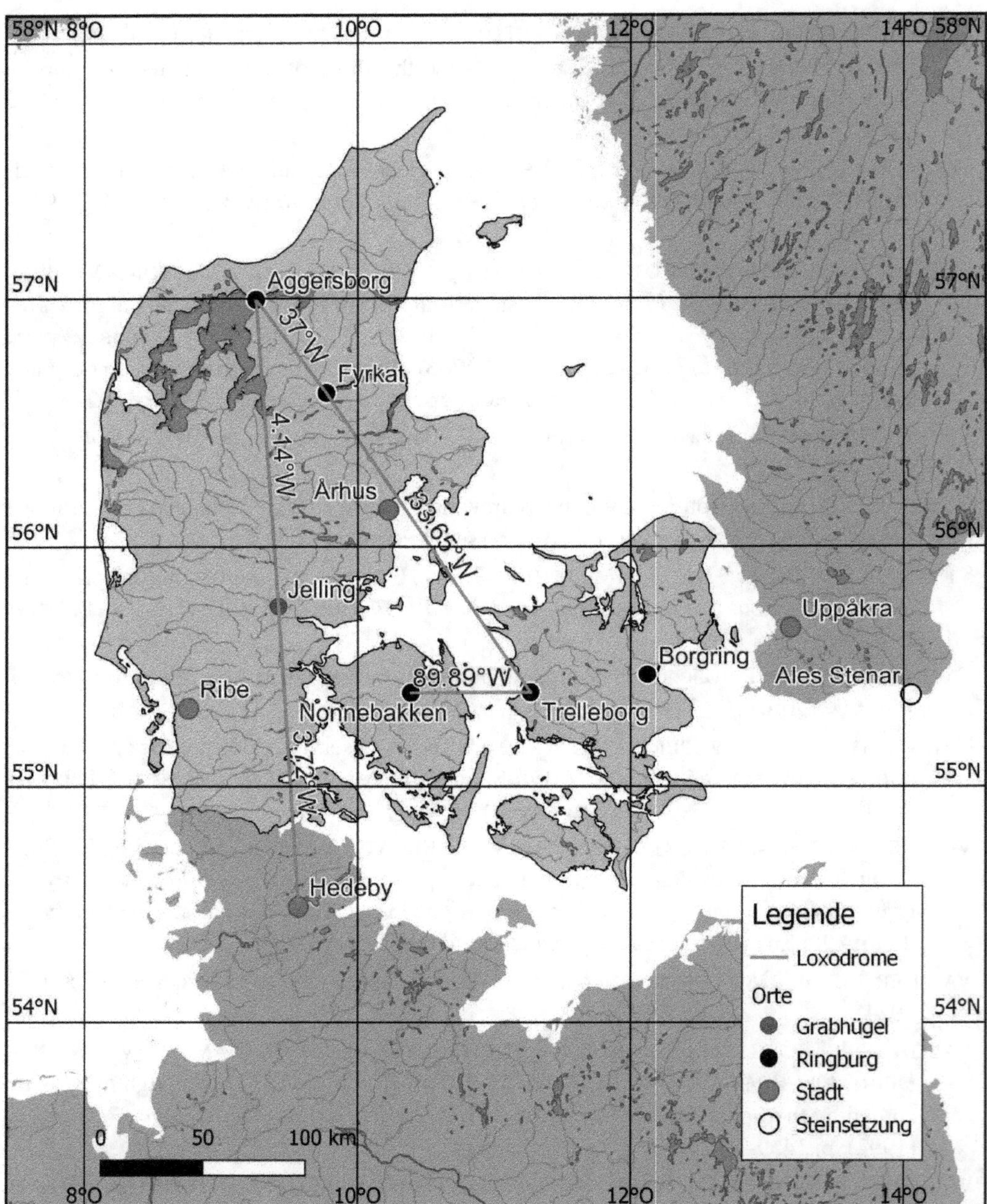

Abbildung 13.1:
Landkarte von Dänemark um 950 n. Chr. in Mercatorprojektion.
Die Verbindungslinien zwischen den Orten stellen Loxodrome
mit den jeweiligen Kurswinkeln dar.

(Graphik: Andreas Fuls)

Die Trelleborgen in Dänemark: Eine astronomisch-geometrische Siedlungsplanung der Wikinger

Andreas Fuls (TU Berlin)

Abstract: The Trelleborgen in Denmark: An astronomical-geometric settlement plan of the Vikings

Around 980 AD, five ring fortresses with a circular rampart and houses arranged at right angles to each other were built in Denmark. They date to the time of King Harald Gormsson (called Bluetooth, Danish king from 958 to 987 AD) and his son Sven Tveskæg ("Forkbeard", Danish king from 987 to 1014 AD).

The strictly geometrical arrangement in the ground plans of the ring fortresses as well as their geographical position in relation to each other suggests that they were based on a uniform construction plan. This raises the question of whether their approximate orientation to the north is coincidental or related to the observation of the rising and setting of the Sun and Moon. Two of the ring fortresses, Trelleborg and Nonnebakken, are located at almost the same latitude, which is known for having a right angle between the solstice points on the horizon. Two other ring fortresses, Fyrkat and Aggersborg, lie on an almost perfect great circle as seen from Trelleborg. This can also be demonstrated between the trading centre of Hedeby, the royal seat of Jelling and Aggersborg, so that large-scale settlement planning can be assumed.

The prototype and eponym of the ring fortresses is Trelleborg on the island of Zealand. It was the first of the ring fortresses to be excavated by Poul Nørlund between 1934 and 1941. His detailed report allows a precise astronomical and geometrical analysis of the house ground plans and the reconstruction of the original building plan with the longitudinal axes of the houses. Seen from the centre of the site, the construction plan shows a symmetry associated with the rising and setting directions of the Sun at the equinox and summer and winter solstices.

Zusammenfassung

Um 980 n. Chr. entstanden in Dänemark fünf Ringburgen mit einem kreisförmigen Ringwall und zueinander rechtwinklig angeordneten Häusern im Inneren. Ihre Datierung fällt in die Zeit von König Harald Gormsson (genannt Blauzahn, dänischer König von 958 bis 987 n. Chr.) und dessen Sohn Sven Tveskæg („Gabelbart", dänischer König von 987 bis 1014 n. Chr.).

Die streng geometrische Anordnung in den Grundrissen der Ringburgen sowie ihre geographische Lage zueinander lässt vermuten, dass ihnen ein einheitlicher Konstruktionsplan zu Grunde liegt. Dabei stellt sich die Frage, ob ihre ungefähre Ausrichtung nach Norden zufällig ist oder mit der Beobachtung von Auf- und Untergängen der Sonne und des Mondes in Zusammenhang steht. Zwei der Ringburgen, Trelleborg und Nonnebakken, befinden sich auf nahezu der gleichen geographischen Breite, welche dafür bekannt ist, dass zwischen den Sonnenwendpunkten am Horizont ein rechter Winkel vorliegt. Zwei andere Ringburgen, Fyrkat und Aggersborg, liegen von Trelleborg aus gesehen auf einem nahezu perfekten Großkreis. Dies kann auch zwischen dem Handelsplatz Haitabu, dem Königssitz Jelling und Aggersborg nachgewiesen werden, so dass von einer großräumigen Planung ausgegangen werden kann.

Der Prototyp und Namensgeber der Ringburgen ist Trelleborg auf der Insel Seeland. Sie wurde als erste der Ringburgen zwischen 1934 und 1941 von Poul Nørlund ausgegraben. Sein detaillierter Bericht erlaubt eine genaue astronomische und geometrische Analyse der Hausgrundrisse und die Rekonstruktion des ursprünglichen Bauplans mit den Längsachsen der Häuser. Vom Mittelpunkt der Anlage aus gesehen weist der Konstruktionsplan eine Symmetrie auf, die mit Auf- und Untergangsrichtungen der Sonne zur Tag- und Nachtgleiche sowie Sommer- und Wintersonnenwende in Verbindung steht.

13.1 Die Wikingerzeit in Dänemark im 10. Jahrhundert

Das dänische Königreich erstreckte sich im 10. Jahrhundert n. Chr. über das heutige Dänemark, nördliche Teile von Schleswig-Hosstein sowie Skåne in Südschweden. Unter den bedeutenden Städten existierten Ribe, Hedeby (Haithabu), Århus und Uppåkra.

Jelling war damals der Sitz der Könige und das Zentrum des Königreichs. Zwischen den beiden Hügeln, deren Konstruktion an die Wallanlagen der Ringburgen erinnern, liegen zwei Runensteine, die als „Geburtsurkunde" von Dänemark gelten. (Tab. 13.1).

13.2 Die Ringburgen der Wikinger

Vor 90 Jahren wurde die erste Ringburg in Dänemark entdeckt und zwischen 1934 und 1941 von Poul Nørlund ausgegraben. Bis heute sind noch vier weitere Ringburgen entdeckt worden: Fyrkat, Aggersborg, Nonnebakken und Borgring. Sie besitzen

Tabelle 13.1:
Überblick zur dänischen Geschichte

Zeit	Ereignis
935–945	König Gorm der Alte: Königssitz in Jelling, Haithabu erobert.
945–985	König Harald Blauzahn: Reichseinheit geschaffen.
948	Bistümer Ripen, Århus und Schleswig gegründet.
960	Taufe von König Harald Blauzahn.
ab 980	Verfall Haithabus: Halbkreisförmiger Schutzwall erbaut.
980/981	Trelleborg gebaut.
985–1014	König Sven Gabelbart erobert England.
988	Heiligtum Odins in Odense erwähnt.
1036	Aggersborg zerstört.
1066	Schlacht von Haisting, Wandteppich von Bayeux.

alle einen ähnlichen Grundriss mit einer besonderen Symmetrie und Rechtwinkligkeit zwischen ihren Häusern. Dies führte zu der Vermutung, dass sie als Militärlager dienten. Die Datierung von Trelleborg (980/981 n. Chr. erbaut) legt die Vermutung nahe, dass sie für die Eroberung von England 994 durch Sven Gabelbart gebaut wurden. Andere Überlegungen und besonders die archäologischen Funde ließen den Verdacht aufkommen, dass es sich um Zwingburgen handelte, die als Stützpunkte an strategisch günstigen Stellen auf den dänischen Inseln verteilt von Harald Blauzahn erbaut wurden [Klose (1982)].

Bei allen Burgen tritt ein gemeinsames Schema im Grundriss hervor: Zwei rechtwinklig aufeinander stehende Hauptwege mit 4 Toren im Wall. In den dadurch entstandenen Vierteln stehen 4 Häuser parallel zu einen Hauptweg im Karree, bei Aggersborg ist diese Anordnung durch zwei weitere Karrees pro Viertel ergänzt (Abb. 13.2, Tab. 13.2).

Es stellt sich die Frage, warum die Häuser so symmetrisch und rechtwinklig zueinander angeordnet sind und nicht z. B. radial wie in Ismantorp/Gotland oder Lembecksburg/Föhr? Außerdem zeigen die 4 Tore immer ungefähr nach den vier Himmelsrichtungen. Dies brachte mich auf die Idee, die Ringburgen nach astronomischen Beobachtungen hin zu untersuchen.

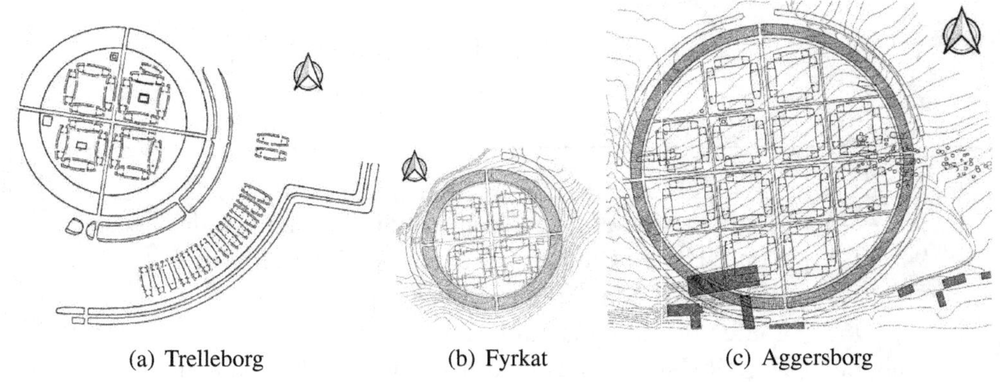

(a) Trelleborg (b) Fyrkat (c) Aggersborg

Abbildung 13.2:
Grundrisse der Ringburgen Trelleborg, Fyrkat und Aggersborg
(Pläne nach geographisch Norden orientiert).

(Graphik: Andreas Fuls, modifiziert nach [Nørlund (1968), Olsen (1970), Jensen (1986)])

13.2.1 Trelleborg

Trelleborg auf der Insel Seeland wurde 1934 bis 1941 von Poul Nørlund ausgegraben. Sein detaillierter Bericht erlaubt eine genaue astronomische und geometrische Analyse der Hausgrundrisse und die Rekonstruktion des ursprünglichen Plans mit den Längsachsen der Häuser.

Es gibt 16 Häuser im Inneren des Rinwalls mit einer Durchschnittslänge von 29.4 Meter. Im Südosten befanden sich darüber hinaus 15 kleinere Häuser (je 26.3 m lang), von denen 13 radial zum Mittelpunkt und 2 parallel zur Ost-Westachse angeordnet waren (Abb. 13.2a, 13.3).

Das geodätische Institut von Kobenhavn hatte bei der Ausgrabung alle Hausecken in einem lokalen Koordinatensystem vermessen. Über Richtungsmessungen zu einigen bekannten Fernzielen wurde das lokale System mit dem Landeskoordinatensystem von 1934 verknüpft [Nørlund (1948), S. 181–184]. Aber erst die Kenntnis der Meridiankonvergenz[1] von 0.74336° erlaubt es den Winkel zwischen der lokalen Y-Achse und der geographischen Nordrichtung zu bestimmen: $3.3973 + 0.74336 = 4.14066°$. Dadurch ist es möglich exakte Azimutwinkel zu berechnen und so zu überprüfen, ob im Grundriss von Trelleborg astronomische Ausrichtungen vorliegen.

1 Pers. Mitteil. des Geoteknisk Instituts von København, 26.08.1998

Tabelle 13.2:
Vergleich der Ringburgen

Ringburg	Breitengrad	Wall-breite	Innerer Wall-durchmesser	Orientierung der Anlage	Anzahl Häuser	Haus-länge
Trelleborg	55° 23' 38.8"	19 m	136 m	9.2° Ost	16	29.4 m
					+15	26.3 m
Nonnebakken	55° 23' 31"	?	120 m	?	?	?
Borgring	55° 28' 10.8"	11 m	120 – 123 m	19° Ost	?	?
Fyrkat	56° 37' 23.5"	13 m	120 m	3.2° West	16	28.7 m
Aggersborg	56° 59' 40"	11 m	240 m	6.6° West	48	32.0 m

Abbildung 13.3:
Modell von Trelleborg

(Foto: Andreas Fuls)

13.2.2 Fyrkat

Die Ringburg Fyrkat liegt auf einer Landspitze im Onsild-Flußtal in der Nähe des Mariager Fjords. Seit 1950 wurde es unter Leitung des Museumsinspektors C. G. Schulz bis zu dessen frühen Tod 1958 ausgegraben.

Aufgrund der Ausgrabungen ist bekannt, dass Fyrkat nicht dauerhaft bewohnt war, denn typische Wohnplatzfunde wie Gefäßscherben, Gebeine und Werkzeuge sind in nur geringen Mengen gefunden worden [Olsen (1970), S. 12]. Nach den Ausgrabun-

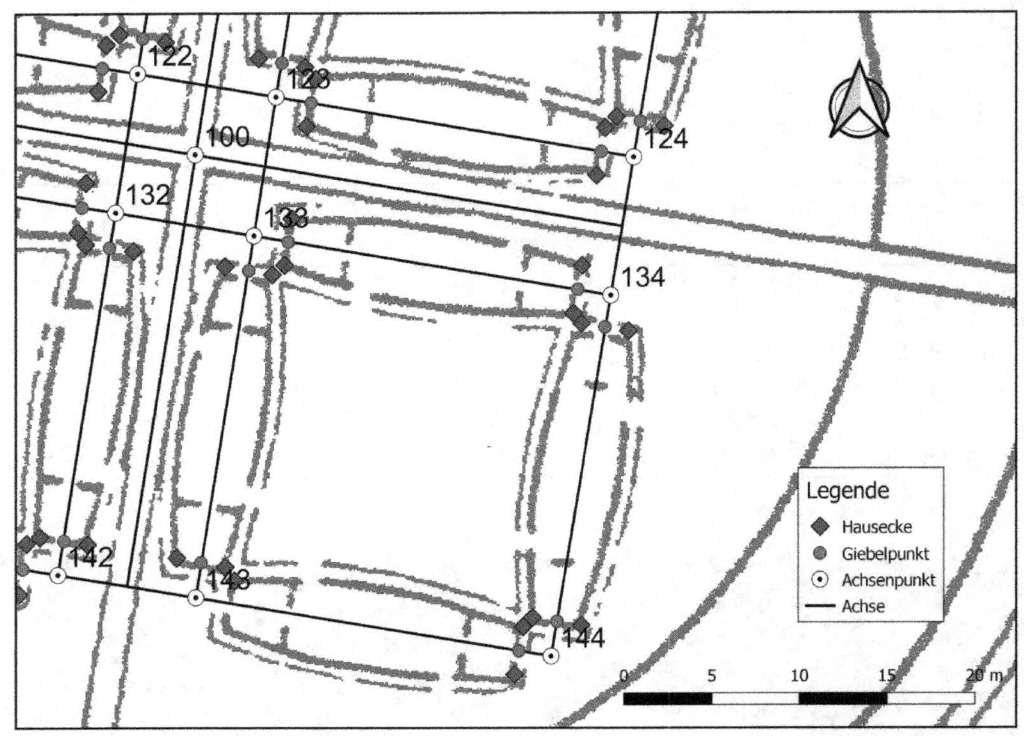

Abbildung 13.4:
Ausschnitt aus dem Grundriss von Trelleborg des südöstlichen Viertels und dem Ringwall mit Osttor. Die Achsen sind gegenüber geographisch Norden um 9.1 bis 9.5 Grad gedreht.

(Graphik: Andreas Fuls)

gen wurde der Ringwall und der Burggraben wieder hergestellt und die ehemals hölzernen Pfostenlöcher der Häuser im Boden mit Beton markiert.

Im Gegensatz zu den Häusern von Trelleborg haben die Langhäuser von Fyrkat zwei gegenüberliegende Eingänge an den Längsseiten mit je einen Windfang. Schmale Bohlen um die Häuser herum waren schräg angeordnet und stützten so Wand und Dach ab. Dadurch ergibt sich eine andere Rekonstruktion der Holzhäuser als in Trelleborg zuvor angenommen worden war (Abb. 13.5). Nicht alle Häuser dienten zum Wohnen, denn es wurde auch eine Schmiede und ein Haus mit viel Getreide, welches vermutlich als Speicher diente, gefunden [Olsen (1970), S. 6–8].

Abbildung 13.5:
Rekonstruierte Häuser in Trelleborg und Fyrkat.
Die Rekonstruktion in Trelleborg wird heutzutage nicht mehr als korrekt angesehen.

(Fotos: Andreas Fuls)

13.2.3 Aggersborg

Die Lage von Aggersborg am Limfjord ist besonders gut zur Kontrolle des Schiffsverkehrs geeignet. Bereits seit dem 8. Jahrhundert gab es dort eine Siedlung mit vielen kleinen Hütten und einigen größeren Häusern.[2] Direkt über die ältere Siedlung wurde am Ende des 10. Jahrhunderts die Ringburg errichtet. Sie bestand aber nicht lange, denn bereits 1036 wird es von aufrührerischen Vendelbewohnern zerstört.

13.2.4 Nonnebakken

Auf der Insel Fyn befand sich eine weitere Ringburg. Sie wird aufgrund des gleichnamigen Stadtteils in Odense Nonnebakken genannt. Die Stadt Odense überdeckt das ehemalige Gebiet der Ringburg, so dass nur an wenigen Stellen Ausgrabungen durchgeführt werden konnten.

13.2.5 Borgring

Hochauflösende LiDAR Daten und radiometrische Bodenuntersuchungen brachten 2013 eine weitere Ringburg auf Seeland zum Vorschein. In einem Kataster von 1682 wurde sie noch als Borgring bzw. Borre ring („Burgring“) bezeichnet, war aber danach in Vergessenheit geraten [Goodchild, Holm & Sindbæk (2017)].

Der Ringwall hat einen inneren Durchmesser von ca. 120 bis 123 Meter und eine Breite von 10 bis 11 Meter [Kristiansen (2022)]. Zwei AMS-Radiokarbondatierungen aus dem nördlichen Torbereich der Anlage ergaben eine Datierung zwischen 893 und 1017 n. Chr. (95% Wahrscheinlichkeitsbereich). Zwei Tore konnten im Norden und im Osten identifiziert werden, welche 90 Grad auseinander liegen. Die Orientierung des Nordtors, welches 4.4 bis 5 Meter breit ist, liegt bei knapp 19 Grad östlich von Nord.

13.2.6 Funktion der Ringburgen

Die archäologischen Funde, besonders in Fyrkat, weisen die Verwendung der Burgen als Kasernen zurück, denn es gibt neben den Wohnräumen auch Arbeitsräume und Lagerräume und auf dem Friedhof findet man auch Frauengräber. Vielmehr scheinen es Zwingburgen gewesen zu sein, die von König Harald Blauzahn gebaut wurden. Die dünne Kulturschicht deutet auf eine kurze Benutzung hin, den Fyrkat wurde wie Trelleborg und Aggersborg durch einen Brand zerstört. Eine ausführliche Diskussion über die archäologisch historische Quellenlage findet sich bei [Weibull (1974)].

---

2 `https://intarch.ac.uk/journal/issue36/brown_toc.html` (letzter Zugriff 13.11.2023).

13.3 Geographische Lage der Ringburgen

Ein Blick auf die Landkarte (Abb. 13.1) zeigt, dass die Lage der Ringburgen nicht zufällig ausgewählt wurde. Sie lagen etwas im Landesinneren, waren aber mit Schiffen erreichbar. Von ihnen aus konnte man wichtige Handelswege auf dem Land als auch auf dem Seeweg kontrollieren.[3]

Neben militärischen und wirtschaftlichen Gründen zur Lage der Ringburgen könnte es aber auch astronomisch-geometrische Gründe gegeben haben, die im Folgenden diskutiert werden. Ausgangspunkt der Überlegung ist der Grundriss von Trelleborg mit den rekonstruierten Achsen, welche aus den Giebelpunkten der Häuser berechnet wurden (vgl. Abb. 13.4). Peilt man vom Mittelpunkt der Anlage in Trelleborg aus genau nach Westen, dann ist dies die Richtung nach Nonnebakken. Eine zweite Peilrichtung ergibt sich über den nordwestlichen Schnittpunkt der Achsen der Innenhäuser. Sie zeigt in die Richtung nach Fykat und Aggersborg. Eine dritte Peilrichtung zwischen zwei Achsenschnittpunkten zeigt nach Borgring (Abb. 13.6a). Somit ist es möglich die für die Navigation wichtige Himmelsrichtung von Trelleborg aus zu den anderen Ringburgen anzugeben.

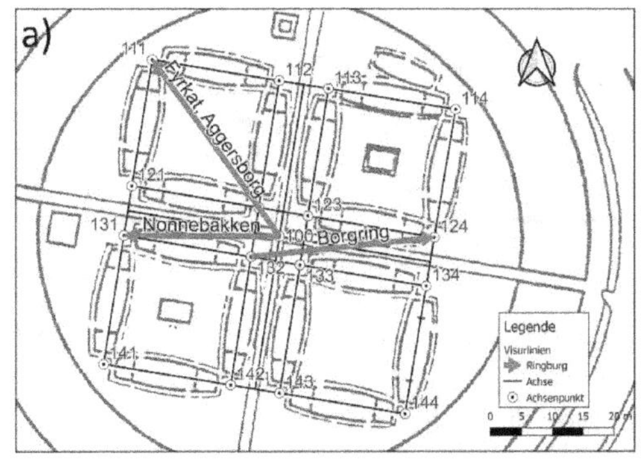

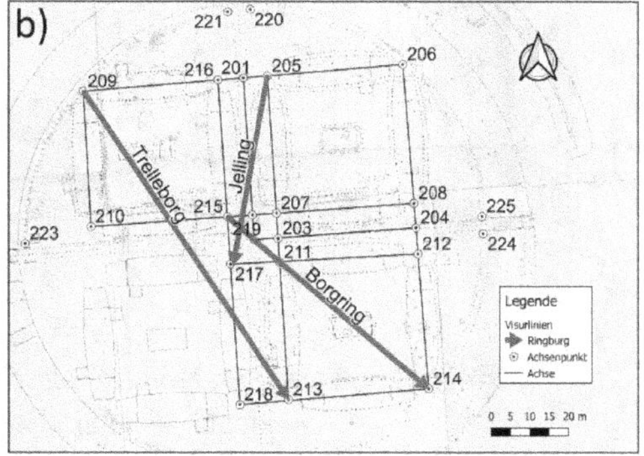

Abbildung 13.6:
Peilrichtungen zu den anderen Ringburgen über Schnittpunkte der Hauptachsen im Grundriss von a) Trelleborg, b) Fyrkat.

(Graphik: Andreas Fuls)

In Fyrkat können ebenfalls Peilrichtungen zwischen den Achsenpunkten nachgewiesen werden, und zwar nach Trelleborg und Borgring. Außerdem gibt es eine Peilrichtung zum Königshof und den Grabhügeln in Jelling (Abb. 13.6b).

3 Karte mit rekonstruierten Handelswegen und Bevölkerungsdichte siehe [Goodchild, Holm & Sindbæk (2017), Figure 1 auf S. 1030]

13.3.1 Der Breitengrad von Trelleborg

Eine gleiche Breite beibehalten, das kannten die Wikinger vom sogenannten Breitensegeln, bei der man die offene See überquerte,indem man immer auf einen Breitenkreis segelte. Diese Praxis fand offenbar auch bei der Auswahl der Lage von Nonnebakken, Trelleborg und Borgring ihre Anwendung, denn alle drei Ringburgen liegen ungefähr auf der gleichen geographischen Breite (Abb. 13.7).

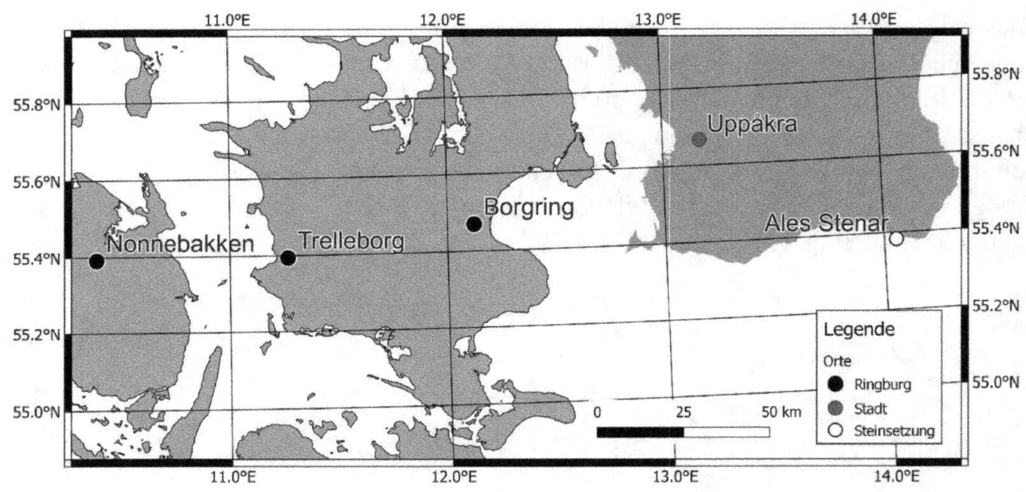

Abbildung 13.7:
Breitengradkarte von Nonnebakken – Trelleborg – Borgring – Ales Stenar.
Sie liegen nahe am Breitengrad von 55.4° Nord.

(Graphik: Andreas Fuls)

Die Eigenschaft der Breite von Trelleborg brachte mich nun dazu, eine sogenannte Sollbreite auszurechnen, bei der der Azimutwinkel zwischen Sommersonnenwende und Wintersonnenwende genau 90° beträgt. Das Azimut für die Deklinaten 23.5714° (Sommersonnenwende) beträgt, bei einer theoretischen Sonnenhöhe von Null Grad, dann genau 45° und und der für die Wintersonnenwende liegt bei 135°. Dieser als Sollbreite bezeichnete Breitengrad liegt bei 55.56°. Trelleborg ist davon 18.6 km entfernt und Nonnebakken liegt 18.8 km südlich davon. Borgring liegt auf einer geographischen Breite von 55.4697° an der Ostküste von Seeland und damit 10.1 km südlich von der Sollbreite.

13.3.2 Grosskreis 1: Trelleborg, Fyrkat und Aggersborg

Vergleicht man die Lage der Ringburgen Trelleborg, Fyrkat und Aggersborg zueinander, dann fällt auf, dass sie ungefähr auf einer Linie liegen (Abb. 13.1). Der Abstand von Fyrkat zu dem Großkreis zwischen Trelleborg und Aggersborg beträgt 1.5 km.

13.3.3 Grosskreis 2: Haithabu, Jelling und Aggersborg

Einen zweiten Großkreis kann man zwischen Haithabu, Jelling und Aggersborg finden (Abb. 13.1). Unter einen Azimut von -3.87° zeigt die Verbindungslinie ungefähr nach Norden. Jelling war damals der Sitz des dänischen Königs und stellte somit das Zentrum im Königreich dar. Es befindet sich ungefähr auf halber Strecke zwischen Haithabu und Aggersborg und liegt nur 363 m vom Großkreis entfernt.

13.4 Astronomie von Trelleborg

Für die Berechnung von astronomischen Visierlinien sind genaue Koordinaten nötig, weswegen nur Trelleborg detailiert untersucht werden konnte. Vom Mittelpunkt der Anlage aus gesehen weist der Konstruktionsplan eine Symmetrie auf, die mit Auf- und Untergangsrichtungen der Sonne zur Tag- und Nachtgleiche sowie zur Sommer- und Wintersonnenwende in Verbindung steht.

Aufgrund des relativ flachen Geländes lassen sich Himmelsereignisse in Horizontnähe beobachten. Das Horizontprofil ist aber nicht überall flach, da das Gelände in Richtung Osten leicht ansteigt. Deswegen wurde ein Horizontprofil aus einem digitalen Oberflächenmodell mit 30 m Horizontalauflösung berechnet, wobei die Höhe des Beobachters 3 m üNN beträgt und von einer Augenhöhe von 1.5 m über Grund ausgegangen wurde. Die Topographie wurde bis zu einer Entfernung von 25 km berücksichtigt, da in noch größerer Entfernung keine weiteren Berge vorhanden sind (Abb. 13.8).

Beim Anblick des rekonstruierten Hauses in Trelleborg kam mir ursprünglich die Idee, dass die Giebel der Häuser ein ideales Visierkreuz darstellen und für die astronomische Beobachtung verwendet werden könnten (Abb. 13.5a). Dazu müsste ein Beobachter auf dem Wall stehen. Allerdings wird die Rekonstruktion des Hauses von Trelleborg heutzutage als veraltet angesehen und die Giebel könnten auch anders gebaut worden sein, z. B. so wie in Fyrkat (Abb. 13.5b).

Ein anderes Problem bei der Suche nach astronomischen Visierlinien ist die große Anzahl von 59 Giebelpunkten. Dabei besteht die Gefahr, dass man zufällig immer irgendwelche Visierlinien finden könnte. Außerdem mussten die Erbauer zuallererst einen Grundriss der Anlage abgesteckt haben, bevor die Häuser gebaut werden konnten. Deswegen werden im Folgenden nur die 16 Achspunkte im Inneren des Ringwalls

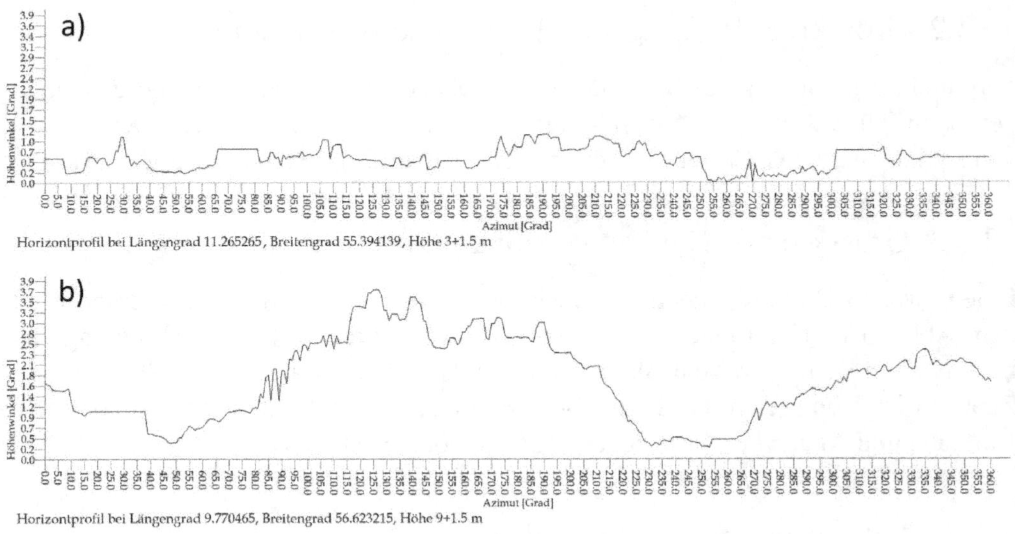

Abbildung 13.8:
Horizontprofil von a) Trelleborg und b) Fyrkat, berechnet aus einem digitalen Oberflächenmodell mit 30 m Horizontalauflösung

(Graphik: Andreas Fuls)

untersucht. Damit soll geklärt werden, ob das System der Achsen auf astronomische Beobachtungen beruht.

In Trelleborg liegen genaue Koordinaten vor [Nørlund (1948), S. 182–183], aus denen sich die Visierlinien der Auf- und Untergänge für Sonne und Mond berechnen lassen. Allerdings muss das lokale Koordinatensystem in das Landessystem transformiert und anschließend die Meridiankonvergenz berücksichtigt werden (S. 268).

Die genaueste Visierlinie von Trelleborg ist von Punkt 143 über den Mittelpunkt hinweg nach Punkt 112 (Abb. 13.9a). Das Azimut von 359.96397° zeigt nahezu exakt nach Norden. Auf eine Entfernung von 88.03 m entspricht das einer Querabweichung von 55 mm. Vom Schwerpunkt der Anlage und Mittelpunkt 100 zum Achsenpunkt 112 beträgt das Azimut sogar 359.99783°, weicht also nur um -0.00217° von Norden ab (Querabweichung von 2 mm auf 44.03 m).

Senkrecht dazu verläuft die Achse 131 – 100 – 124 nahezu Ost-West (Azimut 90.3665° bzw. 270.3665°). Um die Deklination der Sonne zu berechnen, muss für den jeweiligen Höhenwinkel einer Visur, abgeleitet aus dem Höhenprofil, der Einfluss

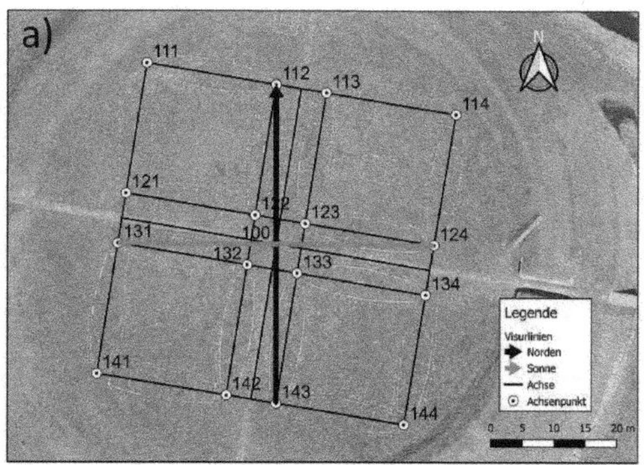

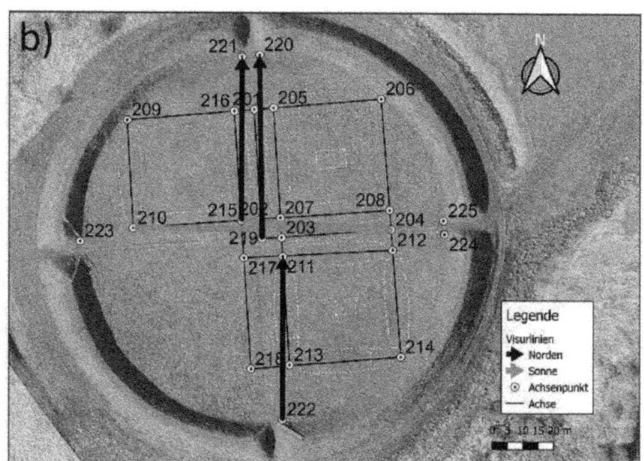

Abbildung 13.9:
Visierlinien zur Nordrichtung sowie Tag- und Nachtgleiche in
a) Trelleborg (Tab. 13.3), b) Fyrkat (Tab. 13.5)

(Graphik: Andreas Fuls)

der Refraktion berücksichtigt werden[4]. Die Refraktion ist allerdings in Horizontnähe nur ungefähr bekannt.

Scheinbare Höhe	Wahre Höhe	Sichtbarkeit
-0.27°	-0.85°	Oberkante Sonne
+0.00°	-0.58°	Mitte der Sonne, Sterne
+0.27°	-0.27°	Unterkante Sonne

Die Achse 131–124 im Grundriss ergibt demnach eine Deklination von -0.09° bei Sonnenaufgang bzw. -0.07° bei Sonnenuntergang (Abb. 13.9, Tabelle 13.3).

Bei Beobachtungen des Mondes in Horizontnähe ist außerdem die Horizontalparallaxe von etwa 1° zu beachten. Mit der wahren Höhe, der Azimutrichtung und der Breite kann nun die Deklination ausgerechnet werden. Je nachdem, ob Oberkante, Mitte oder Unterkante der Sonne beobachtet wurden, erhält man drei Werte für die Deklination. Sie können mit dem Sollwert für das Himmelsobjekt verglichen werden.

Die Schiefe der Ekliptik betrug im Jahr 980 n. Chr. 23.5714°. Dies entspricht der maximalen Deklination der Sonne zur Zeit der Sommersonnenwende. Zur Wintersonnenwende beträgt die minimale Deklination entsprechend -23.5714°.

Für den Mond gibt es zwei markante Ereignisse im Laufe eines Zyklus von 18.61 Jahren, die Große Mondwende bei maximaler bzw. bei minimaler Deklination. Sie werden entsprechend der extremen Azimute bei Auf- und Untergang als nördliche

4 Siehe Steinrücken (2011): Sonnenwenden und Mondwenden.

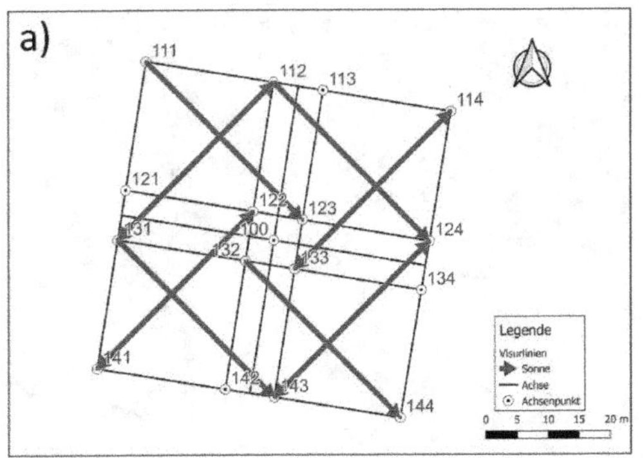

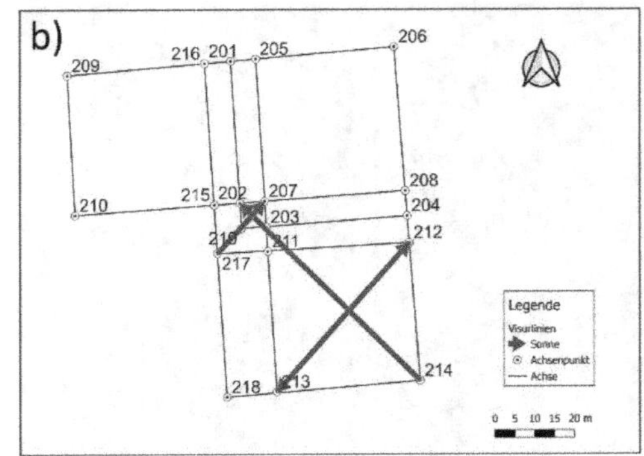

Abbildung 13.10:
Visierlinien zur Sommer- und Wintersonnenwende in
a) Trelleborg (Tab. 13.3), b) Fyrkat (Tab. 13.5)

(Graphik: Andreas Fuls)

bzw. südliche große Mondwende bezeichnet. Die Bahnneigung des Mondorbits zur Ekliptik schwankt im Laufe eines halben Finsternisjahres um ±0.15° und beträgt im Mittel 5.1467° [Ginzel (1906), S. 31, 37]. Die Extremwerte der Deklination für den Mond liegen demnach im Mittel bei ±28.7181°, können aber bis zu ±28.8681° groß werden.

Als kleine Mondwenden werden die Zeitpunkte bezeichnet, an denen die Deklinationswerte des Mondes vom Betrag her ein Minimum erreichen. Die Deklination des Mondes bewegt sich dann im Mittel um ±18.4247° und schwankt zwischen ±18.2747° und ±18.5747°.

Die astronomischen Visierlinien von Trelleborg sind in Tabelle 13.3 und 13.4 zusammengefasst. Hauptaugenmerk bei den astronomischen Visierlinien scheint auf die Sonnenwende gelegt worden zu sein (Abb. 13.10a). Es wurde die Unterkante der Sonne bei Aufgang und bei Untergang die Oberkante, also das allerletzte Aufblitzen der Sonne vor Beginn der Dämmerung, beobachtet.

Eine große Mondwende fand u. a. im Jahr 982 n. Chr. statt, es sind aber keine entsprechenden Visierlinien im Grundriss von Trelleborg erkennbar. Eine kleine Mondwende war z. B. von Ende 972 bis Anfang 974 n. Chr. zu beobachten, also wenige Jahre von dem Bau von Trelleborg um 980/981 n. Chr. Im Achsensystem von Trelleborg sind tatsächlich entsprechende Visierlinien für den Monduntergang zur kleinen nördlichen und zur kleinen südlichen Mondwende gefunden worden, wobei nur Visurlinien über 50 m Länge berücksichtigt sind (Tab. 13.4). Die längste dieser Visurlinien verläuft diagonal über das Achsensystem von Punkt 114 nach 141 und ist 122.8 m lang

Tabelle 13.3:
Visurlinien in Trelleborg zur Sonne.
Beobachtet wurde die Oberkante (o), Mitte (m) bzw. Unterkante (u) der Sonne: Frühlingsanfang (FA), Sommersonnenwende (SSW), Herbstanfang (HA) und Wintersonnenwende (WSW) bei Sonnenaufgang (SA) bzw. Sonnenuntergang (SU). Eine Abweichung zum exakten Zeitpunkt des Ereignisses ist in ganzen Tagen angegeben.

Von	Nach	Azimut [Grad]	Höhe [Grad]	Objekt	Deklination [Grad]	Ereignis
131	112	45.1349	0.26	Sonne (u)	23.65	SSW SA
133	114	45.1395	0.26	Sonne (u)	23.65	SSW SA
141	122	45.1183	0.26	Sonne (u)	23.65	SSW SA
143	124	45.1063	0.26	Sonne (u)	23.66	SSW SA
131	124	90.3665	0.40	Sonne (u)	-0.05	FA/HA SA
124	131	270.3665	0.00	Sonne (u)	-0.02	FA/HA SU
124	131	270.3665	0.00	Sonne (o)	-0.51	FA-1/HA+1 SU
111	123	135.1136	0.55	Sonne (u)	-23.40	WSW±6 SA
112	124	135.2812	0.53	Sonne (u)	-23.49	WSW±5 SA
131	143	135.1372	0.55	Sonne (u)	-23.41	WSW±6 SA
132	144	135.3222	0.52	Sonne (u)	-23.52	WSW SA
112	131	225.1349	0.77	Sonne (o)	-23.57	WSW SU
114	133	225.1395	0.77	Sonne (o)	-23.57	WSW SU
122	141	225.1183	0.77	Sonne (o)	-23.58	WSW SU
124	143	225.1063	0.77	Sonne (o)	-23.58	WSW SU

(Abb. 13.11a). Dabei zeigt die Analyse, dass bei Monduntergang zur südlichen kleinen Mondwende die Mitte der Mondscheibe verwendet wurde, aber zur nördlichen kleinen Mondwende die Oberkante des Mondes angezielt wurde.

13.5 Astronomie von Fyrkat

Bei der Untersuchung des Grundrisses von Fyrkat wurde in gleicher Weise vorgegangen wie bei Trelleborg. Der einzige Unterschied ist, dass Fyrkat im südwestlichen Viertel nicht ausgegraben wurde. Zwei Achsenpunkte in diesen Bereich sind deswegen mit Hilfe der anderen Achsen rekonstruiert worden, wobei auf eine kurze Entfer-

Tabelle 13.4:
Visurlinien in Trelleborg zum Mond. Beobachtet wurde die Oberkante (o), Mitte (m) bzw. Unterkante (u) des Mondes: Nördliche (n) und südliche (s) kleine Mondwende (KMW) bei Mondaufgang (MA) bzw. Monduntergang (MU).

Von	Nach	Azimut [Grad]	Höhe [Grad]	Objekt	Deklination [Grad]	Ereignis
134	143	234.1988	0.63	Mond (m)	-18.41	s KMW MU
150	141	234.2697	0.63	Mond (m)	-18.37	s KMW MU
114	123	234.2764	0.63	Mond (m)	-18.37	s KMW MU
132	141	234.2903	0.63	Mond (m)	-18.36	s KMW MU
124	142	234.2942	0.63	Mond (m)	-18.36	s KMW MU
114	141	234.3172	0.63	Mond (m)	-18.35	s KMW MU
114	132	234.3366	0.63	Mond (m)	-18.34	s KMW MU
112	121	234.3445	0.63	Mond (m)	-18.34	s KMW MU
123	141	234.3468	0.63	Mond (m)	-18.34	s KMW MU
114	150	234.3646	0.63	Mond (m)	-18.32	s KMW MU
113	131	234.3759	0.63	Mond (m)	-18.32	s KMW MU
124	111	302.1743	0.73	Mond (o)	18.45	n KMW MU
144	131	302.1793	0.73	Mond (o)	18.45	n KMW MU

nung zu bekannten Achsenpunkten geachtet wurde, um Extrapolationsfehler gering zu halten.

Dabei ergaben sich einige Visierlinien zur Sonne (Tab. 13.5) und zum Mond (Tab. 13.6). Wie in Trelleborg wurde in Fyrkat die Ober- bzw. Unterkante der Sonne bei Auf- und Untergang am Horizont beobachtet. Bei Mondbeobachtungen scheint nur die Oberkante verwendet worden zu sein. Da das Horizontprofil stellenweise bis zu 3.7° hoch ist, geht die Sonne bzw. der Mond entsprechend später auf bzw. früher unter (Abb. 13.8b).

Die genauesten Richtungen findet man im Grundriss von Fyrkat zur Nordrichtung. Dazu sind auch die Eckpunkte der Tore verwendet worden, welche durch Pfostenlöcher markiert waren und auf einem Kreis um den Mittelpunkt der Anlage liegen (Abb. 13.9b: Punkte 220 bis 225). Die Richtung vom Mittelpunkt der Anlage aus nach Punkt 220 an der inneren östlichen Ecke des Nordtores liegt nur 0.28° westlich von Nord.

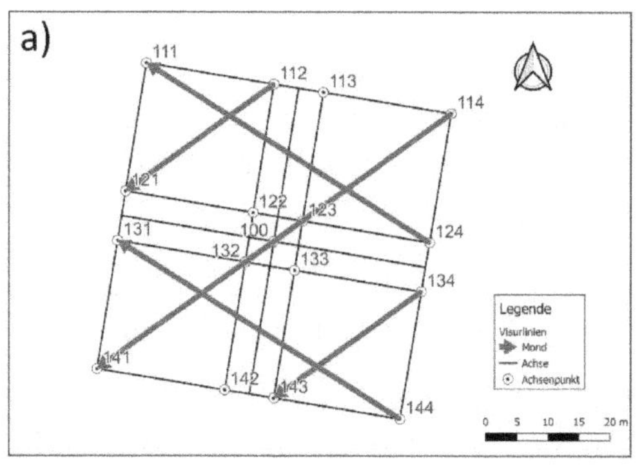

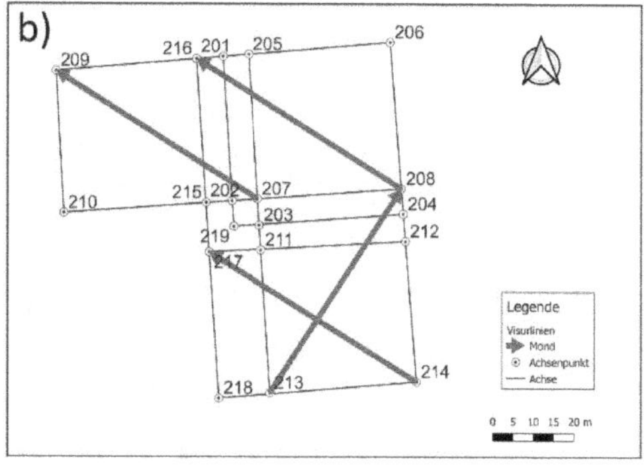

Abbildung 13.11:
Visierlinien zur den Mondwenden in
a) Trelleborg (Tab. 13.4), b) Fyrkat (Tab. 13.6)

(Graphik: Andreas Fuls)

Tabelle 13.5:
Visurlinien in Fyrkat zur Sonne.
Beobachtet wurde die Oberkante (o) bzw. Unterkante (u) der Sonne: Frühlingsanfang (FA), Sommersonnenwende (SSW), Herbstanfang (HA) und Wintersonnenwende (WSW) bei Sonnenaufgang (SA) bzw. Sonnenuntergang (SU). Eine Abweichung zum exakten Zeitpunkt des Ereignisses ist in ganzen Tagen angegeben.

Von	Nach	Azimut [Grad]	Höhe [Grad]	Objekt	Deklination [Grad]	Ereignis
217	207	42.6093	0.51	Sonne (o)	23.66	SSW SA
217	219	44.0480	0.50	Sonne (o)	23.57	SSW SA
215	204	93.2681	1.94	Sonne (u)	-0.21	FA/HA SA
208	218	221.1287	1.00	Sonne (u)	-23.68	WSW SU
207	219	221.2002	1.00	Sonne (u)	-23.65	WSW SU
212	213	221.8178	0.85	Sonne (u)	-23.56	WSW SU
204	210	270.1687	0.98	Sonne (o)	0.34	FA+1/HA-1 SU
214	202	314.3424	1.84	Sonne (o)	23.75	SSW SU

Noch exakter nach Norden zeigen die Visuren von Punkt 222 nach Punkt 211 (Azimut 0.02°) und von Punkt 215 nach Punkt 221 (Azimut 0.05°).

Tabelle 13.6:
Visurlinien in Fyrkat zum Mond. Beobachtet wurde die Oberkante (o) des Mondes: Nördliche (n) kleine Mondwende (KMW) bzw. große Mondwende (GMW) bei Mondaufgang (MA) bzw. Monduntergang (MU).

Von	Nach	Azimut [Grad]	Höhe [Grad]	Objekt	Deklination [Grad]	Ereignis
213	208	32.7791	1.05	Mond (o)	28.82	n GMW MA
214	217	301.9497	1.66	Mond (o)	18.69	n KMW MU
208	216	302.1807	1.65	Mond (o)	18.80	n KMW MU
207	209	302.2188	1.65	Mond (o)	18.82	n KMW MU

13.6 Konstruktion von Trelleborg

Um den Grundriss der Häuser von Trelleborg festzulegen, war es notwendig zuerst die Achsen der Innenhäuser festzulegen. Dazu sollen zwei mögliche Konstruktionsmethoden diskutiert werden. Die erste Methode basiert auf konstante Abstände und rechte Winkel und die zweite Methode auf astronomische Visierlinien.

13.6.1 Das verwendete Längenmaß in Trelleborg

N. E. Nørlund schlug vor, dass sich im Grundriss von Trelleborg konstante Abstände wiederfinden, die ein einheitliches Maß von 29.33 cm verwendeten [Nørlund (1948), S. 186–188]. Er leitete dies aus der mittleren Länge der Innenhäuser mit je 100 Fuß und dem Abstand der Giebelmittelpunkte zu den Achsen von 12 Fuß ab, und zwar unter der Voraussetzung, dass alle Achsen zueinander im rechten Winkel bzw. parallel konstruiert wurden. Der Ringwall hätte einen inneren Radius von 234 Fuß sowie eine Dicke von 60 Fuß. Die verwendete Maßeinheit entspricht in etwa dem römischer Fuß von 29.57 cm.

Der Konstruktionsvorschlag entspricht in etwa dem, was wir über die Konstruktion römischer Militärlager wissen. Zur Absteckung von rechten Winkel wurde von den römischen Legionären eine Groma verwendet.

Die Außenhäuser würden dann 90 Fuß lang sein und deren Achsen zeigen radial auf einen gemeinsamen Punkt, der in der Nähe des Schwerpunktes der Innenhäuser liegt.

Berechnet man allerdings die Länge der Häuser in Fyrkat, dann schwanken die Werte zwischen 28.3 m und 29.0 m und liegen im Durchschnitt bei 28.6 m. Dies entspräche 97.7 Fuß von 29.33 cm Länge und es stellt sich die Frage, warum ein solch krummes Fußmaß verwendet werden sollte.

Radial angeordnete Lagerhäuser und ein kreisförmiger Ringwall sind allerdings im römischen Militärwesen unbekannt. Um die Konstruktionsmethode und auch das Längenmaß zu überprüfen, wurde eine Helmerttransformation der Giebelpunkte durchgeführt. Der Maßstab ist tatsächlich 0.29334 m groß, bei einem Drehwinkel von -13.4796°. Allerdings beträgt der mittlere Gewichtseinheitsfehler nach der Ausgleichung bei ±0.3830 Fuß bzw. ±0.112m. Dies ergibt sich aus den Restklaffungen zwischen Soll- und Ist-Koordinaten, welche zwischen 0.21 und 0.85 Fuß liegen. Ein mittlerer Fehler von fast einen halben Fuß erscheint recht groß zu sein und es stellt sich die Frage, ob nicht eine andere Konstruktionsmethode in Frage kommt.

13.6.2 Astronomische Konstruktion des Grundrisses

Nach dem Festlegen des Mittelpunktes in Trelleborg (Punkt 100) wurde zuerst die Nord- und Südrichtung festgelegt. Die Endpunkte 112 und 143 liegen jeweils 150.09 bzw. 150.01 Fuß (à 29.334 cm) vom Mittelpunkt entfernt.

Die Richtungen zur Tag- und Nachtgleiche ergibt die Punkte 124 und 131, welche jeweils 149.68 bzw. 149.95 Fuß vom Mittelpunkt entfernt liegen. Punkt 124 könnte aber auch von Punkt 143 bei Sonnensonnenwende im Abstand von 211.02 Fuß abgesteckt worden sein. Von 112 nach 124 sind es 212.73 Fuß, eine dritte Variante, falls die Richtung zum Sonnenaufgang bei Wintersonnenwende zu Grunde liegt.

Vergleicht man die Länge der Visierlinien untereinander und wendet die Cosinus Quantogramm Methode an, dann ergibt sich als vermutliches Längenmaß, welches am besten zu allen Visierlängen passt, eine Strecke von 8.768 m. Das entspricht in etwa 30 Fuß.

13.7 Teppich von Bayeux

Der Teppich von Bayeux beschreibt die Eroberungsgeschichte von England durch die Normannen im Jahr 1066. Szene 45 zeigt den Bau eines Kastells nach der Landung der Normannen in England, um vor Angriffen geschützt zu sein, direkt vor der Schlacht bei Haistings (Abb. 13.12). Der Begleittext in lateinischer Sprache lautet:
ISTE IVSSIT VT FODERETVR CASTELLVM AD HESTENGA
(„Der da befahl, dass bei Hastings eine Befestigung errichtet werde“).

Im ersten Teil der Szene sind vier Männer mit Schaufeln dargestellt, die zwischen zwei Fahnenträgern stehen. Dieser Teil der Szene wird oftmals als Prügelei interpre-

Abbildung 13.12:
Darstellung des Baus eines Kastells auf dem Teppich von Bayeux (Szene 45)

(Foto: Official Bayeux Tapestry digital representation, Eleventh century.
Credentials: City of Bayeux, DRAC Normandie, University of Caen Normandie,
CNRS, ENSICAEN; Pictures: 2017 – La Fabrique de patrimoines en Normandie)

tiert. Bei genaueren Hinsehen sieht man allerdings, wie der dritte Mann von links mit den Augen schräg nach oben schaut, möglicherweise, um eine Peilung vorzunehmen, während die zweite Person über die beiden Schaufeln schaut, welche kreuzförmig gehalten werden. Im Kontext des Baus eines Kastells erscheint mir deswegen dieser Teil der Szene eher das Ausrichten der Burganlage darzustellen. Wonach gepeilt wurde ist nicht dargestellt. Der eigentliche Bau ist dann im zweiten Teil der Szene zu sehen. Innerhalb des Kastells steht das Wort CAESTRA, welches einfarbig in Rot gestickt ist, um es von dem anderen Text mit abwechselnd roten und schwarzen Buchstaben abzugrenzen. Dessen Bedeutung ist unsicher, bezeichnet vermutlich das Kastell.[5]

13.8 Zusammenfassung

Die Ringburgen Nonnebakken, Trelleborg und Borgring liegen nahe an einen Breitengrad, bei dem die Auf- und Untergangsrichtungen zu Sommer- und Wintersonnenwende untereinander einen rechten Winkel bilden. Sie liegen nur wenige Kilometer

5 Zur Verwendung des Wortes CAESTRA siehe z. B. in der gnomischen Dichtung *Maximes II*: ... Ceastra beoð feorran gesyne, ... („Die Städte [Burgen?] sind von fern zu sehen [?], ...") s. [Göller (1971), S. 64–65])

davon entfernt. Die Intension die Ringburgen mit Sonnenbeobachtungen im Beziehung zu setzen, lässt sich an den Grundrissen von Trelleborg und Fyrkat nachweisen. Bei den anderen Ringburgen konnte aufgrund fehlender Daten keine Detailanalyse der Grundrisse durchgeführt werden.

Die Untersuchung des Grundrisses von Trelleborg hat gezeigt, dass die Achsenpunkte so ausgewählt wurden, dass man von ihnen aus gesehen die Sonne bei Auf- und Untergang zu den Sonnenwenden und der Tag- und Nachtgleiche beobachten hatte. Wie die genaue Nordachse damals festgelegt wurde ist nicht näher bekannt, da der Polarstern im 10. Jahrhundert zu weit von geographisch Nord abwich.

Peilrichtungen zu den anderen Ringburgen im Grundriss von Trelleborg ermöglichten eine Orientierung für die Navigation über große Entfernungen. Eine historische Bestätigung, dass im 10. und 11. Jahrhundert es unter den Wikingern bzw. Normannen üblich war Kastelle nach bestimmten Richtungen zu orientieren findet man auf dem Teppich von Bayeux dargestellt.

13.9 Literatur

[Åkerlund (1974)] ÅKERLUND, HARALD: *Alnar och Fot berättar om forntida bosättning i Väst- och Nordeuropa – och i Amerika – av folk från Främre Orienten. En metrologisk studie.* [Elle und Fuß erzählen uns von der antiken Besiedlung West- und Nordeuropas – und Amerikas – durch Menschen aus dem Nahen Osten. Eine metrologische Studie]. Lidingö: Författaren 1974 (schwedisch).

[Ginzel (1906)] GINZEL, KARL FRIEDRICH: *Handbuch der mathematischen und technischen Chronologie. Band 1.* Leipzig: J. C. Hinrichs'sche Buchhandlung 1906.

[Göller (1971)] GÖLLER, KARL HEINZ: *Geschichte der altenglischen Literatur.* Berlin: Erich Schmidt Verlag (Grundlagen der Anglistik und Amerikanistik, Band 3) 1971.

[Goodchild, Holm & Sindbæk (2017)] GOODCHILD, HELEN; HOLM, NANNA & SØREN M. SINDBÆK: *Borgring: the discovery of a Viking Age ring fortress.* In: *Antiquity* **91** (2017), 358, S. 1027–1042.

[Ingvorsen (o. J.)] INGVORSEN, LEIF: *Jelling in der Wikingerzeit.* Jelling: Bogtrykkeris Forlag o. J.

[Jensen (1986)] JENSEN, LARS BO: *Aggersborg og omegn i vikingetiden.* [Aggersborg und seine Umgebung in der Wikingerzeit]. Løgstør: Limfjordsmuseet 1986 (dänisch).

[Klose (1982)] KLOSE, OLAF: *Handbuch der historischen Stätten. Dänemark.* Stuttgart: Kröner Verlag 1982.

[Kristiansen (2022)] KRISTIANSEN, SØREN M.; STOTT, DAVID; CHRISTIANSEN, ANDERS VEST; HENRIKSEN; PETER STEEN; JESSEN, FISCHER MORTENSEN, CATHERINE MORTEN JESPER; PEDERSEN, SINDBÆK, BJERGSTED SØREN MICHAEL & JENS ULRIKSEN: Non-destructive 3D prospection at the Viking Age fortress Borgring, Denmark. In: *Journal of Archaeological Science: Reports* **42** (2022), S. 1–11.

[Nørlund (1948)] NØRLUND, N. E.: Trelleborgs opmaaling og den benyttede længdeenhed. [Trelleborgs Vermessung und die benutzte Längeneinheit]. Kapitel IX. In: NØRLUND, POUL: *Trelleborg*. København: Nordiske Forlag (Nordiske Fortidsminder, Band IV) 1948, S. 181–188 (dänisch).

[Nørlund (1968)] NØRLUND, POUL: *Trelleborg*. København: Nationalmuseet (Nationalmuseets blå bøger) 1968 (englisch).

[Olsen (1970)] OLSEN, OLAF: *Fyrkat. Ein Wikingerlager in Jütland.* København: Nationalmuseet 1970.

[Olsen (1977)] OLSEN, OLAF: *Fyrkat – En jysk vikingeborg. I: Borgen og bebyggelsen af Olaf Olsen og Holger Schmidt. II: Oldsagerne og gravpladsen af Else Roesdahl.* København: Det Kgl. nordiske Oldskriftselskab (Nordiske Fortidsminder, Serie B, Band 3) 1977 (dänisch).

[Pásztor & Roslund (2001)] PÁSZTOR, EMÍLIA & CURT ROSLUND: Orientation of Danish geometrical Viking fortresses. In: BELMONTE, JUAN A. (ed.): *Astronomy and Cultural Diversity.* Oxford VI. and SEAC 99, Belgrade, Serbia: La Laguna 2001, S. 65–69.

[Weibull (1974)] WEIBULL,CURT: *Die dänischen Trelleburgen.* Göteborg: Kungl. Vetenskaps- och Vitterhets-Samhället 1974.

Abbildung 14.1:
Rundbogennische mit Sonnenloch im Nordturm des Erfurter Doms.
Der Blick durch das Sonnenloch ist versperrt durch den angebauten gotischen Chor.

(Foto: Burkard Steinrücken)

Das Lichtereignis in der Nordturmkapelle im Erfurter Mariendom

Burkard Steinrücken (Recklinghausen)

Abstract: The Light Event in the North Tower Chapel in Erfurt Cathedral

In the north tower of the Erfurt Cathedral there is a Chapel, dedicated to the Virgin Mary, with a round arched niche and a round light opening (Fig. 14.1) aligned with the axis of the Romanesque cathedral from 1150–1160. This opening was already covered by the addition of the Gothic choir around 1350, so that the original light event, the incidence of sunlight through the round opening and the backlighting of the head of a Marian figure, can no longer be observed since then. The figure of Mary from around 1160 has been preserved together with the altar retable that surrounded the round-arched niche and is exhibited in a new location in Erfurt Cathedral [Möller 1996]. The cosmic symbolism of the altar retable underlines the intended effect of relating the Sun to the Virgin Mary (Fig. 14.2). The north tower chapel, like the entire Romanesque cathedral, is aligned to the sunrise around August 15 – *Assumption Day* [Sareik (1986)]. In this contribution, the geometry of the complex, its orientation to the horizon line formed by the Ettersberg between Erfurt and Weimar, is re-analysed and the relevant sunrise paths in the construction period of the Romanesque cathedral and the Gothic choir extension are visualized (Fig. 14.3). The question of interest is whether the alignment of the cathedral in Erfurt to the sunrise on the Assumption of Mary was done by observation or by calculation, and whether conclusions can be drawn about one of these possibilities from the analysis of the accuracy.

The Erfurt north tower chapel is also relevant with regard to a possibility of comparison with the similarly designed round-arched niche in the sacellum of the *Externsteine* [Steinrücken 2018], which was and is interpreted in a long, controversial and ideologically colored interpretation tradition on the one hand as a pre-Christian celestial sanctuary and on the other hand as a Christian hermitage.

Abbildung 14.2:
Marienfigur und Altarretabel der Rundbogennische von 1160;
jetzt ausgestellt im Erfurter Dom

(Foto: Burkard Steinrücken)

Zusammenfassung

Im Nordturm des Erfurter Doms befindet sich eine Marienkapelle mit einer Rundbogennische und einer runden, in der Achse des romanischen Doms von 1150–1160 orientierten Lichteinfallsöffnung (Abb. 14.1). Durch den Anbau des gotischen Chores ist diese Öffnung bereits in der Zeit um 1350 verdeckt worden, so dass das ursprüngliche Lichtereignis, der Einfall des Sonnenlichtes durch die runde Öffnung und die Hinterleuchtung des Kopfes einer Marienfigur seitdem nicht mehr beobachtet werden kann. Die Marienfigur aus der Zeit um 1160 ist, zusammen mit dem Altarretabel, das die Rundbogennische umgab, erhalten und an neuer Stelle im Erfurter Dom ausgestellt [Möller 1996]. Die kosmische Symbolik des Altarretabels unter-

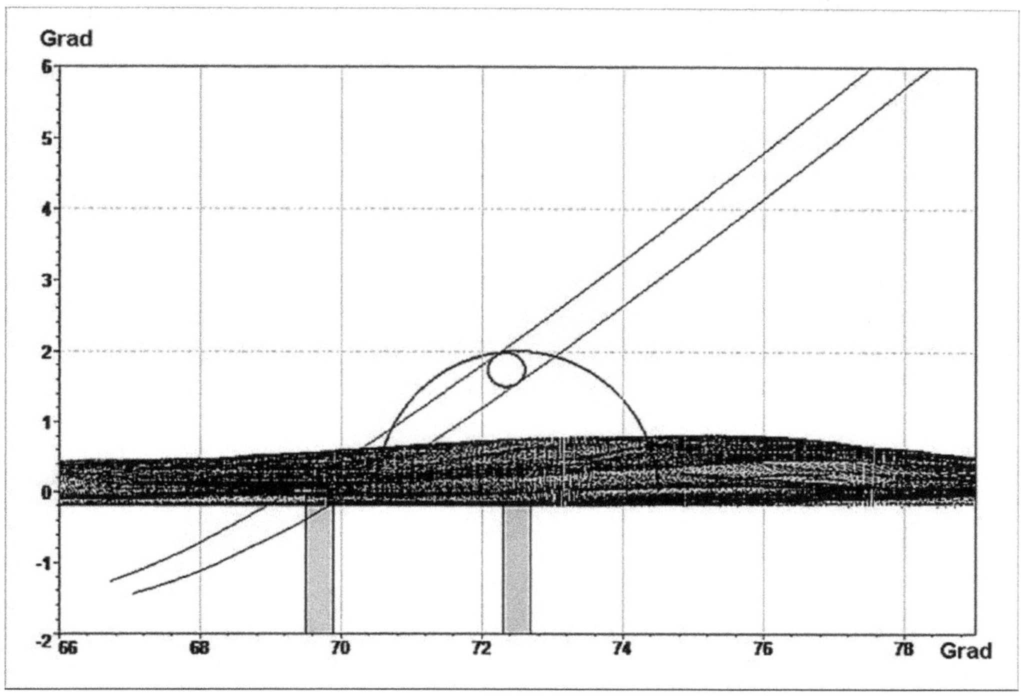

Abbildung 14.3:
Simulation des Sonnenaufgangs am 15. August 1150 (Mariä Himmelfahrt) bei der Betrachtung durch das Sonnenloch

Aufgetragen sind die Horizontlinie und die Bahn der aufgehenden Sonne im Horizontsystem (Elevation gegen Nordazimut). Der Ettersberg (bei Nordazimuten von ca. 70° bis 78°) stört das Ereignis nicht. Eine gute Ausleuchtung der gesamten Wandung des Sonnenlochs (und damit der Effekt eines Heiligenscheins um das Haupt der Marienfigur) wird bei Sonnenständen erzielt, die nicht um mehr als 2° von der Mittelachse abweichen (roter Halbkreis). Die Ausrichtung des Sonnenlochs (angezeigt durch den rosa Balken unterhalb der Horizontlinie; auch Ausrichtung des romanischen Doms) beträgt $72,5^\circ \pm 0,2^\circ$, ermittelt aus Luftbildkarten der Thüringer Landesvermessung im Internet. Der graue Balken links daneben markiert die Ausrichtung des später angebauten gotischen Chors von $69,7^\circ \pm 0,2^\circ$.

(Grafik: Burkard Steinrücken)

streicht die beabsichtigte Wirkung einer Inbeziehungsetzung der Sonne mit der Gottesmutter Maria (Abb. 14.2). Die Nordturmkapelle wie auch der gesamte romanische Dom ist ausgerichtet auf den Sonnenaufgang in den Tagen um den 15. August – *Mariä Himmelfahrt* [Sareik (1986)]. In diesem Beitrag werden die Geometrie der Anlage, ihre Ausrichtung auf die Horizontlinie, die durch den Ettersberg zwischen Erfurt und Weimar gebildet wird, neu analysiert und die relevanten Sonnenaufgangsbahnen in der Bauzeit des romanischen Doms und des gotischen Choranbaus visualisiert (Abb. 14.3). Dabei ist die Frage von Interesse, ob die Domausrichtung in Erfurt auf den Sonnenaufgang an Mariä Himmelfahrt durch Beobachtung oder durch Berechnung erfolgte und ob sich aus der Analyse der Genauigkeit Rückschlüsse auf eine dieser Möglichkeiten gewinnen lassen.

Die Erfurter Nordturmkapelle ist auch von Relevanz hinsichtlich einer Vergleichsmöglichkeit mit der ähnlich gestalteten Rundbogennische im Sazellum der *Externsteine* [Steinrücken 2018], die in langer kontroverser und ideologisch gefärbter Deutungstradition einerseits als vorchristliches Gestirnsheiligtum als auch andererseits als christliche Eremitage gedeutet wurde und wird.

14.1 Literatur

MÖLLER, ROLAND: Zur Farbigkeit mittelalterlicher Stuckplastik. In: EXNER, MATTHIAS (Hg.): *Stuck des frühen und hohen Mittelalters – Geschichte, Technologie, Konservierung.* München (ICOMOS – Hefte des Deutschen Nationalkomitees; Bd. 19) 1996.

SAREIK, UDO: Angewandte Astronomie im Mittelalter: Die Lichtöffnungen am Erfurter Dom und an der Klosterkirche zu Veßra. In: *Die Sterne* **62** (1986), 5, S. 284–292.

STEINRÜCKEN, BURKARD: Archäoastronomie der Externsteine – Neuunterschung der mutmaßlichen astronomischen Peilungen, Analyse der Forschungsliteratur und archäoastronomische Deutungsmöglichkeiten. In: EIKERMANN, LARISSA, HAUPT, STEFANIE, LINDE, ROLAND & MICHAEL ZELLE (Hg.): *Die Externsteine zwischen wissenschaftlicher Forschung und völkischer Deutung.* Münster: Aschendorf Verlag (Veröffentlichungen der Historischen Komission für Westfalen; 31) 2018, S. 223–266.

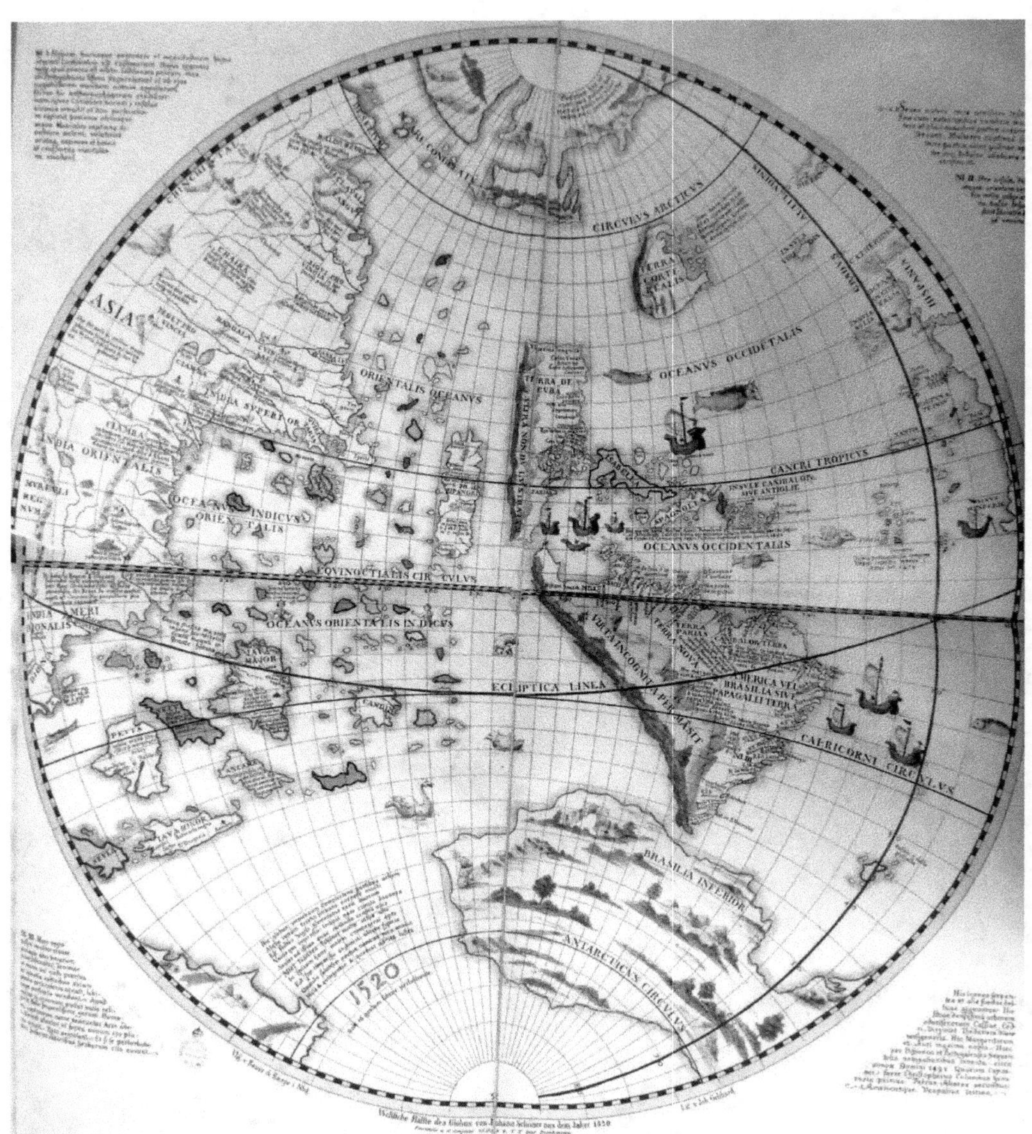

Abbildung 15.1:
Globus von Johannes Schöner (1515), Westliche Hemisphäre

(CC)

Kosmische Lebenswelten auf Himmels- und Erdgloben in Weimar und Gotha

Harald Gropp (Heidelberg)

Abstract: Cosmic life worlds on celestial and terrestrial globes in Weimar (and Gotha)

This paper is a follow-up to my talk in Gilching 2019 [Gropp 2020] on cosmovision and imaginary celestial worlds. Here the focus is on globes in the first third of the 16th century. The most important globe maker was Johannes Schöner (1477–1547) [Maruska 2008]. The connection to Thuringia is obvious where important globes are in possession of the *Stiftung Klassik Weimar, Herzogin Anna Amalia Bibliothek*, and of Castle Friedenstein in Gotha.

In 2023, we celebrate the 500th anniversary of the 1523 globe, which had been unknown for a long time, and was first "postulated" in 1881 [Wieser 1881]. The Schöner globe of 1515 (today copies are in Weimar and Frankfurt am Main) is the oldest printed terrestrial globe still in existence. In Weimar, there is also a pair of Schöner globes (1533). The pair of a terrestrial and a celestial globe shows the cosmic world of man and its dramatic change 500 years ago. In particular, the importance of Amerigo Vespucci's travels voyages will be considered [Omodeo 2020].

Last but not least, a globe in Thuringia is discussed, namely the marble globe of Gotha [Horn 1976], not much investigated in papers, difficult to attribute, but almost certainly more or less 500 years old. An attribution to Johannes Schöner does not seem very probable.

Zusammenfassung

In gewisser Weise als Fortsetzung meines Vortrags von Gilching 2019 [Gropp 2020] über Kosmovision und imaginäre Himmelswelten stehen diesmal im Fokus die Lebenswelten auf Globen im ersten Drittel des 16. Jahrhunderts. Der wichtigste Globenhersteller war Johannes Schöner (1477–1547) [Maruska 2008, Wolfschmidt 1978]. Für diesen Vortrag ist eine Verbindung zu

Thüringen der Anlass, da sich wichtige Globen in Weimar (*Stiftung Klassik Weimar, Herzogin Anna Amalia Bibliothek*) und in Gotha (Schloss Friedenstein) befinden.

Im Jahre 2023 feiern wir das 500. Jubiläum des Globus von 1523, der lange verschollen war und erstmals 1881 „postuliert" wurde [Wieser 1881]. Der Schönerglobus von 1515 (heute Weimar und Frankfurt am Main) ist der älteste gedruckte Erdglobus, der noch existiert. Ausserdem befindet sich in Weimar ein Globenpaar von Schöner (1533). Das Paar eines Erd- und eines Himmelsglobus zeigt in besonderer Weise die kosmische Lebenswelt der Menschen und ihre dramatische Veränderung vor 500 Jahren. Insbesondere wird untersucht, wieweit Amerigo Vespuccis Reisen dabei wichtig sind [Omodeo 2020].

Last but not least wird ein Globus in Thüringen diskutiert, nämlich der Marmorglobus von Gotha [Horn 1976], wenig beachtet, kaum besprochen, schwierig zuzuordnen, mit grosser Sicherheit aus der Zeit vor 500 Jahren. Eine Zuschreibung zu Johannes Schöner scheint mir wenig wahrscheinlich.

15.1 Literatur

GROPP, HARALD: Phantastische Inseln und imaginäre Himmelswelten – Kosmovision auf dem Okeanos. In: WOLFSCHMIDT, GUDRUN (Hg.): *Himmelswelten und Kosmovisionen –Imaginationen, Modelle, Weltanschauungen.* Proceedings der Tagung der Gesellschaft für Archäoastronomie in Gilching 2019. Hamburg: tredition (Nuncius Hamburgensis; Band 51) 2020, S. 264–279.

HORN, WERNER: *Die alten Globen der Forschungsbibliothek und des Schloßmuseums Gotha.* Gotha (Veröffentlichungen der Forschungsbibliothek Gotha; 17) 1976.

MARUSKA, MONIKA: *Johannes Schöner – Leben und Werk eines fränkischen Wissenschaftlers an der Wende vom 15. zum 16. Jahrhundert.* Dissertation, Universität Wien 2008.

OMODEO, PIETRO: *Amerigo Vespucci: The Historical Context of his Explorations and Scientific Contribution.* Venezia: Edizioni Ca'Foscari (Knowledge Hegemonies in the Early Modern World 1) 2020.

WIESER, FRANZ: *Magalhães-Strasse und Austral-Continent auf den Globen des Johannes Schöner.* Innsbruck: Verlag der Wagner'schen Universitaets-Buchhandlung (Beiträge zur Geschichte der Erdkunde im XVI. Jahrhundert) 1881.

WOLFSCHMIDT, GUDRUN: Johann Schöner – ein fränkischer Geograph und Astronom. In: *Sterne und Weltraum* **17** (1978), S. 86–90.

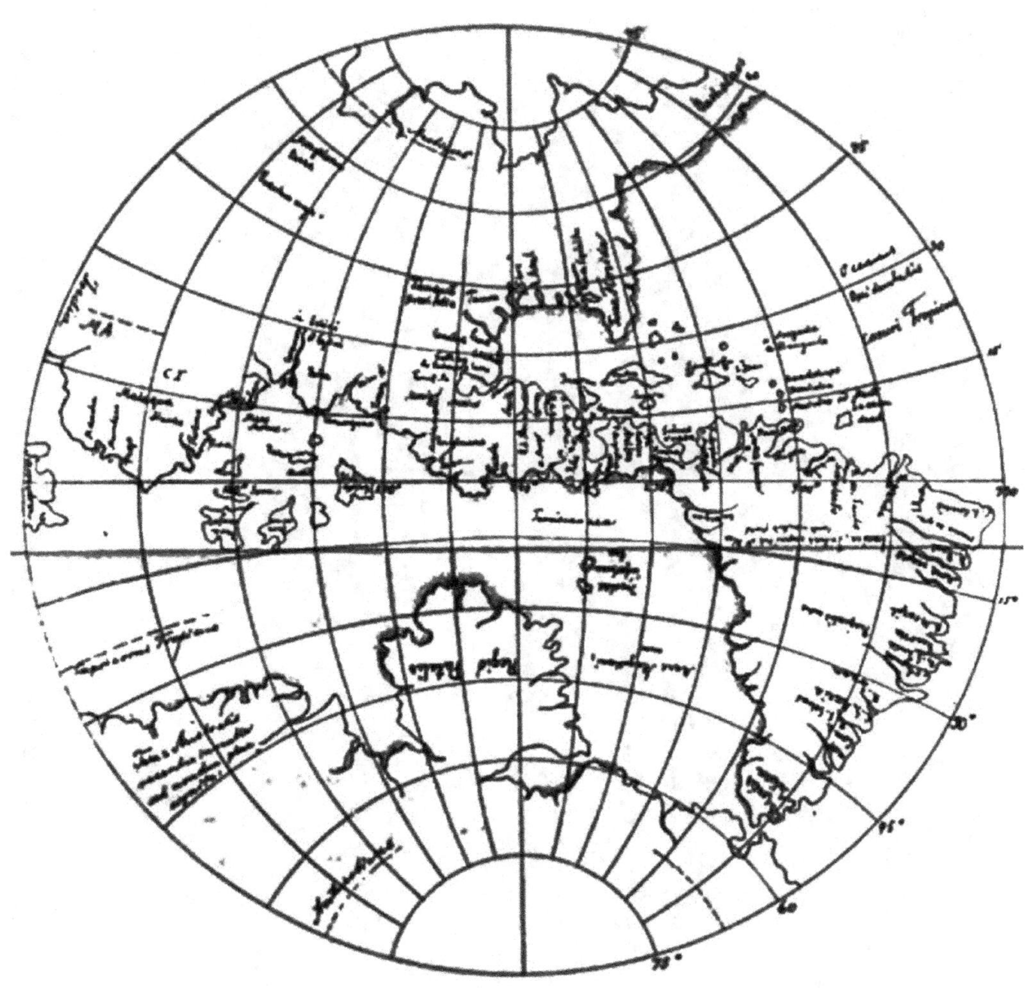

Abbildung 15.2:
Weimarer Globus von Johannes Schöner (1533)

Abbildung 16.1:
Prunkkronleuchter im Blauen Salon

(Foto: Louis Held, um 1912)

Ist das Römische Haus im Park an der Ilm in Weimar astronomisch verankert? Teil II: Vertiefung der archäoastronomischen Befunde

Ettore Ghibellino (Weimar)

„Let all rays, bundles of rays, clusters of rays, ray brushes and twines of rays continue to flow if one wants to attain the pure perception of the phenomenon."'
Goethe: Theory of Colours.

Abstract: Is the Roman House in the Park on the Ilm in Weimar astronomically anchored? Part II: Deepening the archaeoastronomical findings

The sub-discipline of archaeoastronomy is usually associated with the astronomical anchoring of buildings in prehistoric and ancient times. However, there is much to suggest that this knowledge was never really lost among architects and civil engineers, but was rather passed on. This is borne out, for example, by the orientation towards the east of church buildings over thousands of years. For the first time, the question of whether this ancient knowledge was also used at the beginning of the modern age will be examined using the example of the Roman House in the Park on the Ilm in Weimar (1791–1798) built by Johann Wolfgang von Goethe.

Zusammenfassung

Mit der Teilbereichsdisziplin der Archäoastronomie verbindet man gewöhnlich die astronomische Verankerung von Bauwerken in der Prähistorik und in der Antike. Viel spricht indes dafür, dass dieses Wissen unter Architekten und Bauingenieuren nie wirklich verloren ging, vielmehr weiter tradiert wurde. Dafür spricht zum Beispiel die Ostung von Kirchenbauten über Jahrtausende. Erstmals wird der Frage nachgegangen, ob auch zu Beginn der Moderne dieses antike Wissen Verwendung fand am Beispiel des von Johann Wolfgang von Goethe erbauten Römischen Hauses im Park an der Ilm in Weimar (1791–1798).

> *„Man lasse alle Strahlen, Strahlenbündel, Strahlenbüschel, Strahlenpinsel, Strahlenzwirn nur immerhin fahren, wenn man zu der reinen Anschauung des Phänomens gelangen will.“*[1]
> Goethe: Farbenlehre.

16.1 Vorgeschichte

Im Jahre 2019 durfte der Verfasser anlässlich der Tagung der Gesellschaft für Archäoastronomie in Gilching seine Studie „Das Römische Haus als ‚Geheimster Wohnsitz'“ (Weimar 2020) vorstellen, genauer: den Teil, der die astronomische Verankerung behandelt.[2]

Die Grundannahme ist, dass das Römische Haus auf drei Ebenen astronomisch verankert ist: Das Gebäude ist geostet, die Sonne steigt bei Aufgang am 28. August dem Anschein nach aus dem Schornsteinkopf auf. In der Cella im Erdgeschoss wird im Frühling und im Sommer eine Lichtsymphonie vor dem Gemälde ‚*Anna Amalia in Rom*' aufgeführt: Ein durch große Spiegel potenzierter Kronleuchter orchestriert Brechungen der Sonnenstrahlen als funkenstobender Lichtreigen. Der untere Durchgang stellt schließlich eine monumentale Sonnenuhr mit Öffnungen im Opaion-Maßstab des Pantheons dar. Markante Effekte sind die Sonnenwenden sowie die Äquinoktien. Der zweite Teil der Studie wurde ausgelassen: Die poetische Verankerung des Römischen Hauses, wo anhand unabhängiger Quellen nach dem Kontext der astronomischen Verankerung am Bauwerk, deren Funktion sowie der Verifizierbarkeit der archäoastronomischen Interpretation gefragt wird.[3]

---

1 Weimarer Ausgabe. Weimar 1906, Paralipomena zum didaktischen Teil II, 5, S. 61.

2 Abgedruckt in: Ghibellino: Goethes Testament ist mit Sonnenstrahlen geschrieben: Das Römische Haus im Park an der Ilm zu Weimar. Teil I: Archäoastronomische Befunde, 2020, S. 206–221.

3 Zur Methodik siehe Fuls: Archäoastronomische Methodik bei Baudenkmälern, 2018, S. 158, S. 160.

16.2 Tagung in Weimar 2023

Zur Sommersonnenwende vom 21. bis 25.06.2023 fand die Jahrestagung der *Gesellschaft für Archäoastronomie* in Weimar statt. Erstmals konnte die Genehmigung erwirkt werden, an drei Tagen am frühen Morgen zum Sonnenaufgang in das Gebäude zu gelangen und den Verlauf der Sonnenstrahlen In der Cella im Erdgeschoss zu beobachten. Beide anderen Lichtinszenierung können von außen beobachtet werden.

Entsprechend konnte im Sonnenobservatorium im unteren Durchgang die auftreffende Lichtstrahlfläche bei Sonnenaufgang genau in der linken Hälfte der Westwand unterhalb des Musentanzes beobachtet werden (Bild 16.2).

Abbildung 16.2:
Sonnenobservatorium im unteren Durchgang:
Beobachtung der auftreffenden Lichtstrahlfläche bei Sonnenaufgang genau in der linken Hälfte der Westwand unterhalb des Musentanzes, Beobachtung vom 21. Juni 2023 bei Sonnenaufgang

(© Ettore Ghibellino)

Bis zum Herbstäquinoktium wandert sie bis zur rechten Hälfte. Zur Wintersonnenwende wird die Sonnenstrahlfläche durch die Südöffnung ohne antike Schattenattribute projiziert und trifft wieder links auf. Bis zum Frühlingsäquinoktium wandert der Lichtkegel bei Sonnenaufgang wieder genau in die rechte Hälfte zurück.

Eine Sonnenuhr zeigt als Instrument verschiedene Größen an, die von der Himmelslage der Sonne abhängig sind. Damit sind Höhe und Tiefe der Halle so bemessen, dass die Äquinoktien jeweils auf Beginn und Ausklang der Sonnenlichtinszenierung im Tempel weisen. Die Westwand gibt damit in der begehbaren Sonnenuhr Auskunft über den Stand des jährlichen Erdenlaufs um die Sonne (Bild 16.3).

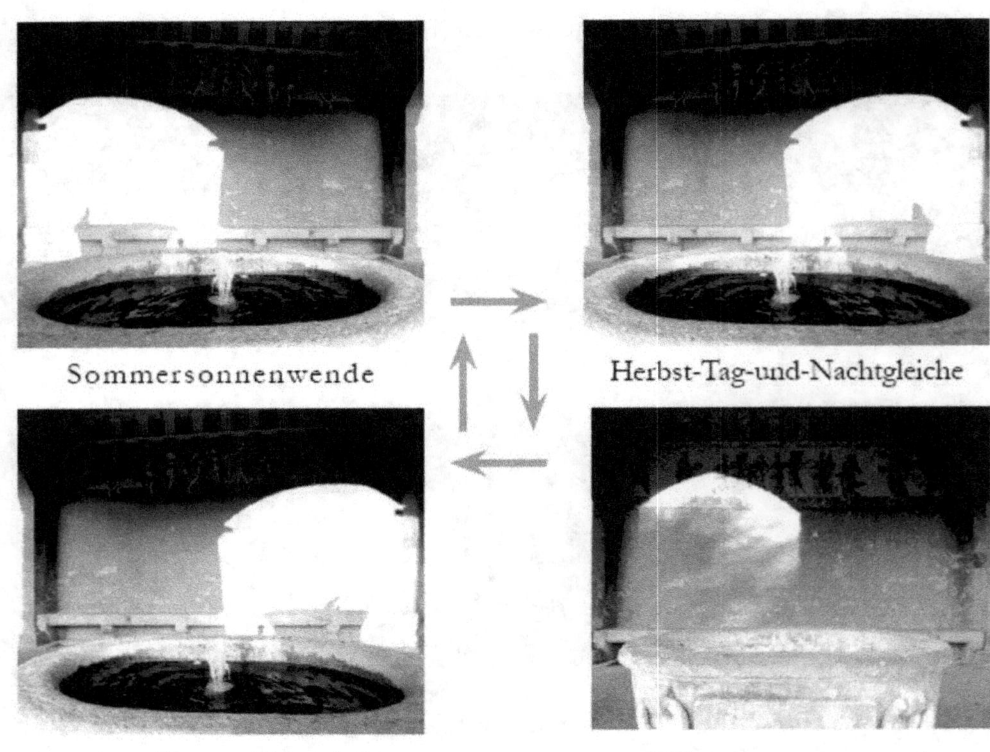

Abbildung 16.3:
Sonnenlichtinszenierung im Tempel zur Zeit der Äquinoktien und Sonnenwenden: Die Westwand spiegelt den jährlichen Umlauf der Erde um die Sonne.

(© Ettore Ghibellino: Das Römische Haus als „Geheimster Wohnsitz“, Weimar 2020)

16.3 Erstmalige Beobachtung des Verlaufs der Sonnenstrahlen in der Cella im Erdgeschoss zur Sommersonnenwende 2023 bei Sonnenaufgang

Die Vertiefung der archäoastronomischen Befunde war indes vorrangig für die erstmalige Beobachtung des Verlaufs der Sonnenstrahlen in der Cella im Erdgeschoss zur Sommersonnenwende 2023 bei Sonnenaufgang (Bild 16.4).

16.3.1 Erwartungen

Wenn das Römische Haus bereits zwei wahrscheinliche Lichtinszenierung am Schornsteinkopf und im unteren Durchgang ausweist, stellt sich die Frage nach einer dritten im Prunkraum im Erdgeschoss. Der Blaue Salon weist die Massen auf, die der antike Architekt Vitruv für die Cella in einem Tempel vorgibt.[4] Die Cella ist nach antikem Verständnis der innere Hauptraum des Tempels: der Raum der Gottheit. Hier ist die Aufstellung eines Abbildes der Gottheit zu erwarten und tatsächlich wählte Goethe ein repräsentatives Gemälde von Herzogin Anna Amalia (1739–1807) in Rom. Aus einem Foto von um 1912 wissen wir, dass ein Prunkkronleuchter im Blauen Salon hang, aus der Inventarliste von 1797 lernen wir, dass es mit Böhmischen Kristallglas behangen war.

Konnte Goethe eine dritte Lichtinszenierung mit Hilfe dieses Kronleuchters geplant haben? Im Bereich des Fachgebiets der Optik arbeitete Goethe auf höchstem Niveau an der Theorie des Lichts, des Schattens und der Farben und operierte auf Grundlage von Sonnenstrahlen. Ein Zitat aus seiner Farbenlehre zeigt wie ausgeklüngelt er hierbei vorging:

> *„In der Mitte zweier dünnen Bretter machte ich runde Öffnungen, ein drittel Zoll groß, und in den Fensterladen eine viel größere. Durch letztere ließ ich in mein dunkles Zimmer einen breiten Strahl des Sonnenlichtes herein, ich setzte ein Prisma hinter den Laden in den Strahl, damit er auf die entgegengesetzte Wand gebrochen würde, und nahe hinter das Prisma befestigte ich eines der Bretter dergestalt, daß die Mitte des gebrochnen Lichtes durch die kleine Öffnung hindurchging und das übrige von dem Rande aufgefangen wurde.“*[5]

Auf der anderen Seite schreibt man über Kristallkronleuchter:

---

4 Herzog, Doris: Das Römische Haus und die zeitgenössische Architekturtheorie, 2001, S. 61, unter Einbezug des Vestibüls als „vermittelnder Raum zwischen Innen und Außen“.

5 Goethe: Sechster Versuch. Zur Farbenlehre: Polemischer Theil, 1890, II, 2, S. 70.

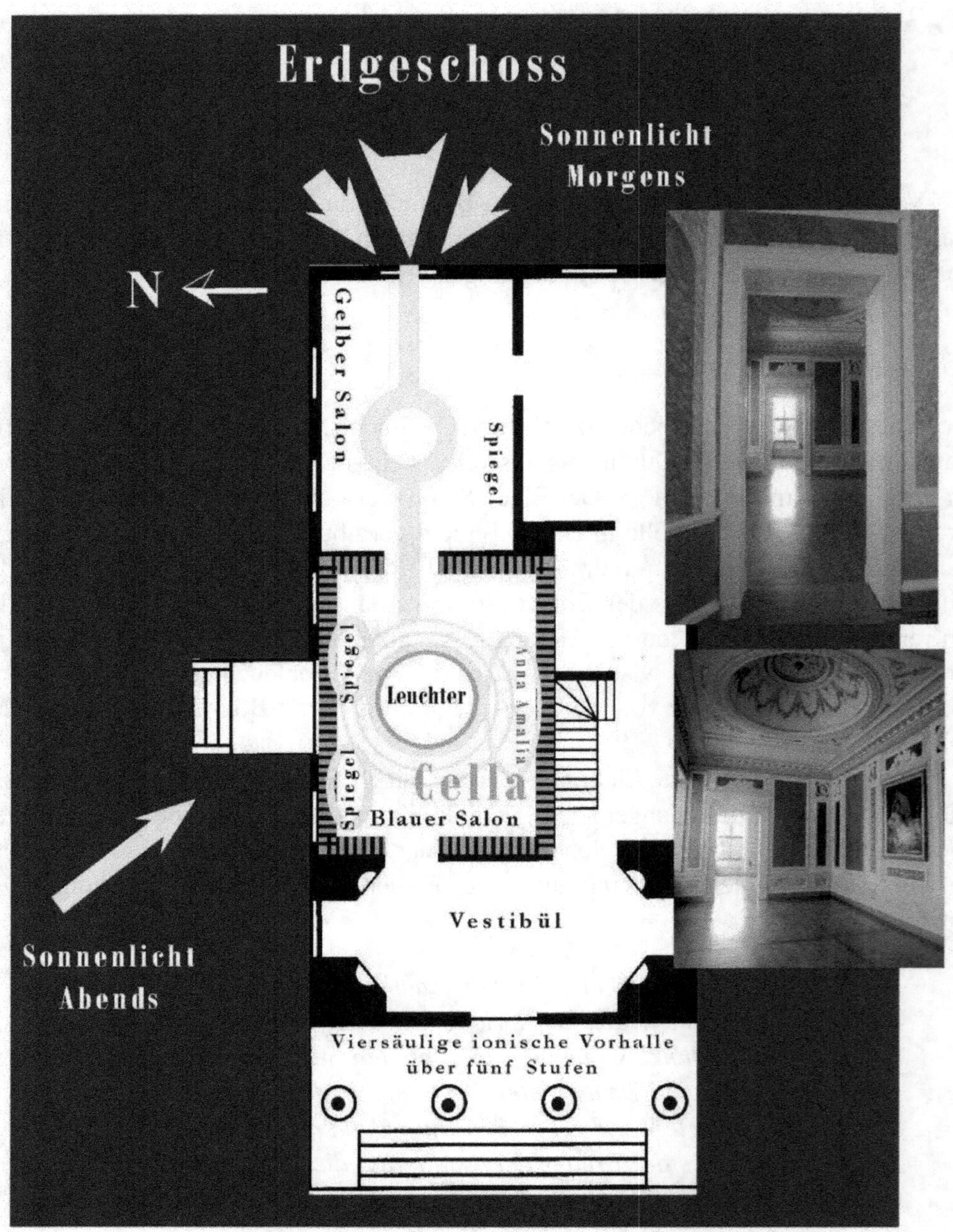

Abbildung 16.4:
Möglicher Verlauf der Sonnenstrahlen in der Cella im Erdgeschoss
zur Sommersonnenwende 2023 bei Sonnenaufgang

(© Ettore Ghibellino: Das Römische Haus als „Geheimster Wohnsitz", Weimar 2020)

> *„Durch Experimente, Berechnungen und Schaltungen können wir die Wirkung des Lichts planen. Der Kristallluster ist hier in seiner Idee ein Meisterwerk.“*[6]

Dies gilt – soweit ersichtlich – nur für künstliche Lichtquellen wie Kerzen, Öl oder Gas, in der Literatur fand sich (noch) kein Beispiel für die Einbindung der Sonne in einer Sonnenlichtinszenierung. Aufgrund von Goethes Voraussetzungen wie Fähigkeiten, ist dies in unserem Fall nahezu anzunehmen.

Der Kronleuchter im Blauen Salon

Dank des Bildes um 1912 sowie einer Zeichnung und Beschreibung im *Journal des Luxus und der Moden*, Ausgabe 12 von 1797 (Bild 16.5) kann das Modell des Kronleuchters identifiziert werden, es kommt von der Chursächsischen Spiegelfabrik in Friedrichsthal bei Dresden.

Das Modell bestätigt auch ein Entwurf mit Blick auf die Nordwand im Blauen Salon vom Innenarchitekten des Römischen Hauses Christian Friedrich Schuricht von 1794 (Bild 16.6).[7]

Von diesem Modell sind mindestens zwei ausgezeichnet erhaltene Exemplare bekannt.[8]

Am 17.08.2023 konnte der Verfasser Leuchterfragmente im Zentraldepot der Klassikstiftung Weimar besichtigen, darunter der vermeintliche Kronleuchter im Blauen Salon.[9] Der Zustand ist nahezu zerstört, dennoch wurde berichtet, dass Pläne erörtert werden, den Kronleuchter wieder herzurichten.

Die halbrunden Spiegelkronleuchter im Blauen Salon

Aus der Inventarliste von 1797 (Bild 16.8) wissen wir, dass die zwei großen Spiegel an der Nordwand im Blauen Salon mit halbrunden Spiegelkronleuchter versehen waren („[...] *mit je einem halbrunden Kronleuchter mit geschliffenen Quasten und Berloquen sowie einem vergoldeten Bügel [...]*“).[10]

6 Rath/Holey: Möbel der Lüfte – Der Kristallluster in Europa, Weitra 2020, S. 52.

7 Möller, Frank C.: The Chursächsische Spiegelfabrik at Friedrichsthal in the Electorate of Saxony: Rediscovery of a Forgotten Glass Factory and its Products. In: Además de revista on line de artes decorativas y diseño (2022), No. 8, S. 188.

8 In Kopenhagen im Museum *The David Collection*, Möller (2022), S. 189, hier abgebildet als Bild 16.7a, sowie im Behnhaus Museum in Lübeck, siehe Käthe Klappenbach: Kronleuchter mit Behang aus Bergkristall und Glas sowie Glasarmkronleuchter bis 1810, Berlin 2001, S. 49, hier Bild 16.7b.

9 Klassik Stiftung Weimar, Inventarnummer: Kg-2020/46, Kg-2020/47, Kg-2020/48.

10 Hauptstaatsarchiv Weimar / Landesarchiv – Freistaat Thüringen, Bausachen B 8991a Bl. 5v.

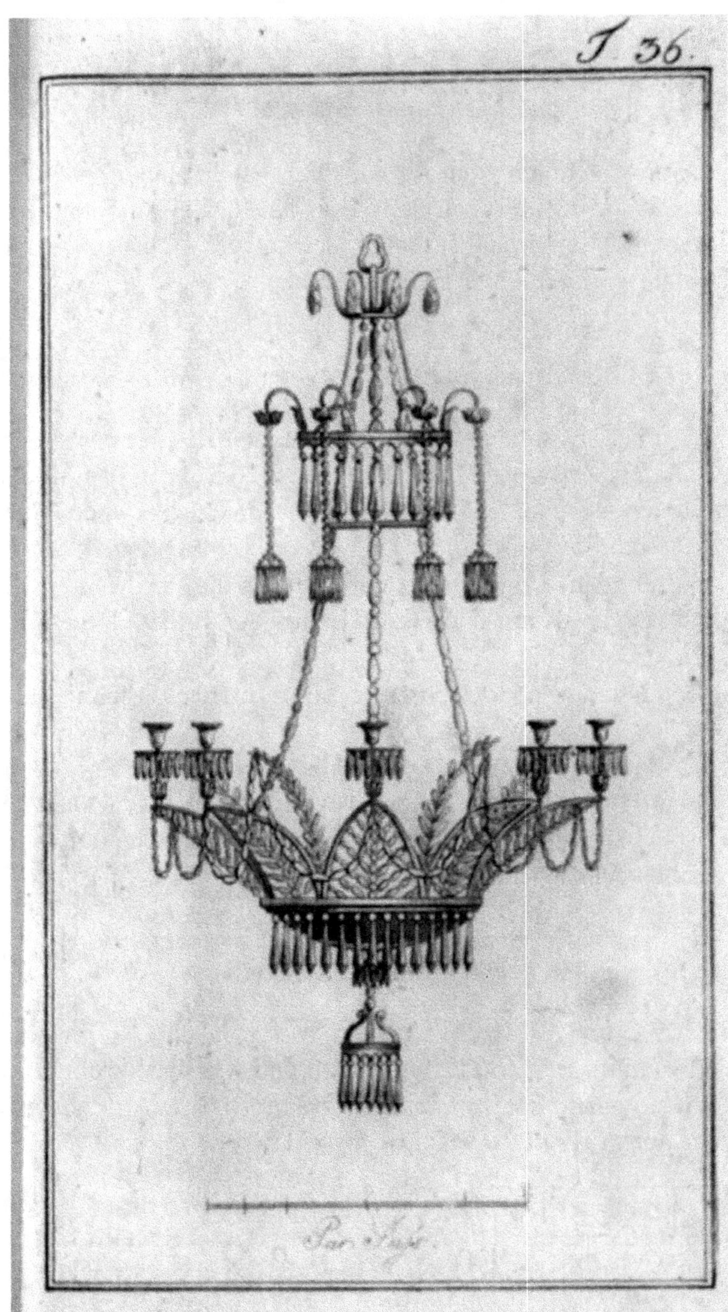

Abbildung 16.5:
Modell des Kronleuchters, Chursächsische Spiegelfabrik in Friedrichsthal bei Dresden

(Journal des Luxus und der Moden, Ausgabe 12 von 1797)

Abbildung 16.6:
Nordwand im Blauen Salon
vom Innenarchitekten des Römischen Hauses Christian Friedrich Schuricht (1794)

(Möller, Frank C.: *The Chursächsische Spiegelfabrik* (2022), No. 8, S. 188)

Im Entwurf vom Innenarchitekten Schuricht sieht man rechts eine Abbildung (Bild 16.6). Im *Journal des Luxus und der Moden*, Ausgabe 12 von 1797 (Bild 16.9), sieht man das Modell.

Bei der oben erwähnten Besichtigung (Fußnote 9) im Zentraldepot der Klassikstiftung Weimar waren zwei halbrunden Spiegelkronleuchter in einem recht guten Zustand zu sehen, die der Beschreibung und den Zeichnungen aus dem Blauen Salon entsprechen.

16.3.2 Sonnenstrahlenbeobachtung im Blauen Salon 2023

Die ersten Sonnenstrahlen am 21.06.2023 Uhr treffen um 05:45 Uhr in der linken Ecke des Gelben Salons auf (Bild 16.10). Damit spielt der Kaminspiegel etwa als Reflektionsfläche in den Blauen Salon keine Rolle. Die Sonnenstrahlen wandern nach rechts und wenn sie etwa eine Stunde später im Bereich der Tür zum Blauen Salon eintreffen, sind sie zu hoch, um auf Höhe des Kronleuchters zu gelangen.

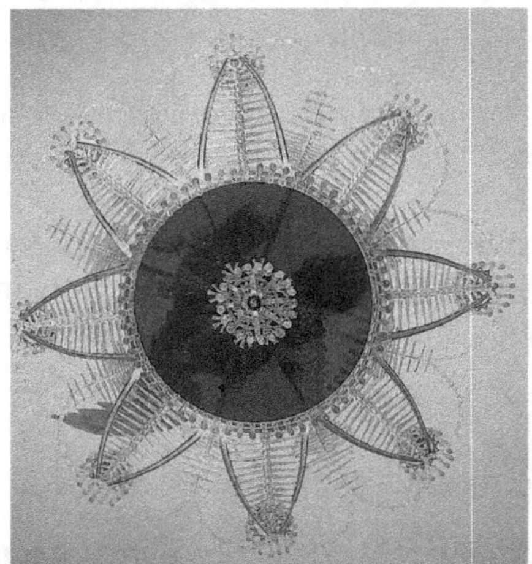

Abbildung 16.7:
Oben: Achtkerzen Kronleuchter aus der Chursächsischen Spiegelfabrik, spätes 18. Jahrhundert
Unten: Kronleuchter von unten aus dem Behnhaus in Lübeck

(Die David Collection, Kopenhagen (Inv.-Nr. L10), Foto: Pernille Klemp. Foto: Käthe Klappenbach)

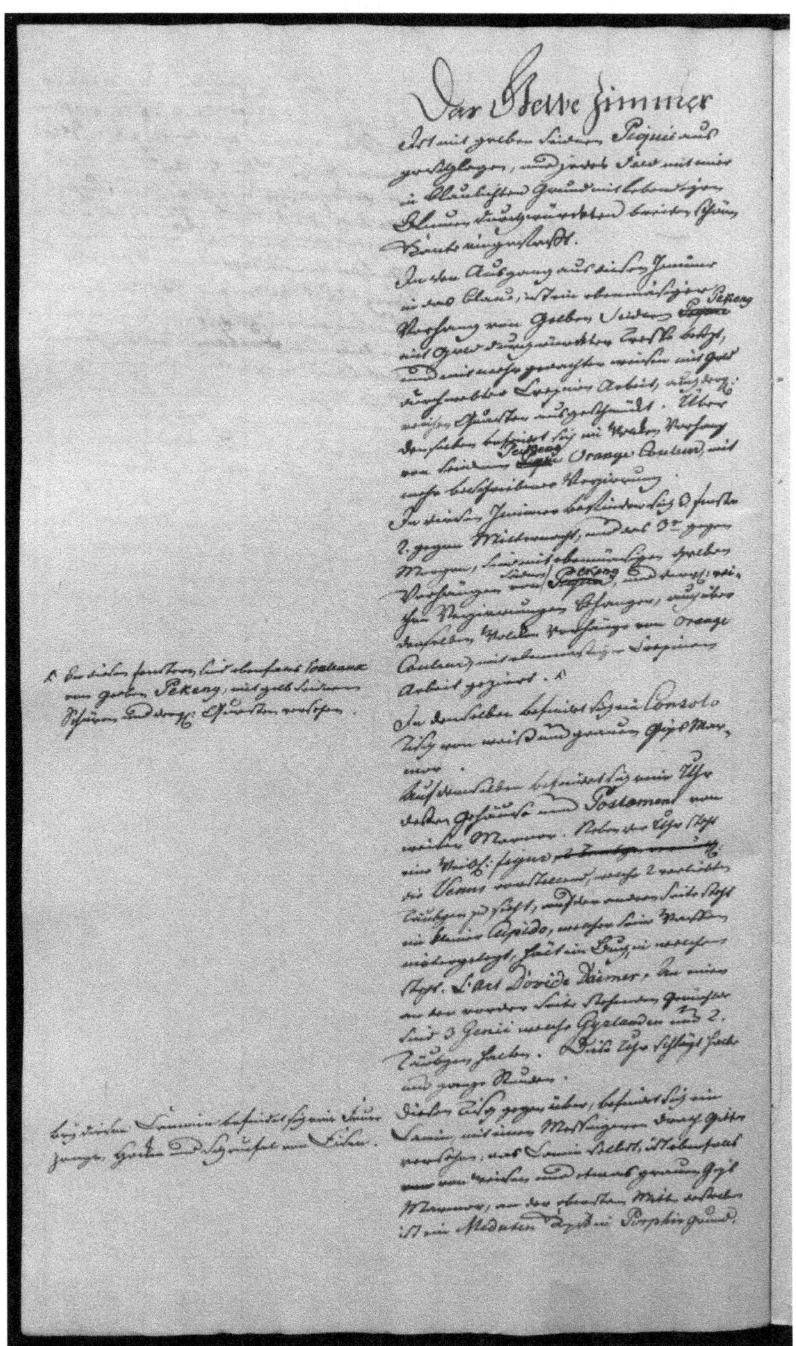

Abbildung 16.8:
Inventarliste des Römischen Hauses von 1797

(Hauptstaatsarchiv Weimar / Landesarchiv - Freistaat Thüringen, Bausachen B 8991a Bl. 5v)

Abbildung 16.9:
Spiegel an der Nordwand im Blauen Salon mit halbrunden Spiegelkronleuchter

(*Journal des Luxus und der Moden*, Ausgabe 12 von 1797)

Abbildung 16.10:
Erste Sonnenstrahlen am 21.06.2023 Uhr,
die um 05:45 Uhr in der linken Ecke des Gelben Salons auftreffen.

(Beobachtung vom 21. Juni 2023 bei Sonnenaufgang, © Ettore Ghibellino)

Wenn die Sonne am 06.04. gen Sommersonnenwende in die Achse des Römischen Hauses eintritt und am 05.09. gen Wintersonnenwende wieder austritt, dürften die Sonnenstrahlen mindestens vier Wochen in gleicher Stärke wie im Bild 16.10 in den Blauen Salon eintreffen und damit auf den Prunkkronleuchter (Bild 16.11).

Zur Zeit unserer Beobachtungen fand eine Ausstellung statt und vor dem Ostfenster lag eine Glasvitrine.

Die auf das Glas prallende Sonnenstrahlen wurden in das Blaue Zimmer gelenkt und man konnte zufälligerweise im kleinen Maßstab von Minikronleuchter die Brechungen der Sonnenstrahlen als funkenstobender Lichtreigen beobachten (Bild 16.13).

So half der Zufall, den zu einem anderen Zeitpunkt erwarteten, im Blauen Salon stattfindenden Lichteffekt, doch noch zu beobachten (Bild 16.14a und 16.14b).

Abbildung 16.11:
Beobachtung vom 21. Juni 2023 um 06:44 Uhr, die Sonnenstrahlen im Blauen Salon werden aufgrund der Höhe der Sonne durch den Türrahmen begrenzt.

(© Ettore Ghibellino)

Abbildung 16.12:
Sonnenhöhe in der Achse des Römischen Hauses bei Sommersonnenwende am 23. Juni 2016, April bis September Azimut: 78°36′

(© Ettore Ghibellino: Das Römische Haus als „Geheimster Wohnsitz“, Weimar 2020)

16.4 Ausblick

„Kompaß und Pol-Stern, Zeitenmesser / Und Sonn und Mond verstehst du besser“, dichtete Goethe in Zahme Xenien VI (1827). Die Wahrscheinlichkeit, mit dem Römischen Haus die erste astronomische Verankerung eines Bauwerkes am Beginn der Moderne nachzuweisen, ist hoch. Ein entsprechender Antrag ist bereits formuliert und wird hier im Anhang wiedergegeben. Mit der Ausleihe eines der Prunkkronleuchters der Spiegelmanufaktur Dresden sowie der Aktivierung der vorhandenen halbrunden Spiegelkronleuchtern lassen sich im Blauen Salon valide Messungen durchführen.

Abbildung 16.13:
Ettore Ghibellino und Helfer mit selbstgebastelten „mobilen“ Kronleuchter, davor die Glasvitrine, am 21. Juni 2023 bei Sonnenaufgang

(© Ettore Ghibellino)

Eingebettet mit Messungen am Schornsteinkopf und im unteren Durchgang wird anhand der Daten eine genaue Aussage möglich, ob das Römische Haus astronomisch verankert ist. Von da aus kann der hier ausgelassene zweite Teil meiner Studie *„Das Römische Haus als ‚Geheimster Wohnsitz‘“* (Weimar 2020) mit dem Titel *„Poetische Verankerung“* angegangen werden. Quellen und Kunstwerken werden untersucht, darunter *‚Faust. Eine Tragödie‘*, und im Kontext der archäoastronomischen Interpretation des Römischen Hauses im Park an der Ilm zu Weimar gestellt.

Abbildung 16.14:
Beobachtung des Lichteffekts am 21. Juni 2023 um 06:53 Uhr
mit selbstgebastelten, kleinen Kronleuchter im Blauen Salon:
Die Brechungen der Sonnenstrahlen als funkenstobender Lichtreigen
können bereits im kleinen Maßstab beobachtet werden.

(© Ettore Ghibellino)

Anhänge

16.5 Fragen an den Verfasser anlässlich der Tagung in Weimar

Von den vielen Fragen sei nur eine ausgewählt, die vom Doyen der Archäoastronomie Clive Ruggels:[11] *„Wie sind überhaupt zwei Lichtinszenierungen gleichzeitig am Schornsteinkopf und im unteren Durchgang möglich?"*
Diese Frage leitete ich an den führenden Experten für die Orientierung von Bauwerken Christian Wiltsch[12] weiter, der sie dankenswerterweise wie folgt beantwortete:

„[...] *neben einer Orientierung des Kaminkopfes des römischen Hauses (kann) im unteren Bereich etwas ganz anderes realisiert werden* [...] *Das ist eigentlich gar nicht so schwierig. Die Arkade außen bildet die Kontur, deren Schattenwurf genutzt werden kann. Die Himmelsrichtungen, aus denen die Sonne an den gedachten Tagen zu den gedachten Zeiten scheint, sind bekannt. Ich kann also eine Richtung festlegen, an der sich die Grenze Licht-Schatten befindet. Ich kann nun hierzu an der zurückliegenden Wand die Tiefe dieser Wand so festlegen, dass ich dadurch ein Merkzeichen bekomme. Es gibt genügend freie Variablen, die Abmessungen so fest zu legen, dass solche Inszenierungen möglich sind.*

Ob und welche Inszenierungen dann gewollt sind, muss zum einen durch Beobachtung gewonnen werden und idealerweise durch Zitate belegt werden.

Neben der Horizontalprojektion kann auch die Lichteinfallshöhe zu bestimmten Tagen und Zeiten in den Fokus der Planung gelangen. Es ist letztlich ein Spiel mit geometrischen Vorgaben, das zeichnerisch auf dem Reißbrett angestellt werden kann, um dann aus dem Rahmenkonstrukt der gefundenen Markierungen die Architektur zu entwickeln. Variablen sind die Abstände der Wände, die Dicke von Wänden und Pfeilern, Höhenlagen usw.

Ob Goethe so konstruiert hat, ist meines Wissens nicht schriftlich hinterlegt. Also kann hier nur die Beobachtung Anhaltspunkte geben, aus denen sich zunächst ein sinnvolles Konzept ergeben müsste, das nicht nach Willkür ausschaut, zu dem dann in seinen Werken nach Hinweisen gesucht werden müsste. Dass das römische Haus mit seiner Orientierung den Sonnenaufgang zu Goethes Geburtstag darstellt, wo der erste sichtbare Lichtstrahl den Kamin axial trifft, lässt sich nachrechnen und beobachten, wie von Ihnen geschehen. Der nächste Schritt, ob es eine Beziehung oder zumindest Liebeserklärung an die Herzogin gab, müsste auf gleicher Basis, idealerweise ihr Ge-

11 Vgl. nur Clive L. N. Ruggles (ed.): Handbook of Archaeoastronomy and Ethnoastronomy, 3 Bände, 2014. Boutsikas, McCluskey & Steele: Clive Ruggles and the Development of Cultural Astronomy, 2021, S. 1–7.

12 Wiltsch: Das Prinzip der Heliometrie im Lageplan mittelalterlicher Kirchen, 2014.

burtstag und evtl. Sonnenuntergang (Venus-Abendstern-Prinzip) am Gebäude gezeigt werden.

In gleicher Weise ist es mit der Arkade im Untergeschoss, wo die Lichtspiele zu einem sinnvollen Konzept zusammengesetzt werden müssten, denn wenn es nur einen einzigen Lichteffekt gibt, könnte es tatsächlich noch Zufall sein. Wenn aber markante Effekte zu markanten Tagen (Sonnwenden, Äquinoktien) dargestellt werden können, ist eine entsprechende Komposition nicht abwegig.

Fazit: Völlig unabhängige Inszenierungen am Kaminkopf (Gebäudeachse) und auf der Terrasse im unteren Geschoss sind durchaus möglich, müssten aber ein in sich nachvollziehbares Gesamtkonzept darstellen, das aufzuzeigen ist und nicht aus Einzelmomenten besteht, die eher zufällig sein könnten.

[...] *ich habe eine Skizze gefertigt (Bild 16.15), die ein bisserl die Gestaltungsmöglichkeiten und Variationen erläutert, die möglich sind, Lichtinszenierungen zu planen, hier am Beispiel zweier Rundsäulen, die einen Azimut von 75° haben für den Lichteinfall der aufgehenden Sonne zu den Sonnenwenden auf dem 51. Breitengrad.*

Bei den alten Kirchen war der Zwischenraum zwischen den Säulen das Ostfenster, und die Projektionsfläche das Kruzifix über dem Altar. Mit der aufkommenden wissenschaftlicheren Betrachtung solcher Phänomene seit der Renaissance und Aufklärung beginnen in der Architektur deutlich mehr solcher Überlegungen, die sich heute in der Regel nicht als geplante Lichtspiele nachweisen lassen, weil immer die Planungsidee nicht schriftlich fixiert wurde. Goethe mit seiner bekannten Forschung zu Licht und Farbe wird sich mit diesen Gedanken befasst haben.

Zusätzlich kann in den (hier breiten Raum zwischen den Säulen) auch noch ein Architekturelement gestellt werden, etwa eine Skulptur, die einen Schattenwerfer darstellen kann, wo z. B. eine ausgestreckte Hand, eine Lanzen- oder Fahnenspitze usw. einen Schattenwurf auf der Wand dahinter produziert, wo zu bestimmten Tageszeiten und definierten Tagen ganz bestimmte Dinge angezeigt werden können. Solche Schattenzeiger können gefühlt zwar mittig stehen, aber der genaue Punkt muss es nicht sein. Klassisches Beispiel ist eine Figur, die mittig steht, wo aber die Spitze einer Lanze (etwa Georg beim Drachentöten) sich deutlich ausmittig befindet. Die Gestaltungsmöglichkeiten sind also unbegrenzt.“

16.6 Forschungsantrag Römisches Haus: Wissenschaftlicher Nachweis archäoastronomischer Verankerungen

Weimar, 04.11.2023

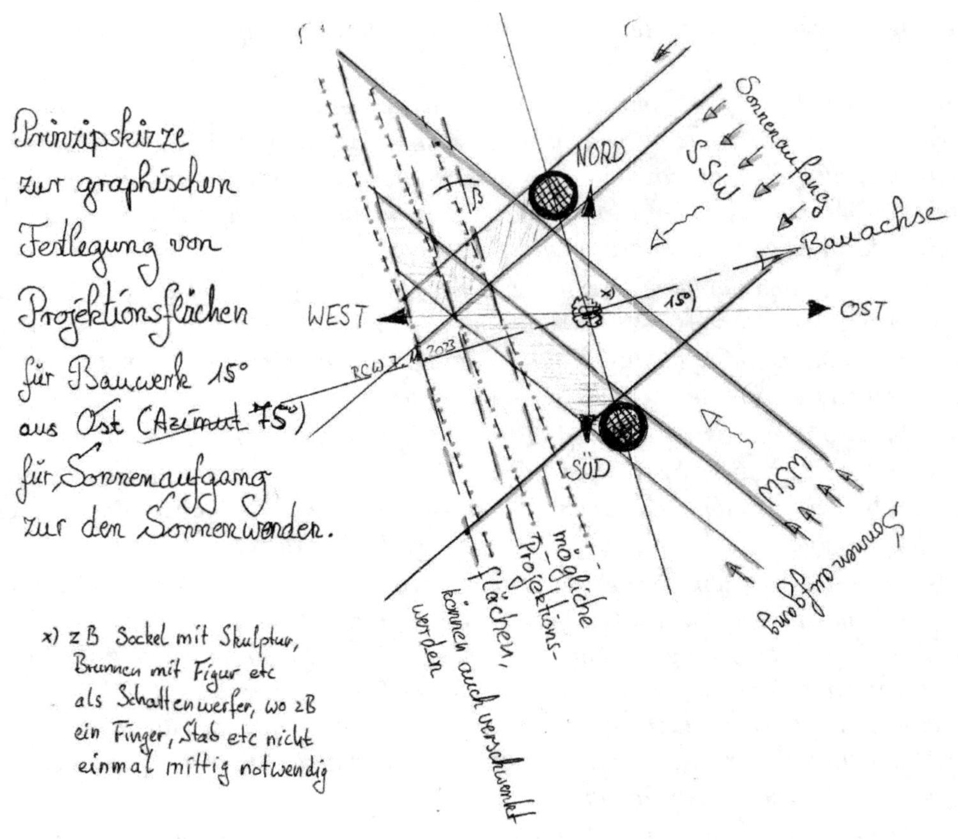

Abbildung 16.15:
Planung von Lichtinszenierungen,
hier am Beispiel zweier Rundsäulen, die einen Azimut von 75° haben für den Lichteinfall der aufgehenden Sonne zu den Sonnenwenden auf dem 51. Breitengrad

(Skizze von Christian Wiltsch)

Gesellschaft für Archäoastronomie (GfA) e.V.

Projektkoordination: Dipl. Des. M. Peuschel (Weimar)

„Fragt man mich, ob es in meiner Natur sei, die Sonne zu verehren, so sage ich abermals: Durchaus! Denn sie ist gleichfalls eine Offenbarung des

Abbildung 16.16:
Römisches Haus in Weimar

(Georg Melchior Kraus: Weimar, das Römische Haus. o.O. 1798)

Höchsten, und zwar die mächtigste, die uns Erdenkindern wahrzunehmen vergönnt ist."[13]

Präambel

Das Römische Haus an der Ilm, ist ein unbestrittener Höhepunkt der Schaffenskraft des Universalgenies Goethes. Hoch inspiriert setzte er nach seiner Italienreise den Auftrag seines Fürsten um, mit einem Herrschaftsentwurf die semantische Leerstelle innerhalb der von Karl Augusts Großvater angelegten architektonischen Repräsentationsachse im Park an der Ilm zwischen Weimarer Stadtschloss und Lustschloss Belvedere umzudeuten und diese neu zu füllen.

13 Zitat: Goethe, J. W.: Gespräche. Mit Johann Peter Eckermann, 11. März 1832.

Das Römische Haus hatte als Musen und Lusttempel für seinen Fürsten Carl August, die volle Aufmerksamkeit Goethes. Zum ersten Mal konnte Goethe hier von Grund auf Architektur planen und nach den höchsten Regeln des eigenen Wissens gestalten. Mit dem Römischen Haus holte er seine Verehrung für die Antike und das ewige Rom nach Weimar. Hier fließt sein umfassendes Wissen über die Antike, Optik, Architektur, Poesie, Astronomie, etc. in einem für die Ewigkeit angelegten, architektonischen Entwurf zusammen.

Ganz entsprechend dem umfassenden Wirkungsanspruch eines Universalgenies, lag die Vermutung nahe, dass Goethe hier sein Wissen und Können auf vielen bisher unbekannten Ebenen einfließen ließ. Er hatte auch einen praktischen Bezug zur Astronomie und machte selbst Himmelsbeobachtungen durch Teleskope (auch zusammen mit Friedrich Schiller). So war ihm sicher spätestens nach seiner Italienreise und dem Besuch des Pantheons wie auch anderer römischer Kirchen, vor allem die astronomische Sonnen-Ausrichtung der repräsentativsten Gebäude bewusst. Immer wieder orientierten klassische Baumeister ihre Bauwerke nach Vorgängen am Himmel, häufig nach dem Lauf der Sonne. Dieser ermöglichte eine sehr genau zu datierende Wiederkehr spezieller Lichtprojektion an den Bauwerken selbst. Nach einer Hypothese bescheinen im Pantheon an jedem 21. April Sonnenstrahlen das Eingangsportal und markieren die Gründung der Stadt Rom, was jährlich viele Besucher anlockt und in ihren Bann zieht.

Die Antragsteller kennen die Hypothesen über archäoastronomische Verankerungen des Römischen Hauses auf mindestens drei Ebenen mit der Sonne. *Dieses Phänomen ist durch eine ordentliche Messreihe wissenschaftlich zu überprüfen.*

Eine mögliche Bestätigung der Hypothesen könnte das Römische Haus an der Ilm zu einem Juwel und einer architektonischen Glanzleistung am Beginn der Moderne erheben. Eine mögliche Widerlegung der Hypothesen würde niemandem schaden und dennoch Gewissheit bescheren. Der Nachweis einer (und falls auch nur teilweisen) archäoastronomischen Verankerung eröffnet viele Möglichkeiten für weiterführende Programme universeller Naturbeobachtungen, Bildung und Tourismus. Die Öffentlichkeitsarbeit der Klassik Stiftung wird bereits während und auf jeden Fall nach der Messreihe (unabhängig vom wissenschaftlichen Ergebnis) von der Aktion profitieren.

Wissenschaftliche Fragestellungen

Folgende Beobachtungen archäoastronomischer Verankerungen sind zu überprüfen:

1. Warum ist das Römische Haus aus der exakten Ostrichtung um 12° nach Norden gedreht? Was ist an der Behauptung Ghibellinos dran, dass die Sonne an jedem 28. August dem Geburtstag Goethes allem Anschein nach aus dem Schornsteinkopf aufsteigt? Wo muss man stehen, damit das der Fall ist, ist der Beobachter-

standort für diesen Effekt in irgendeiner Weise ausgezeichnet? Ist es wirklich nur am 28. August der Fall und damit eine geniale Planleistung Goethes?

2. Ist der untere Durchgang eine Sonnenuhr, die mittels natürlicher Lichtprojektionen an der Westwand die Sommersonnenwenden sowie die Tag- und Nachtgleichen anzeigt?

3. Ist der Blaue Salon im Erdgeschoss eine Cella nach den Maßangaben für Tempelbau des römischen Architekten Vitruv, in der das Zusammenspiel von Kristallkronleuchtern und Spiegeln im Frühling und Sommer eine Lichtsymphonie aufgeführt wird?

Umsetzung

1. und 2. sind in ersten Schritten dokumentiert und belegt. Doch sind genaue Messungen notwendig, die durch Außenkameras mit Zeitschaltuhren vorgenommen werden können, die direkt auf die Westwand ausgerichtet sind. Eine weitere Kamera mit Zeitschaltuhr ist im Blauen Salon auf der Nordseite mit Weitwinkelobjektiv aufzustellen, um möglichst viel vom Raum zu erfassen. Die Zeitschaltuhren stellen sicher, dass die Aufnahmen nur zum Zeitpunkt des Sonnenauf- und -untergangs gemacht werden und (Datenschutz) keine Gäste im Haus aufnehmen.

3. Den Nachweis einer möglicherweise durch Kronleuchter und Spiegel aufgeführten Lichtsymphonie ermöglicht nur ein praktisches Experiment. Die Klassik Stiftung verfügt über die Reste des Prunkkronleuchters. Nach einem gemeinsamen Termin konnten wir feststellen, dass sich der Prunkleuchter und die zwei Halbkronleuchter für die Spiegel in einem stark restaurierungsfähigen Zustand befinden (Inventarnummer: Kg-2020/46, Kg-2020/47, Kg-2020/48). Die Ausleihe eines historisch dokumentierten und identischen Kronleuchters der Spiegelmanufaktur Dresden, kann von Mai – August 2024 im Museum Behnhaus Drägerhaus in Lübeck, als auch bei der The David Collection Museum in København angefragt werden (siehe Bild 16.7a). Im Messzeitraum sind die Fensterläden im Römischen Haus zum Sonnenauf- bzw. Sonnenuntergang zu öffnen.

–

Die Messreihen werden in Kooperation mit der Universität Jena gespeichert, mit KI-gestütztem Image Processing ausgewertet und allgemeingültig dokumentiert. Die Umsetzung der Messungen kann nicht invasiv und ünsichtbaräm und im Römischen Haus eingebettet werden oder auf Wunsch/ bei Bedarf in der Öffentlichkeitsarbeit der Klassik Stiftung gewinnbringend eingesetzt werden.

Dazu werden verschiedene Methoden (Messverfahren, Übertragung und Auswertung der Bilder etc.) in Zusammenarbeit aller Beteiligten entwickelt und nach Evaluierung eingesetzt.

Kooperationspartner

Auf der Jahrestagung der Archäoastronomischen Gesellschaft e.V. im Juni 2023 in Weimar, konnte die Unterstützung renommierter Koryphäen in diesem Fachgebiet für die weitere Erforschung des Römischen Hauses im Park an der Ilm gewonnen werden.

Prof. Dr. Clive Ruggles, international führender Experte für Archäoastronomie und Berater der UNESCO zu Themen des Weltkulturerbes.

Dr. Georg Zotti, 2. Vorsitzender der Gesellschaft für Archäoastronomie

Dr. Michael Rappenglück M.A., 1. Vorsitzender der Gesellschaft für Archäoastronomie

Dr. Dr. Susanne M. Hoffmann, Astronomie & Informatik, Universität Jena

Prof. Dr. Gudrun Wolfschmidt, Universität Hamburg, Hamburger Sternwarte.

Zusammenfassung

Das Forschungsprojekt hat das Ziel der ergebnis-offenen Prüfung der archäoastronomischen Bedeutung des Römischen Hauses. Die Forschenden bitten um eine solide Messreihe, auf deren Grundlage entschieden werden kann, ob Goethe Ruf als Universalgenie sich auch auf die bisher wenig erkannte und kaum gewürdigte architektonische Konzeption mit Ausrichtungen des Römischen Hauses nach dem Sonnenstand ausdehnen lässt. Stellt sich diese Hypothese als haltbar heraus, trägt sie zu einer vielfältigen Öffnung des Gesamtwerks Goethes bei und ließe sich auch im Weimarer (Thüringer) Tourismus gut vermarkten.

Eine Forschungskooperation mit der Universität Jena zur Datenspeicherung und mit der GfA zur wissenschaftlichen Einordnung der Befunde bietet die einzigartige Möglichkeit, die vielfältigen Forschungs- und Bildungsprogramme der Klassik Stiftung auf Jahre zu befruchten und in vielen (neuen) Bereichen weiter zu entwickeln; zum Beispiel über mögliche Forschungsarbeiten zu angewandtem antikem Wissen, über optische Grundsätze, die „aufgeführte“ (Goethe) Architektur, Naturbeobachtungen, astronomische Wissenschaften, literarische Einbettungen ...] u.v.a.m.

16.7 Literatur

BOUTSIKAS, EFROSYNI ; MCCLUSKEY, STEPHEN C. & JOHN STEELE: Clive Ruggles and the Development of Cultural Astronomy. In: *Advancing Cultural Astronomy.* Edited by EFROSYNI BOUTSIKAS, STEPHEN C. MCCLUSKEY & JOHN STEELE. Cham: Springer International Publishing (Historical & Cultural Astronomy) 2021.

FULS, ANDREAS: Archäoastronomische Methodik bei Baudenkmälern. In: WOLFSCHMIDT, GUDRUN (Hg.): *Baudenkmäler des Himmels – Astronomie in gebautem Raum und gestalteter Landschaft.* Hamburg: tredition (Nuncius Hamburgensis; Bd. 35) 2018, S. 146–164.

GHIBELLINO, ETTORE: *Goethe und Anna Amalia. Das Römische Haus als ‚Geheimster Wohnsitz'.* Weimar: Anna Amalia und Goethe Stiftung 2020.

GHIBELLINO, ETTORE: Goethes Testament ist mit Sonnenstrahlen geschrieben: Das Römische Haus im Park an der Ilm zu Weimar. Teil I: Archäoastronomische Befunde. In: WOLFSCHMIDT, GUDRUN (Hg.): *Himmelswelten und Kosmovisionen – Imaginationen, Modelle, Weltanschauungen.* Hamburg: tredition (Nuncius Hamburgensis; Bd. 51) 2020, S. 206–221.

HERZOG, DORIS: Das Römische Haus und die zeitgenössische Architekturtheorie. In: BEYER, ANDREAS (Hg.): *Das Römische Haus in Weimar.* München, Wien: Hanser Verlag (Schriftenreihe des Goethe Nationalmuseums Weimar – Stiftung Weimarer Klassik) 2001, S. 48–62.

KLAPPENBACH, KÄTHE: *Kronleuchter mit Behang aus Bergkristall und Glas sowie Glasarmkronleuchter bis 1810.* Berlin 2001.

MÖLLER, FRANK C.: The Chursächsische Spiegelfabrik at Friedrichsthal in the Electorate of Saxony: Rediscovery of a Forgotten Glass Factory and its Products. In: *Además de revista on line de artes decorativas y diseño* (2022), No. 8.

RATH, PETER & JOSEPH HOLEY: *Möbel der Lüfte – Der Kristallluster in Europa.* Weitra 2020.

RUGGLES, CLIVE L. N. (ed.): *Handbook of Archaeoastronomy and Ethnoastronomy, 3 Vol.* New York, NY: Springer 2014, XXXVI, 2297 Seiten.

WILTSCH, CHRISTIAN: *Das Prinzip der Heliometrie im Lageplan mittelalterlicher Kirchen: Nachweis der Ausrichtung von Kirchenachsen nach Sonnenständen an Kirchweih und Patronatsfest und den Folgen für die Stadtplanung.* Aachen: Shaker (Berichte aus der Geschichtswissenschaft) 2014.

Abbildung 17.1:
Stanley Lloyd Miller (1930–2007) and Harold Clayton Urey (1893–1981):
Experiment in search of the origin of life (1953)

(Photo: Gudrun Wolfschmidt, Museo de Ciencias Naturales, La Plata, Argentina, 2022)

Homo sapiens and the Universe – Where do we come from? Where are we going to?

Jaak Aru (Tartu, Estland)

Abstract

According to the current knowledge human beings can not be sure whether life started from the components available on the Earth or the primary living units were seeded by cosmic objects like meteorites, and further evolutionary process. We also do not have the full picture what possibilities has Mankind to leave the planet Earth if it is full contaminated and no more suitable for larger living organisms like mammals. The aim of this contribution is to give a short overview of the current concepts of biological teleportation of life today. Both aspects are discussed – seeding of life to earth from other planets, and the current status of the art of teleportation of life to other planets.

Zusammenfassung: Der Homo sapiens und das Universum. – Woher kommen wir? Wohin gehen wir?

Nach heutigem Kenntnisstand kann der Mensch nicht sicher sein, ob das irdische Leben aus den auf der Erde vorhandenen Bestandteilen entstanden ist oder ob Leben durch kosmische Objekte wie Meteoriten entstand und einen weiteren evolutionären Prozess. Wir haben auch kein vollständiges Bild davon, welche Möglichkeiten die Menschheit hat, den Planeten Erde zu verlassen, wenn er völlig verseucht und für größere Lebewesen wie Säugetiere nicht mehr geeignet ist. Ziel dieses Beitrags ist es, einen kurzen Überblick über die heutigen Konzepte der biologischen Teleportation von Leben zu geben. Es werden beide Aspekte diskutiert – die

„Aussaat“ von Leben auf die Erde, von anderen Planeten kommend, sowie der aktuelle Stand der Kunst der Teleportation von Leben auf andere Planeten.

17.1 Introduction

The aim of this presentation to give a short overview of the current concepts of teleportation of life. Short overview of both aspects – seeding of life to earth from another planets and current status of the art of teleportation of life to other planets – is given.

17.2 Where do we come from?

The question how did life evolve on the planet Earth has not been finally solved. The two substantial characteristics of the life are capacity to store information and self reproduction.

17.2.1 Miller-Urey-Experiment (1953)

The best known theory how life originated on the planet Earth is that – Life emerged from a primordial soup.

In 1952, Stanley Miller (1930–2007) performed a famous experiment with Harold Urey (1893–1981). Urey was Professor of Chemistry at the University of Chicago from 1945 to 1958. His field of research was the separation of isotopes. His most important achievement during his time as Associate Professor at Columbia University in New York was the discovery of deuterium (1931), for which he was awarded the Nobel Prize in Chemistry *for his discovery of heavy hydrogen* in 1934. The ratio of deuterium to helium is of great importance for the course of primordial nucleosynthesis.[1] Miller and Urey injected ammonia, methane and water vapor into an enclosed glass container to simulate what were then believed to be the conditions of Earth's early atmosphere (Fig. 17.1). Then they passed electrical sparks through the container to simulate lightning. Amino acids, the building blocks of proteins, soon formed – origin of life.

1 Wolfschmidt: Kosmochemie – Chemische Elemente im Kosmos – Meteoriten, Sterne, Kosmologie, 2022, here p. 118–119.

17.2.2 Panspermia – Life is seeded by comets or meteors

The cosmological aspect of the arousal of information-binding polymers is described in the theory of *panspermia or pseudo-panspermia*. This theory suggests that – Life is seeded by comets or meteors.

Abbildung 17.2:
Artist's concept of meteors impacting ancient Earth
Some scientists think such impacts may have delivered water
and other molecules useful to emerging life on Earth.

(Credits: NASA's Goddard Space Flight Center Conceptual Image Lab)

The panspermia hypothesis suggests that many of the small *organic molecules* used for life originated in space, and were distributed to planetary surfaces. Evidence for pseudo-panspermia includes the discovery of organic compounds such as *sugars*,[2] *amino acids, and nucleobases* in meteorites and other extraterrestrial bodies.[3]

At least two key aspects of the *panspermia theory* have been studied during man's exploration of space:

Survival in transit
: The survival of microorganisms has been studied extensively using both simulated facilities and in low Earth orbit. A large number of microorganisms have been selected for exposure experiments, both human-borne microbes (significant

2 Steigerwald: First Detection of Sugars in Meteorites Gives Clues to Origin of Life, 2019.
3 Kaufmann, Marc: Did Life on Earth Come From Mars? (2013).

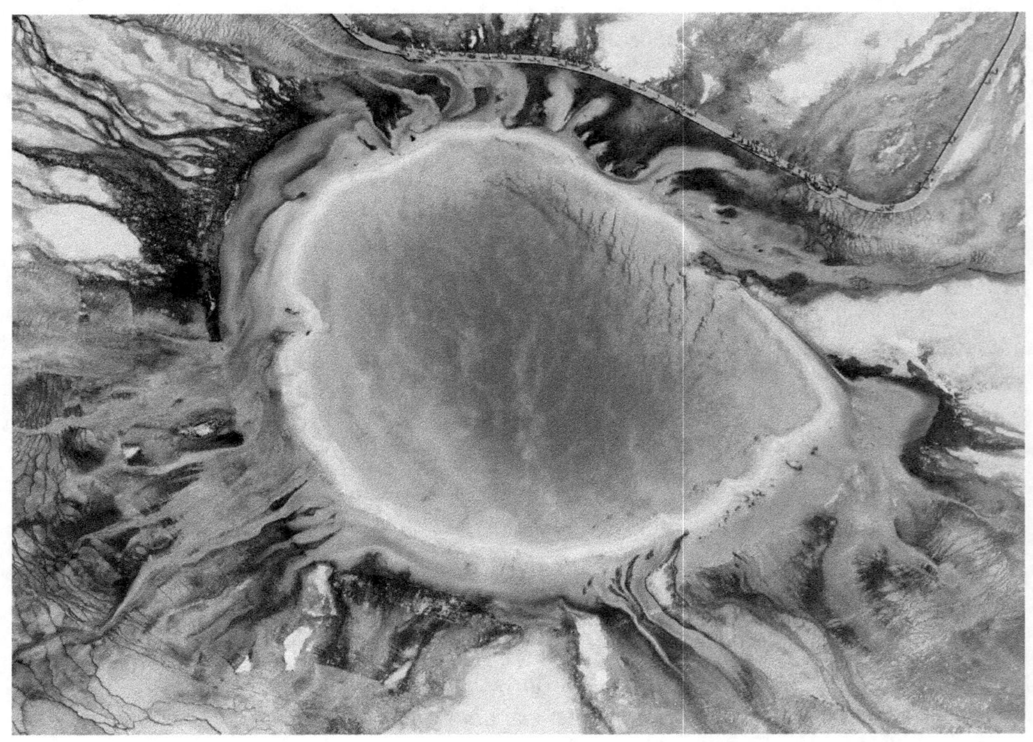

Abbildung 17.3:
The bright colors are produced by thermophiles, a type of extremophile that thrives at relatively high temperatures (41–122°C)
Aerial image of Grand Prismatic Spring, Yellowstone National Park

(Credits: CC4, Carsten Steger)

for future crewed missions) and *extremophiles*[4] (significant for determining the physiological requirements of survival in space).

Atmospheric entry

To test whether microbes on or within rocks could survive hypervelocity entry through Earth's atmosphere. As an example *Bacillus subtilis spores*[5] inoculated onto granite domes were twice subjected to hypervelocity atmospheric transit by launch to a ∼120 km altitude on an Orion two-stage rocket. The spores survived

4 An extremophile is an organism that is able to live or to thrive in extreme environments like extreme temperature, radiation, salinity, or pH level. Cf. Stetter: History of discovery of the first hyperthermophiles, (2006).

5 Errington & Aart: Microbe Profile: Bacillus subtilis, (2020).

on the sides of the rock, but not on the forward-facing surface that reached 145°C. As photosynthetic organisms must be close to the surface of a rock to obtain sufficient light energy, atmospheric transit might act as a filter against them by ablating the surface layers.

17.2.3 Life in the universe – Life on another planets

Another interesting aspect of the possible origins of life in the universe are covered by the studies looking for life on another planets.

The idea of such studies is to find chains on molecules with the potential of carrying information and capable on self-reproduction. For example *phosphine*[6] – a chemical compound made of one phosphorous atom surrounded by three hydrogen atoms (PH_3) – may indicate evidence of life if found in the atmospheres of small rocky planets like our own. An international team of scientists last year claimed to have detected phosphine in the atmosphere of Venus. This discovery raised the prospect of the first evidence of life on another planet-albeit the primitive, single-celled variety.

17.3 Where are we going?

The more interesting and important question today is – Where are we going?

Our planet Earth is suspected not to have enough space and food the for rapidly growing population of *homo sapiens*. Scientific and governmental communities are looking for perspectives to transfer life to other planets of the solar system and elsewhere. The best example in this field are NASA activities.[7]

17.3.1 How NASA plans to build on other planets – Biological teleportation

A very crucial part of the future of human life on other planets[8] is the possibility of *biological teleportation.*[9]

6 Zapata Trujillo et al.: Computational Infrared Spectroscopy of 958 Phosphorus-Bearing Molecules, (2021).

7 Drake, Nadia: Why we explore Mars – and what decades of missions have revealed, 2021. Nadia Drake is the daughter of Frank Donald Drake (1930–2022), director of the Arecibo Observatory in Puerto Rico from 1966 to 1968, searched for extraterrestrial intelligence (SETI), based on the the Drake equation (1961, Project Ozma).

8 Schwab: Here's how NASA plans to build on other planets, 2018.

9 See e. g. Alexander: Biological Teleporter Could Seed Life Through Galaxy, (2017).

The most eminent and productive author in this field is Craig Venter.[10] A short description of his concept is as follows. There are at least two substantial technical steps required to complete teleportation of life to other planets.

First. Converting the analog genetic code into digital code. An analog to digital converter (ADC) converts an analog signal into digital form. An embedded system uses the ADC to collect information about the external world (data acquisition system.) The input signal is usually an analog data, and the output is a binary number.

Second step is described as Digital-to-Biological Converter. The DBC is Venter's attempt to transfer and manufacture life.[11] The DBC downloads DNA files (from the internet) and prints the code using the four chemical bases of DNA-adenosine, guanine, thymine, and cytosine (A, G, T, C). The machine is capable of building proteins from the genetic code (printing biological hardware, so to speak), bringing it one step closer to building living cells from scratch.

The progress in this field of biological teleportation is described in varios short reports by Craig Venter and NASA.

17.3.2 Mohava Desert Experiment

Classical is the Mohava desert experiment:
In November 2013 Venter's team completed a teleporting test in the Mojave desert – the closest environment to Mars on Earth – with NASA. A sample of algae that can be grown under clear quartz because the light penetrates the rock and the algae is protected from exposure to the heat was taken. The DNA genome sequence of the collected algae out in the desert was sent up to the Cloud. In Venter's labs in La Jolla,[12] this sequence was downloaded and finally the algae genome was reconstructed in the computer in automated fashion ("Digital Biological Converter").

The next step planned is the teleportation of life to the satellite systems orbiting the planet Earth.

17.3.3 Aim of such Interventions

What is the aim of such interventions today?

First.

Creating enviroment for man on other planets. Seeding microorganisms and

10 Venter: Synthetic Biology, 2011.
11 Venter: Life at the Speed of Light: From the Double Helix to the Dawn of Digital Life, 2013.
12 Venter: Biological Teleportation, 2014.

Abbildung 17.4:
3D printing buildings in space
(Credits: NASA/Autodesk)

other environmentally important microorganisms in the future destinations of homo sapience is crucial for being supplying astronauts with water, oxygen and other substantial chemical components.

Second.

Protecting man on other planets. Space X's Elon Musk wants to *colonize Mars* with modules where earthlings can live. Teleporting technology is the number one way those individuals will get new information, new treatments of diseases that will occur on the planet. Teleportation technology enables just email to Mars a new antibiotic or a new vaccine. Important reservation is that teleportation of multicellular organisms (mammals, trees) is not possible today.

17.4 Summary

According to the current knowledge human beings can not be sure whether life started from the components available on the Earth or the primary living units were a result of teleportation from other objects in the universe and further evolutionary process.

Abbildung 17.5:
DuAxel Rover in the Mojave Desert in California
(Credit: NASA/JPL-Caltech/J. D. Gammell)

We also do not have the full picture what possibilities has Mankind to leave the planet Earth if it is full contaminated and no more suitable for larger living organisms like mammals. The aim of this presentation was to give a short overview of the concepts of biological teleportation today.

17.5 Literatur

ALEXANDER, BRIAN: *Biological Teleporter Could Seed Life Through Galaxy.* MIT Technology Review (Aug 2, 2017), `http://cdn.technologyreview.com/s/608388/biological-teleporter-could-seed-life-through-galaxy/`.

DRAKE, NADIA: Why we explore Mars – and what decades of missions have revealed. In: *Nationalgeographic.com*, Science, (May 4, 2021), `https://www.nationalgeographic.com/science/article/mars-exploration-article`.

ERRINGTON J. & L. T. AART: Microbe Profile: Bacillus subtilis: model organism for cellular development, and industrial workhorse. In: *Microbiology* **166** (May 2020), 5, p. 425-427.

KAUFMANN, MARC: Did Life on Earth Come From Mars? In: *Nationalgeographic.com* (September 6, 2013), https://www.nationalgeographic.com/science/article/130905-mars-origin-of-life-earth-panspermia-astrobiology.

SCHWAB, CATHERINE: *Here's how NASA plans to build on other planets,* (July 31, 2018), NASA, https://www.fastcompany.com/90210735/heres-our-first-glimpse-at-martian-architecture.

STEIGERWALD, WILLIAM: *First Detection of Sugars in Meteorites Gives Clues to Origin of Life*, NASA, (Nov. 18, 2019), https://www.nasa.gov/news-release/first-detection-of-sugars-in-meteorites-gives-clues-to-origin-of-life/.

STETTER, K.: History of discovery of the first hyperthermophiles. In: *Extremophiles* **10** (2006), 5, p. 357–362.

VENTER, CRAIG J.: *Synthetic Biology.* NASA's Ames Research (Audiovisual), 2011, YouTube: https://www.youtube.com/watch?v=KTzG_HIUu9c.

VENTER, CRAIG J.: *Life at the Speed of Light: From the Double Helix to the Dawn of Digital Life.* London: Penguin 2013.

VENTER, CRAIG J.: *Biological Teleportation.* UC San Diego, (2014), YouTube: https://www.youtube.com/watch?v=DVL1nL3SU6I.

WOLFSCHMIDT, GUDRUN: Kosmochemie – Chemische Elemente im Kosmos – Meteoriten, Sterne, Kosmologie. In: WOLFSCHMIDT, GUDRUN (Hg.): *Kosmochemie – Geschichte der Entdeckung und Erforschung der chemischen Elemente im Kosmos zum 150. Jubiläum des Periodensystems der Elemente. Cosmochemistry – History of Discovery and Research of Chemical Elements in the Cosmos – on the Occasion of the 150th Anniversary of the Periodic Table of the Elements (PSE, 1869).* Hamburg: tredition (Nuncius Hamburgensis; Band 50) 2022, p. 66–135.

ZAPATA TRUJILLO, JUAN C. ET AL.: Computational Infrared Spectroscopy of 958 Phosphorus-Bearing Molecules. In: *Frontiers in Astronomy and Space Sciences (FASS), Section Astrochemistry* **8** (2021), https://doi.org/10.3389/fspas.2021.639068.

Abbildung 17.6:
Mensch im Kosmos (Collage)

Collage (Michael A. Rappenglück) aus Vorlagen von Georg Zotti (Stonehenge), CC (der Denker), Michael A. Rappenglück (Kykladenidol), (Deep Field, James-Webb Telescope, NIRCam: Galaxienhaufen SMACS 0723, Sternbild Fliegender Fisch [Volans], 4,6 Milliarden Lichtjahre entfernt, credit: NASA, ESA, CSA, STScI).

Programm der Tagung *Der Mensch im Kosmos*

18.1 Tagung Gesellschaft für Archäoastronomie in Weimar, 21.–25. Juni 2023

Tagungsort /Tagungsraum:
Kultur- und Jugendzentrum mon ami, Goetheplatz 11, Weimar

Weimar, Mittwoch / Wednesday, 21. Juni 2023

Anreise.

Weimar, Donnerstag / Thursday, 22. Juni 2023

05:30–08:30 Uhr Optional: Beobachtungen des Sonnenaufgangs im Römischen Haus im Park an der Ilm. Im unteren Durchgang die Sonnenuhr beobachten und im Erdgeschoss – erstmals! – die bisher nur theoretisch vermutete Lichtinszenierung mit Kronleuchter. Frühstück in den Hotels. /
Option: Observations of the sunrise in the Roman House in the park on the Ilm. Observing the sundial in the lower passage and on the ground floor – for the first time! – the previously only theoretically suspected light show with chandelier. Observation of the sunrise in the Roman House in the Park on the Ilm. Breakfast in the hotels.

10:15–10:25 Uhr Begrüßung
zur *Tagung Gesellschaft für Archäoastronomie*
durch den 1. Vorsitzenden Dr. Michael A. Rappenglück

10:25–11:05 Uhr Dr. Georg Zotti, Wien:
Die Stehenden Steine von Callanish I: Eine öffentliche Datenpublikation für Stellarium / The Standing Stones of Callanish I: A Public Data Release for Stellarium

11:05–11:20 Uhr Kaffeepause / Coffee break

11:20–12:20 Uhr Prof. em. Dr. Clive Ruggles, University of Leicester, UK:
Stonehenge and the major standstill moon / Stonehenge und die große Mondwende

12:20–14:20 Uhr Mittagspause / Lunch break

Abbildung 18.1:
Group photo in Weimar

(Foto: Gudrun Wolfschmidt)

14:20–15:00 Uhr Prof. Dr. Siegfried Hess & Dr. Thomas Storch, Berlin:
Engraved Stones from burial mounds in Franconia point to observations of the Sun, of Orion and of the Pleiades / Gravierte Steine aus fränkischen Grabhügeln weisen auf Beobachtungen der Sonne, des Orion und der Plejaden hin

15:00–17:00 Uhr Stadtführung durch Weimar / Guided tour through Weimar

19:00–20:00 Uhr Dr. Michael A. Rappenglück M.A., Gilching:
Der Mensch in der Welt: Kulturelle Kosmologie einst und jetzt / Human being in the world: Cultural Cosmology once and now

20.00 Uhr Einkehr / End of the evening in Versilia am Frauenplan, Frauentorstraße 17.

Weimar, Freitag / Friday, 23. Juni 2023

05:30–08:30 Uhr Optional: Beobachtungen des Sonnenaufgangs im Römischen Haus im Park an der Ilm. Im unteren Durchgang die Sonnenuhr beobachten und im Erdgeschoss – erstmals! – die bisher nur theoretisch vermutete Lichtinszenierung mit Kronleuchter. Frühstück in den Hotels.
Option: Observations of the sunrise in the Roman House in the park on the Ilm. Observing the sundial in the lower passage and on the ground floor – for the first time! – the previously only theoretically suspected light show with chandelier. Observation of the sunrise in the Roman House in the Park on the Ilm. Breakfast in the hotels.

10:30–11:10 Uhr Dr. Burkard Steinrücken, Westfälische Volkssternwarte, Recklinghausen:
Die Nordturmkapelle im Erfurter Dom / The North Tower Chapel in Erfurt Cathedral

11:10–11:45 Uhr Poster-Session/Kurzvortrag:

Dr. Aarv Jaak, Estland / Estonia:
Homo sapiens and the Universe – Where do we come from? Where are we going to? / Der Homo sapiens und das Universum – Woher kommen wir? Wohin gehen wir?

Hermann Wenzel, München:
Der kretisch-minoische *Diskos von Phaistos*, ein herausragendes Artefakt der Archäoastronomie auf Basis einer dreiteiligen Kosmologie: „Geist-Leben-Sache".
The Cretan-Minoan Phaistos Disc, an outstanding artefact of archaeoastronomy based on a three-part cosmology: „Spirit-Life-Matter".

11:45–12:25 Uhr Dipl.-Math. Harald Gropp, Heidelberg:
Kosmische Lebenswelten auf Himmels- und Erdgloben in Weimar (und Gotha) / Cosmic life worlds on celestial and terrestrial globes in Weimar (and Gotha)

12:35–14:30 Uhr Mittagspause / Lunch break

14:30–15:10 Uhr Klaus Albrecht, Kassel:
Im zyklischen Rhythmus des Kosmos liegt das Schicksal der Menschen – Die bildhafte Zeit im „Heydnischen" Grab von Göhlitzsch-Halle/Leuna / In the cyclical rhythm of the cosmos lies the fate of man – The pictorial time in „Heydnian" grave of Göhlitzsch-Halle/Leuna

15:10–15:50 Uhr Dr. Andreas Fuls, Berlin:
Die Trelleborgen in Dänemark: Eine astronomisch-geometrische Siedlungspla-

nung der Wikinger / The Trelleborgen in Denmark: An astronomical-geometric settlement plan of the Vikings

15:50–16:00 Uhr Kaffeepause / Coffee break

16:00–17:00 Uhr Jahresversammlung der GfA / General Assembly of the GfA

19:00–20:30 Uhr Public invited lecture / Öffentlicher Gastvortrag:
Prof. em. Dr. Clive Ruggles, University of Leicester, UK:
Ancient Astronomies – Ancient Worlds / Alte Astronomien – Alte Welten

20:30 Uhr Abendessen / End of the evening in Versilia am Frauenplan, Frauentorstraße 17.

Weimar, Samstag / Saturday 24. Juni 2023

09:20–10:00 Uhr Dr. Anna Paule, Linz, Austria:
Mensch, Mythos, Kosmos. Sonne und Sonnenfinsternisse im Alten Orient (2. und 1. Jt. v. Chr.) / Human being, myth, cosmos. The sun and solar eclipses in the Ancient Orient (2nd and 1st millennia BC)

10:00–10:40 Uhr Dr. Jörg Bäcker, Göttingen:
Rätsel der alten Astronomie. // Enigmas of ancient astronomy

10:40–10:50 Uhr Kaffeepause / Coffee break

10:50–11:30 Uhr Dr. Ettore Ghibellino, Weimar:
Ist das Römische Haus im Park an der Ilm in Weimar astronomisch verankert? / Is the Roman House in the Park on the Ilm in Weimar astronomically anchored?

11:30–14:00 Uhr Mittagspause / Lunch break

14:00–16:30 Uhr Besichtigung des Ur- und Frühgeschichtsmuseums, Weimar / Visit to the Museum of Prehistory and Early History, Weimar

16:30–19:00 Uhr Wanderung durch die Stadt bis zum archäologischen Freigelände Weimar-Ehringsdorf (wichtige Altsteinzeitfunde); Aufenthalt im Freigelände; Rückfahrt in die Stadt mit Bus /
Walk through the town to the Weimar-Ehringsdorf archaeological site (important Palaeolithic finds); stay at the site; return to the town (by bus).

20.00 Uhr Abendessen / End of the evening in Versilia am Frauenplan, Frauentorstraße 17.

Abbildung 18.2:
Goethe diktiert in seinem Arbeitszimmer dem Schreiber John.
Ölgemälde von Johann Joseph Schmeller (1796–1841), 1831

(Anna Amalia Bibliothek, Weimar, KGe/00742, CC)

Weimar, Sonntag / Sunday, 25. Juni 2023

09:00–09:40 Uhr Roland Gröber, Leverkusen:
Höhenheiligtum am Pfitscher Sattel, Bedeutung und Entwicklung / High altitude shrine on the Pfitscher Sattel, significance and development

09:40–10:20 Uhr Albrecht Ploum, Belgium:
Ikonographie, Wissenschaft und Blitzfiguren // Iconography, Science and Lightning Figures

10:20–11:00 Uhr Dr. Michael A. Rappenglück, Gilching:
Zwischen Welten und Wirklichkeiten: Öffnungen, Passagen und Wege in der Kosmologie und Kosmopraxis / Between Worlds and Realities: Openings, Passages and Pathways in Cosmology and Cosmopraxis

11:00–11:15 Uhr Kaffeepause / Coffee break

11:15–12:30 Uhr Abschlussdiskussion und Abschluss der Tagung; Abreise / Final discussion and closing of the conference; departure.

Abbildung 18.3:
Weimar City Walk in historical old town and group photo in Weimar

(Foto: Gudrun Wolfschmidt)

Autoren

Dr. Jaak Aru (Tartu, Estland)

Tartu, Estland/Estonia
E-Mail: `jaakaru@hotmail.com`

Klaus Albrecht (Naumburg)

Jahrgang 1948, Studium der Kunstpädagogik und -geschichte an der Kunsthochschule Kassel 1969–1974, Berufschullehrer im Tischlerhandwerk 1982 bis 2010; ab 1996 Beschäftigung mit Inhalten der Archäologie und Astronomie, Veröffentlichungen und Vorträge im Rahmen der Archäo-Astronomie zu neolithische Kammergräber und Kunst, zu Kretischen Tempeln, zu Maltesischen Tempeln, zu indischer Astronomie und Astrologie, zu mexikanischen Tempelanlagen.

Hufeisenstr. 10, 34311 Naumburg-Altendorf
E-Mail: `KAlbrecht@t-online.de`

Dr. Jörg Bäcker (Gummersbach)

Study of general linguistics, sinology (Chinese culture, classical and modern Chinese), Central Asian studies (Mongolian studies, Manchu), Slavistic studies, ethnology, comparative religions. Astronomy as a hobby since my schooldays.
Ph.D. in classical sinology on a Neo-Confucian thinker of the Song dynasty (of the 11th century AD), and Mongolian and Manchu studies. Habilitationsschrift (thesis of professorship) on Khitan ethnogenesis, mythology and customs (still unpublished).

Co-worker in the DFG project (Sonderforschungsbereich) Central Asia and China / Tibet / India 1982–1993; 1996–2010 supported by the *Stifterverband für die Deutsche Wissenschaft* (union of German foundations of science) working on the oral traditions of China's and Russia's minority peoples; contributions to the *Enzyklopädie für vergleichende Erzählforschung* ("Enzyklopädie des Märchens", Encyclopedia of Narrative Research) on shamanism, the cosmic tree, visionary literature, Tungus and Daghur peoples, ancient Chinese tales etc.; on the world view and star lore of Manchu shamanism ("The Shaman's Sky"); on the role of astronomy in the mythology and early migrations of homo sapiens; undeciphered scripts and languages of China, and other subjects.

Some publications:

Mandschurische Göttinnen und iranische Teufel. Die Mandschu-Weltentstehungsmythen als Kultursynthesen. Wiesbaden: Harrassowitz 1997.

Sur l'origine des signes cycliques chinois. In: *Approches critiques de la mythologie chinoise.* Ed. par Charle Le Blanc et Rémi Mathieu. Montréal: Les presses de l'université de Montréal 2007, p. 51–85.

Qidan zongjiao cihui yu A'ertai yuxi minzu yuyan wenhua. 契丹宗教词汇与阿尔泰语系民族语言文化 (The religious vocabulary of Khitan and the cultures and languages of the Altaic people). In: 中国多文字时代的历史文献研究 (*Researches on Historical Records in China's Periods of Muliple Scripts*). Ed. by Hongyin Nie 聂鸿音 & Bojun Sun 孙伯君。. Beijing: Shehui kexue wenxian chubanshe 社会科学文献出版社 2010, p. 418–426.

Reconstructing Man's Earliest Creation Myths. On E. J. Michael Witzel: *The Origins of the World's Mythologies.* Oxford, New York etc.: Oxford University Press 2012 (665 pps.) In: Fabula 58, Heft 1–2, p. 146–156.

Franz-Schubert-Str. 53, 51643 Gummersbach
E-Mail: yao.bai@t-online.de, yerugokittan@gmail.com

Alistair Carty (UK)

Carty's company *Archaeoptics* was the first 3D laser scanning service specialising in Cultural Heritage and pioneered the use of scanners in archaeology. Operating primarily between 2000–2006 AD, Archaeoptics scanned over 500 heritage objects and monuments worldwide including Callanish, Stonehenge, the Mary Rose, Cutty Sark and St. Paul's Cathedral. Archaeoptics now offer 3D consultancy services and 2D/3D data archival services and recently published their entire historic archive of 3D scan data.

https://archaeoptics.co.uk/
E-Mail: . . . .

Dr. Andreas Fuls (Berlin)

(Jahrgang 1964) studierte Vermessungswesen an der Technischen Fachhochschule Berlin. Seit 1988 arbeitet er an der *Technischen Universität Berlin* am *Institut für Geodäsie und Geoinformationstechnik.* Im Jahr 2006 promovierte er im Fach *Geschichte der Naturwissenschaften, Technik und Mathematik* an der Universität Hamburg zu dem Thema *„Die astronomische Datierung der klassischen Mayakultur (500–1100 n. Chr.)“.*

Seine derzeitigen Forschungsschwerpunkte liegen in der Archäoastronomie, der Epigraphik und der Entzifferung von antiken Schriftsystemen. Innerhalb der Forschungsprojekte sind neue mathematische Methoden entwickelt worden, um archäoastronomische und epigraphische Daten zu analysieren.
Buchpublikationen:
FULS, ANDREAS: *A Catalog of Indus Signs.* Berlin: Eigenverlag (Mathematica Epigraphica; Band 4) 2023.
FULS, ANDREAS: *Corpus of Indus Inscriptions.* Berlin: Eigenverlag (Mathematica Epigraphica; Band 3) 2022.
FULS, ANDREAS: *Deciphering the Phaistos Disk and other Cretan Hieroglyphic Inscriptions – Epigraphic and Linguistic Analysis of a Minoan Enigma.* Hamburg: tredition (Mathematica Epigraphica; Band 1) 2019.
FULS, ANDREAS: *Die astronomische Datierung der klassischen Mayakultur (500–1100 n. Chr.): Implikationen einer um 208 Jahre verschobenen Mayachronologie.* Dissertation an der Universität Hamburg. Norderstedt: Books on Demand 2007.
WELLS, BRYAN & ANDREAS FULS: *The Correlation of the Modern Western and Ancient Maya Calendars.* Monograph No. 5, Early Site Research Society (West) 2000.

Institut für Geodäsie und Geoinformationstechnik, Sekr. KAI 2-2
Technische Universität Berlin (TUB),
Kaiserin-Augusta-Allee 104–106, D-10553 Berlin
Privat: `http://www.archaeoastronomie.de/`, `http://www.epigraphica.de/`,
Dienst: `https://www.tu.berlin/gis/ueber-uns/team/andreas-fuls`
E-Mail: `andreas.fuls@tu-berlin.de`.

Dr. jur. Ettore Ghibellino (Weimar)

geboren 1969 im württembergischen Waiblingen, aufgewachsen in Deutschland und Italien. Studium der Rechtswissenschaften in Tübingen, Belfast, Speyer und Rom als Stipendiat der *Studienstiftung des Deutschen Volkes*. Befähigung zum Richteramt aus München, Meister der Rechte (MJur) aus Oxford sowie zum Thema internationale Rechtsvereinheitlichung in Bayreuth promoviert (summa cum laude).
Seit 2001 lebt Ghibellino als Goethe-Forscher in Weimar, seine Doppelbiographie über Goethe und Anna Amalia erregt Aufsehen. Ghibellino veranstaltet interdisziplinäre Tagungen zum Forschungsfeld *„Anna Amalia und Goethe"* sowie zu zeitgenössischen Dichtern und Komponisten.

Publikationen (Auswahl)
Goethe und Anna Amalia: Eine verbotene Liebe. Weimar (1. Auflage) 2003, (5. Auflage) 2020.
Goethe und Anna Amalia: Das Römische Haus als „Geheimster Wohnsitz". Weimar 2020.

Rothäuserbergweg 2 b, 99425 Weimar
https://annaamalia-goethe.de/wordpress/
E-Mail: ghibellino@annaamalia-goethe.de, AnnaAmaliaGoethe@web.de

Roland Gröber (Leverkusen)

(*1941): Studium Dipl.-Ing. für Nachrichtentechnik an der Technischen Hochschule München. 35 Jahre in leitender Funktion bei der Bayer AG in Leverkusen.
Wissenschaftliches Interesse an der Astronomie (Mitglied *Sternfreunde Köln*), Archäologie (Teilnahme an mehreren Ausgrabungen) und an der Archäoastronomie (Mitglied der *Gesellschaft für Archäoastronomie*).
Publikationen: Das Bergheiligtum am Pfitscher Sattel bei Meran. Schalensteine und astronomische Beobachtungen in der Kupferzeit (ca. 3200 v. Chr.). Leverkusen 2016. Siehe auch Literatur, S. 126.

Dresdenerstraße 2, 51373 Leverkusen
E-Mail: rgroeber@gmx.de

Dipl.-Math. Harald Gropp (Heidelberg)

Harald Gropp forscht über Mathematik (vor allem Configurationen, Graphentheorie, Kombinatorik) und Geschichte der Mathematik und der Astronomie (unter anderem nichteuropäische Kulturen und Archäoastronomie).

Henkel-Teroson-Str. 20, 69123 Heidelberg
E-Mail: d12@ix.urz.uni-heidelberg.de

Prof. Dr. Siegfried Hess (TU Berlin, Röttenbach)

born 1940 in Hof/Saale (Northern Bavaria) is Professor for Theoretical Physics. He retired in 2007 but is still engaged in research projects at the Technical University (TU) Berlin, together with younger colleagues.
In 1959 he was enrolled as student of Mathematics and Physics at the University Erlangen, he also attended lecures in Chemistry and Astronomy. After his Diploma in Physics in Erlangen in 1964, he spent a year as graduate student of Chemical Engineering in Minneapolis, Minnesota, USA. He got his PhD (Dr.rer.nat.) in 1967, was postdoc in Leiden, Holland, and made his Habilitation for Physics in 1970 at the University Erlangen-Nürnberg, where he later became Professor for Theoretical Physics. In 1984 he accepted a call to a chair for Theoretical Physics at the TU Berlin. Over the years, he also worked in research centers and at universities in Toronto

(Canada), again in Leiden (Holland), in Boulder (Colorado, USA), in Grenoble (France), in Canberra (Australia) and in Santa Barbara (California, USA). He is a corresponding member of the "Accademia Peloritana dei Pericolanti" in Messina (Sicily, Italy).
He published more than 250 articles in refereed scientific journals and as contributions to proceedings and to books, mainly on Statistical Physics of non-equilibrium phenomena in Soft Matter, such as gases, liquids, liquid crystals (LC) and polymeric fluids. Among his books is the popular-scientific work "Opa, was macht ein Physiker?", Wiley-VCH 2014.
From his long lasting interest in Astronomy and Archeology emerged, in 2018, the article *"Prehistoric Engraved Stones from Northern Bavaria hint at Observations of the Sun and of Stars"* which, so far, received more than 600 "reads" at ResearchGate. The recent collaboration with Thomas Storch provided an extra stimulus for activities in Archaeo-Astronomy.

Institut für Theoretische Physik,
Technische Universität Berlin, EW 7-1
Hardenbergstr. 36, 10623 Berlin, Germany
E-Mail: sighess@icloud.com, S.Hess@physik.tu-berlin.de

Dr. Anna Paule (Linz, Österreich)

is an Austrian Bronze/Early Iron Age archaeologist. She graduated from the University of Salzburg, Dept. of Classical and Early Aegean Archaeology, with a master's degree with distinction in 2007. This was followed by doctoral studies at the *Université Aix-Marseille I, Maison Méditerranéenne des Sciences de l'Homme* (MMSH), France. In 2013, she graduated from this university with a doctoral degree in Archaeology (PhD) with First Class Honours. After a period of protracted illness, she became postdoctoral researcher in Prof. Lorenz Rahmstorf's grant at the University of Göttingen, Germany (ERC-2014-CoG WEIGHTANDVALUE; 2018–2020). As the end of this research project coincided with the start of the pandemic, this has been followed by a period of independent, self-financed research and project preparation activities (especially in the field of archaeoastronomy).
Her research interests cover a variety of topics related to Bronze/Early Iron Age Cyprus, the Near East, and Europe. This includes small artefacts (gold jewellery, weighing tools), iconography, first dairy skills, and archaeoastronomy (solar symbols, myths, and eclipses). She is author of ten articles (cf. website; three more articles are in print), but also disposes of practical, hands-on experience in various fields of archaeology, such as archaeological fieldwork in Austria and abroad (e. g., Cyprus and Greece).

Buchenweg 11, 4203 Altenberg bei Linz, Austria
Website: https://www.uni-goettingen.de/de/dr.+anna+paule/599352.html
E-Mail: annapaule.archaeo@hotmail.com

Drs. Albrecht Ploum (Belgium)

degree in psychology (Radboud University, Nijmegen, Netherlands), Independent scholar. Fields of study: History of Science, History of Art, Geomorphology, Biomorphology, Astronomy, Cosmology.
Main work: *Spiegel des Universums* (1996, in German and Dutch). UNESCO library citation: true science. The aim of this book was to investigate some of the shapes shared by terrestrial biological and cosmic structures to try to determine whether resemblances are strictly coincidental or whether they are expressions of a more profound coherence among elements in the universe.

Radboud University Nijmegen, Netherlands
Website: https://ploumresearch.eu/
E-Mail: albrecht.ploum@gmail.com

Dr. Michael Rappenglück, M.A. (Gilching)

geb. 1957, Karlsruhe; 1984 M. A. in Philosophie, Logik, Wissenschaftstheorie, Christliche Philosophie und Theologische Propädeutik (Ludwig-Maximilian-Universität, München);
1998 Dr.rer.nat. in Geschichte der Naturwissenschaften / Geschichte der Astronomie (Ludwig-Maximilian-Universität, München). Leiter der Volkssternwarten der *vhs Gilching* und *vhs Stadt Fürstenfeldbruck*.
Forschungsschwerpunkte: Kulturastronomie (mit Schwerpunkt Ur- und Frühgeschichte), Ethnoastronomie, Ethnomathematik, Urgeschichte der Wissenschaften, Impaktforschung, Symbol- und Mythenforschung.

Buchveröffentlichungen:
Eine Himmelskarte aus der Eiszeit? Ein Beitrag zur Urgeschichte der Himmelskunde und zur paläoastronomischen Methodik, aufgezeigt am Beispiel der Szene in Le Puits, Grotte de Lascaux. Com. Montignac, Dép. Dordogne, Rég. Aquitaine, France 1999.
Rappenglück et al. (ed.): *Astronomy and Power: How Worlds Are Structured.* Proceedings of the SEAC 2010 Conference. BAR International S2794 (2016).
Weitere Informationen siehe Web-Seite: http://www.infis.org.

Bahnhofstr. 1, 82205 Gilching, Germany
Web-Seite: http://www.infis.org
E-Mail: mr@infis.org.

Victor Reijs, MA (Lemelerveld, The Netherlands)

Victor Reijs got his interest in archaeocosmology/archaeoastronomy due to visiting the passage mounds of Newgrange and Knowth, where art and astronomy seems to be interlinked. He

got hooked in trying to interpret these Neolithic art expressions and monuments. In 2016 he achieved an MA in Cultural Astrology and Astronomy.
Broadcastings, over the Internet, of the setting Sun/Moon were performed at Maeshowe (Orkney, Scotland from 1998 to 2018) and Callanish (Lewis, Scotland between 2005 and 2006).
The general principles behind the lunar standstill limit events are explained further in:
`http://www.archaeocosmology.org/eng/majorstandstills2024.htm`

Lemelerveld, The Netherlands
E-Mail: `ma.victor.reijs@gmail.com`

Emma Rennie (Callanish, Isle of Lewis, Scotland)

Emma Rennie is a South African born 3D Graphic Artist and Photographer. She lives on the Isle of Lewis, Scotland, out near the Callanish Standing stones and owns the ruin of a Blackhouse (traditional stone, thatched island home) adjacent to the main site at Callanish. It is currently being prepared for restoration – `https://callanishblackhousetearoom.com`
Over the years, she has become a skilled photographer of the moon, the stones, stars, aurora, zodiacal light, moonbows and so much more. Her work can be seen at
`https://callanishdigitaldesign.com`. Follow Emma for amazing pictures and prints and maybe help with the restoration of the blackhouse!

Callanish, Isle of Lewis, Scotland
E-Mail: `callanishdd@gmail.com`

Prof. Dr. Clive Ruggles (University of Leicester, UK)

Emeritus Professor of Archaeoastronomy

University of Leicester, Leicester, UK
Department: School of Archaeology and Ancient History
`https://le.ac.uk/people/clive-ruggles`
E-Mail: `cliveruggles2010@btinternet.com`

Dr. Burkard Steinrücken (Recklinghausen)

geb. 1965 in Olsberg, Studium der Physik in Aachen, promovierte in Dortmund mit einer Arbeit zur Didaktik der Elementarteilchenphysik, seit 1996 Leiter der Westfälischen Volkssternwarte und des Planetariums in Recklinghausen.
Arbeitsfelder: außerschulische astronomische und naturwissenschaftliche Bildung, Astronomie- und Physikdidaktik, Archäoastronomie.

Vorstandsmitglied im *Initiativkreis Horizontastronomie im Ruhrgebiet e.V.* und in der *Gesellschaft für Archäoastronomie e.V.*.
Online-Veröffentlichungen zu archäoastronomischen Fragestellungen auf
`https://sternwarte-recklinghausen.de/`.

Westfälische Volkssternwarte und Planetarium Recklinghausen
Stadtgarten 6, 45657 Recklinghausen
E-Mail: `steinruecken@sternwarte-recklinghausen.de`

Dipl.-Phys. Thomas Storch (Hemhofen)

studierte Physik an der Friedrich-Alexander-Universität Erlangen-Nürnberg und schrieb dort 1984 seine Diplomarbeit *Spektroskopische Analyse des Doppelsternsystems HD48099*. Danach begann er seine berufliche Tätigkeit in den diagnostischen Bereichen der digitalen Bildverarbeitung von Computer- und Kernspintomographie von Siemens Erlangen/Forchheim. 2014–2020 Mitglied und Co-Autor der internationalen ISO/IEC Joint Working Group 1 für die weltweit gültige Sicherheitsnorm *ISO/IEC 14971 – Risk Management for Medical Devices*.
Seit dem Beginn seiner Altersteilzeit (2020) widmet er sich intensiv der Astrofotografie und der breiten Fortbildung hinsichtlich Astronomie (z. B. monatliche Sternespaziergänge am „Baumwipfelpfad Ebrach"/ Bildbearbeitungskurse in der Astrofotografie). Er präsentiert seine Werke auf Instagramm (> 1000 Follower) und Facebook (> 250 Follower) und zeigt sie auch auf verschiedenen Ausstellungen:
2021–08 Kunst am Rathaus Röttenbach, 2022–03 Hopfenhaus – Röttenbach,
2022–10 Schloss Adelsdorf: „Schönheiten der Nacht",
2022/2023 Dauerausstellung:
„Das Ungesehene sichtbar machen" im Naturerlebniszentrum NEZ/Maasholm.

Hemhofen
E-Mail: `sternenstorch@gmx.de`
Instagram/Facebook: @Sternestorch

Dipl.-Ing. Hermann Wenzel (München)

geb. 1939, Humanistisches Gymnasium Koblenz, Studium Architektur TU München, Klassische Archäologie LMU München. Mitarbeit bei Hans Scharun in Berlin an der Bibliothek Preußischer Kulturbesitz; selbständiges Architekturbüro in München.
Mitherausgeber der Literaturzeitschrift TORSO München-Berlin. Private frühgeschichtliche Forschungen. Schwerpunkt: Numerisch/astronomische Strukturen in Alphabeten und zugehöriger Epigraphik: Minoisch, Griechisch, Lateinisch, Arabisch, Germanisch, Entzifferung des *Diskos von Phaistos*.
Publikationen:

Zeit der Runen in Futhark und Alphabet der Germanen. München, Ravensburg: GRIN 2016.
Astronomie in Lateinischer Epigraphik. https://lisa.gerda-henkel-stiftung.de/astronomie_in_lateinischer_epigraphik?nav_id=6776.
Entzifferung des *Diskos von Phaistos* als mathematisch/astronomisches Kompendium (11 Kapitel) – siehe im Wissenschaftsportal der Gerda-Henkel Stiftung, Düsseldorf, 2010 –2012, https://lisa.gerda-henkel-stiftung.de/entzifferung_des_diskos_von_phaistos?nav_id=1304. Astronomie der Germanen im älteren Futhark: lisa.gerda-henkel-stiftung, Essay 2017, https://lisa.gerda-henkel-stiftung.de/astronomie_der_germanen_im_aelteren_fuark?nav_id=7355.
Astronomie der Germanen im jüngeren Futhark: lisa.gerda-henkel-stiftung Essay 2017.
Die Taufe Dänemarks – Astronomie der Germanen in Futhark und Epigraphik der jüngeren Runen, Essay 2019, https://lisa.gerda-henkel-stiftung.de/die_taufe_daenemarks?nav_id=8086.

TU und LMU München
E-Mail: sullus.hw@gmail.com

Prof. Dr. Gudrun Wolfschmidt (Hamburg)

geb. in Nürnberg, Dissertation *Analyse enger Doppelsternsysteme*, Universität Erlangen-Nürnberg (Remeis-Sternwarte Bamberg), 1. und 2. Staatsexamen (Physik und Mathematik), Gymnasiallehrerin im Freistaat Bayern. 1987–1997 Deutsches Museum in München: Dauerausstellung Astronomie/Astrophysik (Eröffnung 1992, bis 2022) und Assistentin am Forschungsinstitut für Geschichte der Naturwissenschaft und Technik des Deutschen Museums; Lehre und Habilitation *Genese der Astrophysik* (1997) an der Ludwig-Maximilians-Universität in München.
Seit 1997 Professorin für Geschichte der Naturwissenschaften an der Universität Hamburg, Center for History of Science and Technology.
Aktivitäten siehe: https://www.fhsev.de/Wolfschmidt/
Forschungsschwerpunkte:
Astronomie-, Physik-, Wissenschafts- und Technikgeschichte (Frühe Neuzeit und 19./ 20. Jahrhundert), Archäo- und Kulturastronomie, Sternwarten und wissenschaftliche Instrumente.
Buchveröffentlichungen (Auswahl):
Copernicus – Revolutionär wider Willen (1994), *Milchstraße – Nebel – Galaxien. Strukturen im Kosmos von Herschel bis Hubble* (1995), *Popularisierung der Naturwissenschaften / Astronomie* (2000, 2002, 2017), *Sterne weisen den Weg – Geschichte der Navigation* (2009), *Cultural Heritage of Astronomical Observatories* (2009), *Colours in Culture and Science* (2011), *Harmony and Symmetry* (2020), *Astronomy in Culture – Cultures of Astronomy* (2022).
Publikationen – Gesamt-Liste: https://www.fhsev.de/Wolfschmidt/publikat.php.

AG Geschichte der Naturwissenschaft und Technik
Hamburger Sternwarte, Universität Hamburg
Bundesstraße 55 – Geomatikum, 20146 Hamburg

https://www.fhsev.de/Wolfschmidt/GNT/home-wf.htm
E-Mail: gudrun.wolfschmidt@uni-hamburg.de.

Dr.-techn. Dipl.-Ing. Georg Zotti (Wien, Österreich)

Zotti ist Informatiker und Astronom mit Interesse für Anwendungen der Computergraphik für kulturastronomische Forschung und Vermittlung. Nach seinem Doktorat (TU Wien) widmete er sich 2008–2012 an der interdisziplinären Plattform für archäologische Forschung (VIAS, Universität Wien) der archäoastronomischen Analyse jungsteinzeitlicher Kreisgrabenanlagen in Niederösterreich und begann mit der Entwicklung von Software für die Virtuelle Archäoastronomie.
Seither ist er auch freiwilliger Entwickler im quelloffenen Computerplanetarium „*Stellarium*" und arbeitete am *Ludwig Boltzmann Institut für Archäologische Prospektion und Virtuelle Archäologie* in Wien (2012–2023). Er ist 2. Vorsitzender der *Gesellschaft für Archäoastronomie*.

LBI ArchPro
Wien, Österreich
https://homepage.univie.ac.at/georg.zotti/
E-Mail: Georg.Zotti@univie.ac.at

Nuncius Hamburgensis

Beiträge zur Geschichte der Naturwissenschaften

Norderstedt: Books on Demand (nur Bd. 2, 6, 7, 8, 10, 11, 14 und 15)

Hamburg: tredition Verlag  tredition® (alle anderen Bände).

Hg. von Gudrun Wolfschmidt,
AG Geschichte der Naturwissenschaft und Technik,
Hamburger Sternwarte, Universität Hamburg – ISSN 1610-6164

Diese Reihe „Nuncius Hamburgensis" wird gefördert von der Hans Schimank-Gedächtnisstiftung. Dieser Titel wurde inspiriert von „Sidereus Nuncius" und von „Wandsbeker Bote".

`https://www.fhsev.de/Wolfschmidt/GNT/research/nuncius.php`

- Band 1 (2009): *Hans Schimank (1888–1979) Ausgewählte Schriften.* Mit einem Beitrag ‚Hans Schimanks Otto von Guericke' von Fritz Krafft. Bearb. von Timo Engels & Igor Abdrakhmanov. Hg. G. Wolfschmidt.
- Band 2 (2007): Wolfschmidt, Gudrun (Hg.): *Hamburgs Geschichte einmal anders – Entwicklung der Naturwissenschaften, Medizin und Technik – Teil 1.*
- Band 3 (2010): Wolfschmidt, Gudrun (Hg.): *Astronomie in Nürnberg.* Proceedings der Tagung vom 2.–3. April 2005 in Nürnberg anläßlich des 500. Todestages von Bernhard Walther (1430–1504) und des 300. Todestages von Georg Christoph Eimmart (1638–1705).
- Band 4 (2011): Wolfschmidt, Gudrun (Hg.): *Entwicklung der Theoretischen Astrophysik.* Proceedings des Kolloquiums des Arbeitskreises Astronomiegeschichte in der Astronomischen Gesellschaft in Köln am 26. September 2005.
- Band 5 (2026): Wolfschmidt, Gudrun (Hg.): *400 Jahre Physik in Hamburg vom Akademischen Gymnasium über das Staatsinstitut zur Universität.*
- Band 6 (2007): Wolfschmidt, Gudrun (Hg.): *Von Hertz zum Handy – Entwicklung der Kommunikation.* Begleitbuch zur Ausstellung zum 150. Geburtstag von Heinrich Hertz (1857–1894).
- Band 7 (2009): Wolfschmidt, Gudrun (Hg.): *Hamburgs Geschichte einmal anders – Entwicklung der Naturwissenschaften, Medizin und Technik, Teil 2.*

Nuncius Hamburgensis –
Beiträge zur Geschichte der Naturwissenschaften, Band 1
Hans Schimank (1888-1979)
Ausgewählte Schriften
Mit einem Beitrag „Hans Schimanks Otto von Guericke"
von Fritz Krafft
Bearbeitet von
Timo Engels und Igor Abdrakhmanov
Norderstedt: Books on Demand 2009

Nuncius Hamburgensis –
Beiträge zur Geschichte der Naturwissenschaften, Band 2
Gudrun Wolfschmidt (Hrsg.)
Hamburgs Geschichte einmal anders
Entwicklung von Naturwissenschaft, Medizin und Technik
Norderstedt: Books on Demand 2007

Nuncius Hamburgensis –
Beiträge zur Geschichte der Naturwissenschaften, Band 3
Gudrun Wolfschmidt (Hg.)
Astronomie in Nürnberg
tredition

Nuncius Hamburgensis –
Beiträge zur Geschichte der Naturwissenschaften, Band 4
Gudrun Wolfschmidt (Hg.)
Entwicklung der
Theoretischen Astrophysik
tredition

Nuncius Hamburgensis –
Beiträge zur Geschichte der Naturwissenschaften, Band 12
Gudrun Wolfschmidt (Hg.)
Astronomy
in new Wavelengths
Astronomie in neuen Wellenlängen
tredition

Nuncius Hamburgensis –
Beiträge zur Geschichte der Naturwissenschaften, Band 6
Gudrun Wolfschmidt (Hrsg.)
Von Hertz zum Handy
Entwicklung der Kommunikationstechnik
Vision 1930:
Bildtelefon

Nuncius Hamburgensis –
Beiträge zur Geschichte der Naturwissenschaften, Band 7
Gudrun Wolfschmidt (Hg.)
Hamburgs Geschichte
einmal anders
Entwicklung der Naturwissenschaften,
Medizin und Technik, Teil 2

Nuncius Hamburgensis –
Beiträge zur Geschichte der Naturwissenschaften, Band 8
Gudrun Wolfschmidt (Hrsg.)
Prähistorische Astronomie
und Ethnoastronomie
2008

Nuncius Hamburgensis –
Beiträge zur Geschichte der Naturwissenschaften, Band 10
Gudrun Wolfschmidt (ed.)
Heinrich Hertz (1857-1894)
and the Development
of Communication
Proceedings of the Symposium

- Band 8 (2008): Wolfschmidt, Gudrun (Hg.):
Prähistorische Astronomie und Ethnoastronomie. Proceedings der Tagung des Arbeitskreises Astronomiegeschichte in der AG in Würzburg 2007.
- Band 9 (2024): Wolfschmidt, Gudrun (Hg.):
Genese der Astrophysik – The Rise of Astrophysics.
- Band 10 (2008): Wolfschmidt, Gudrun (ed.):
Heinrich Hertz (1857–1894) and the Development of Communication. Proceedings of the International Symposium in Hamburg, Oct., 8–12, 2007.
- Band 11 (2008): Wolfschmidt, Gudrun (Hg.):
Astronomisches Mäzenatentum. Proceedings des Symposiums in der Kuffner-Sternwarte in Wien 2004.
- Band 12 (2024): Wolfschmidt, Gudrun (Hg.):
Astronomie in neuen Wellenlängen – Astronomy in New Wavelength. Proceedings der Tagung des Arbeitskreises Astronomiegeschichte in der Astronomischen Gesellschaft in Würzburg.
- Band 13 (2025): Cura, Katrin:
Alchemie im Deutschen Museum. Bearb. und hg. von G. Wolfschmidt.
- Band 14 (2008): Wolfschmidt, Gudrun (Hg.):
„Navigare necesse est“ – Geschichte der Navigation. Begleitbuch zur Ausstellung 2008/09 in Hamburg und Nürnberg.
- Band 15 (2009): Wolfschmidt, Gudrun:
„Sterne weisen den Weg“ – Geschichte der Navigation. Katalog zur Ausstellung 2008/10 in Hamburg und Nürnberg.
- Band 16 (2012): Wolfschmidt, Gudrun (Hg.):
Simon Marius, der fränkische Galilei, und die Entwicklung des astronomischen Weltbildes.
- Band 17 (2026): Cura, Katrin: *Auf den Leim gehen – Geschichte der Klebstoffe.* Hg. von Gudrun Wolfschmidt.
- Band 18 (2011): Wolfschmidt, Gudrun (Hg.):
Farben in Kulturgeschichte und Naturwissenschaft. Begleitbuch zur Ausstellung in Hamburg 2010–2012.
- Band 19 (2011): Andre Koch Torres Assis und Karl Heinrich Wiederkehr und Gudrun Wolfschmidt: *Weber's Planetary Model of the Atom.*
- Band 20 (2011): Wolfschmidt, Gudrun (Hg.):
Hamburgs Geschichte einmal anders – Entwicklung der Naturwissenschaften, Medizin und Technik, Teil 3.
- Band 21 (2019): Wolfschmidt, Gudrun (Hg.): *Vom Abakus zum Computer – Geschichte der Rechentechnik, Teil 1. Begleitbuch zur Ausstellung, 2015–2018.*
- Band 22 (2011): Wolfschmidt, Gudrun (ed.): *Colours in Culture and Science. 200 Years Goethe's Colour Theory.* Proceedings of the Interdisciplinary Symposium in Hamburg, October 12–15, 2010.

Gudrun Wolfschmidt (Hrsg.)
Astronomisches Mäzenatentum in Europa
2008

Nuncius Hamburgensis –
Beiträge zur Geschichte der Naturwissenschaften, Band 14
Gudrun Wolfschmidt (Hrsg.)
Navigare necesse est
Geschichte der Navigation

Nuncius Hamburgensis –
Beiträge zur Geschichte der Naturwissenschaften, Band 15
Gudrun Wolfschmidt (Hrsg.)
Sterne weisen den Weg
Geschichte der Navigation

Gudrun Wolfschmidt (Hg.)
der fränkische Galilei,
und die Entwicklung
des astronomischen Weltbildes
tredition

Nuncius Hamburgensis –
Beiträge zur Geschichte der Naturwissenschaften, Band 18
Gudrun Wolfschmidt (Hg.)
Farben
in Kulturgeschichte und Naturwissenschaft
science

Nuncius Hamburgensis –
Beiträge zur Geschichte der Naturwissenschaften, Band 19
Andre Koch Torres Assis,
Karl Heinrich Wiederkehr
and Gudrun Wolfschmidt
Weber's Planetary Model
of the Atom
Ed. by Gudrun Wolfschmidt

Gudrun Wolfschmidt (Hg.)
Hamburgs Geschichte einmal anders
Entwicklung der Naturwissenschaften,
Medizin und Technik, Teil 3
tredition

Nuncius Hamburgensis –
Beiträge zur Geschichte der Naturwissenschaften, Band 22
Gudrun Wolfschmidt (ed.)
Colours
in Culture and Science
tredition science

Nuncius Hamburgensis –
Beiträge zur Geschichte der Naturwissenschaften, Band 26
Gudrun Wolfschmidt (Hg.)
Der Himmel über Tübingen
Barocksternwarten -
Landesvermessung -
Hochenergieastrophysik
tredition

- Band 23 (2024): Wolfgang Lange:
Edition des Briefwechsels von Carl Friedrich Gauß (1777–1855) und Johann Friedrich Benzenberg (1777–1846). Hg. von G. Wolfschmidt.
- Band 24 (2014): Wolfschmidt, Gudrun (Hg.):
Kometen, Sterne, Galaxien – Astronomie in der Hamburger Sternwarte.
Zum 100jährigen Jubiläum der Hamburger Sternwarte in Bergedorf.
- Band 25 (2016): Wolfschmidt, Gudrun (Hg.):
Wissen aus 400 Jahren Chemie in Hamburg.
Hamburgs Geschichte einmal anders, Teil 4.
- Band 26 (2026): Eike-Christian Harden: *Concordia Res Parvae Crescunt – Fortschritte in Naturwissenschaft und Technik im Goldenen Zeitalter der Niederlande.* Hg. von Gudrun Wolfschmidt.
- Band 27 (2014): Susanne M. Hoffmann: *lingua sine limitibus – Analysen zur Sprache der Bilder und Bildsprachen, insbesondere zur Kommunikation von Fachinformationen.* Sprachen der Populärdidaktik mit zwei- bis vierdimensionalen Medien an Beispielen der Astronomie. Hg. G. Wolfschmidt.
- Band 28 (2014): Wolfschmidt, Gudrun (Hg.): *Der Himmel über Tübingen. Barocksternwarten – Landesvermessung – Astrophysik.*
Proceedings der Tagung des AKAG in Tübingen 2013.
- Band 29 (2013): Wolfschmidt, Gudrun (Hg.):
Sonne, Mond und Sterne – Meilensteine der Astronomiegeschichte.
Zum 100jährigen Jubiläum der Hamburger Sternwarte in Bergedorf.
- Band 30 (2024): Hans G. Beck: *Astrobecks Sternzeiten. Aus dem Leben des Industrie-Astronomen Hans G. Beck.*
Festschrift zu Ehren von Alfred Jensch und Rolf Riekher.
Bearbeitet und herausgegeben von Gudrun Wolfschmidt.
- Band 31 (2015): Wolfschmidt, Gudrun (Hg.):
Astronomie in Franken. Von den Anfängen bis zur modernen Astrophysik – 125 Jahre Dr. Karl Remeis-Sternwarte Bamberg (1889). Proceedings der Tagung des Arbeitskreises Astronomiegeschichte in der AG in Bamberg 2014.
- Band 32 (2018): Wolfschmidt, Gudrun (Hg.):
Astronomie und Astrologie im Kontext von Religionen. Proceedings der Tagung des Arbeitskreises Astronomiegeschichte in der AG in Göttingen 2017.
- Band 33 (2016): Wolfschmidt, Gudrun (ed.):
Enhancing University Heritage-Based Research.
Proceedings of the XV Universeum Network Meeting, Hamburg, 2014.
- Band 34 (2026): Ewering, Christoph: *Bürgerliches Sammeln im 19. Jahrhundert. Das Museum Godeffroy.* Hg. von G. Wolfschmidt.
- Band 35 (2018): Wolfschmidt, Gudrun (Hg.):
Baudenkmäler des Himmels – Astronomie in gebautem Raum und gestalteter Landschaft. Proceedings der Tagungen der Gesellschaft für Archäoastronomie, 2014 bis 2016.

Nuncius Hamburgensis –
Beiträge zur Geschichte der Naturwissenschaften, Band 39
Elvira Pfitzner
Vom Jakobstab zur Spektralanalyse
Astronomie an der Rostocker Universität
Hg. von Gudrun Wolfschmidt
tredition

tredition
WOLFRAM ZIEGLER
AUS DEM LEBEN
EINES FRANKEN
Dr. AUGUST ZIEGLER (1885-1937)
Pflanzenzüchter in Togo
Rebenzüchter in Bayern

Nuncius Hamburgensis –
Beiträge zur Geschichte der Naturwissenschaften, Band 31
Gudrun Wolfschmidt (Hg.)
Astronomie in Franken
Von den Anfängen bis zur modernen Astrophysik.
125 Jahre Dr. Remeis-Sternwarte Bamberg (1889)
tredition

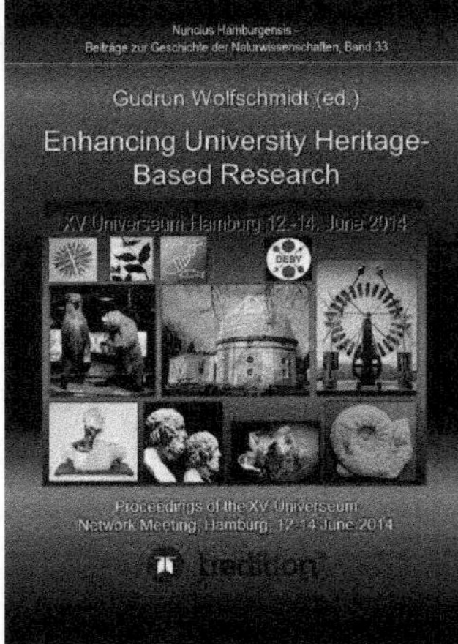
Nuncius Hamburgensis –
Beiträge zur Geschichte der Naturwissenschaften, Band 33
Gudrun Wolfschmidt (ed.)
Enhancing University Heritage-Based Research
XV Universeum Hamburg 12.-14. June 2014
Proceedings of the XV Universeum
Network Meeting, Hamburg, 12-14 June 2014
tredition

Nuncius Hamburgensis –
Beiträge zur Geschichte der Naturwissenschaften, Band 25
Gudrun Wolfschmidt (Hg.)
400 Jahre Chemie in Hamburg
Hamburgs Geschichte einmal anders –
Entwicklung der Naturwissenschaften,
Medizin und Technik, Teil 4
tredition

Nuncius Hamburgensis –
Beiträge zur Geschichte der Naturwissenschaften, Band 43
Carlotta Martini
Zwei Frauenleben für die
Wissenschaft im 18. Jahrhundert
Eine vergleichende Fallstudie
zu Émilie du Châtelet und
Maria Gaetana Agnesi

Nuncius Hamburgensis –
Beiträge zur Geschichte der Naturwissenschaften, Band 27
Susanne M. Hoffmann (Hg.)
Lingua sine limitibus
Analysen zur Sprache der Bilder und Bildsprachen,
insbesondere zur Kommunikation von Fachinformationen
Sprachen der Populärdidaktik
mit zwei- bis vierdimensionalen Medien
an Beispielen der Astronomie
tredition

Nuncius Hamburgensis –
Beiträge zur Geschichte der Naturwissenschaften, Band 40
Jürgen Kost
Wissenschaftlicher Instrumentenbau
der Firma Merz in München (1838-1932)
tredition

Gudrun Wolfschmidt (Hg.)
Festschrift – Proceedings of the
Christoph J. Scriba Memorial Meeting
History of Mathematics
„Mathematik ist eine Bedingung aller exakten Erkenntnis"
Immanuel Kant (1724-1804)
tredition

- Band 36 (2017): Wolfschmidt, Gudrun (ed.):
Festschrift – Proceedings of the Christoph J. Scriba Memorial Meeting – History of Mathematics.
- Band 37 (2020): Wolfschmidt, Gudrun (Hg.):
70 Jahre Observatorium Hoher List. Sieben Jahrzehnte astronomische Beobachtung in der Eifel.
- Band 38 (2018): Wolfschmidt, Gudrun (Hg.):
Astronomie im Ostseeraum – Astronomy in the Baltic. Proceedings der Tagung des Arbeitskreises Astronomiegeschichte in der Astronomischen Gesellschaft in der AG in Kiel 2015.
- Band 39 (2015): Pfitzner, Elvira:
Vom Jakobsstab zur Spektralanalyse – Astronomie an der Rostocker Universität. Bearb. und hg. von Gudrun Wolfschmidt.
- Band 40 (2015): Kost, Jürgen:
Wissenschaftlicher Instrumentenbau der Firma Merz in München (1838–1932). Bearb. und hg. von Gudrun Wolfschmidt.
- Band 41 (2017): Wolfschmidt, Gudrun (Hg.):
Popularisierung der Astronomie. Proceedings der Tagung des Arbeitskreises Astronomiegeschichte in der AG in Bochum 2016.
- Band 42 (2019): Wolfschmidt, Gudrun (Hg.):
Orientierung, Navigation und Zeitbestimmung – Wie der Himmel den Lebensraum des Menschen prägt. Tagung der Gesellschaft für Archäoastronomie in Hamburg 2017.
- Band 43 (2017): Carlotta Martini: *Zwei Frauenleben für die Wissenschaft im 18. Jahrhundert. Eine vergleichende Fallstudie zu Émilie du Châtelet und Maria Gaetana Agnesi.* Bearbeitet und herausgegeben von Gudrun Wolfschmidt.
- Band 44 (2018): Panagiotis Kitmeridis: *Popularisierung der Naturwissenschaften am Beispiel des Physikalischen Vereins Frankfurt.* Überarbeitet und herausgegeben von Gudrun Wolfschmidt.
- Band 45 (2026): Wolfschmidt, G. (Hg.): *Hamburgs Geschichte einmal anders – Entwicklung der Naturwissenschaften, Medizin und Technik, Teil 5.*
- Band 46 (2024): Wolfschmidt, Gudrun (Hg.):
Vom Abakus zum Computer – Geschichte der Rechentechnik, Teil 2. Mit einem Katalog der Instrumente und Modelle in der Sammlung des GNT.
- Band 47 (2018): Irena Kampa: *Die astronomischen Instrumente von Johannes Hevelius.* Hg. von Gudrun Wolfschmidt.
- Band 48 (2020): Wolfschmidt, G. (Hg.): *Maß und Mythos, Zahl und Zauber: Die Vermessung von Himmel und Erde.* Tagung der Gesellschaft für Archäoastronomie in Dortmund 2018.

Nuncius Hamburgensis –
Beiträge zur Geschichte der Naturwissenschaften, Band 24
Gudrun Wolfschmidt (Hg.)
Kometen, Sterne, Galaxien
Astronomie in der Hamburger Sternwarte
tredition

Nuncius Hamburgensis –
Beiträge zur Geschichte der Naturwissenschaften, Band 29
Gudrun Wolfschmidt (Hg.)
Sonne, Mond und Sterne
Meilensteine der Astronomiegeschichte
Zum 100jährigen Jubiläum
der Hamburger Sternwarte in Bergedorf
tredition

Nuncius Hamburgensis –
Beiträge zur Geschichte der Naturwissenschaften, Band 42
Gudrun Wolfschmidt (Hg.)
Orientierung, Navigation
und Zeitbestimmung
Wie der Himmel den Lebensraum
des Menschen prägt
tredition

Nuncius Hamburgensis –
Beiträge zur Geschichte der Naturwissenschaften, Band 38
Gudrun Wolfschmidt (Hg.)
Astronomy in the Baltic
Astronomie im Ostseeraum
Proceedings der Tagung des Arbeitskreises
Astronomiegeschichte in Kiel 2015
tredition

Nuncius Hamburgensis –
Beiträge zur Geschichte der Naturwissenschaften, Band 44
Panagiotis Kitmeridis
Popularisierung der
Naturwissenschaften
am Beispiel des
Physikalischen Vereins Frankfurt
tredition

Nuncius Hamburgensis –
Beiträge zur Geschichte der Naturwissenschaften, Band 32
Gudrun Wolfschmidt (Hg.)
Astronomie und Astrologie
im Kontext von Religionen
Astronomy and Astrology in Context of Religions
tredition

Nuncius Hamburgensis –
Beiträge zur Geschichte der Naturwissenschaften, Band 41
Gudrun Wolfschmidt (Hg.)
Popularisierung
der Astronomie
tredition

Nuncius Hamburgensis –
Gudrun Wolfschmidt (Hg.)
Baudenkmäler des Himmels –
Proceedings der Tagungen der
Gesellschaft für Archäoastronomie
tredition

Irena Kampa
Die astronomischen
Instrumente von
Johannes Hevelius
tredition

- Band 49 (2020): Wolfschmidt, Gudrun (Hg.): *Internationalität in der astronomischen Forschung (18. bis 21. Jahrhundert).* Proceedings der Tagung des Arbeitskreises Astronomiegeschichte in Wien 2018.
- Band 50 (2022): Wolfschmidt, Gudrun (Hg.): *Kosmochemie – Geschichte der Entdeckung und Erforschung der chemischen Elemente im Kosmos.* Zum 150. Jubiläum des Periodensystems der Elemente (PSE, 1869). und anläßlich des 50. Jubiläums der Mondlandung. Proceedings der Tagung des Arbeitskreises Astronomiegeschichte in der AG in Stuttgart 2019.
- Band 51 (2020): Wolfschmidt, G. (Hg.): *Himmelswelten und Kosmovisionen – Imaginationen, Modelle, Weltanschauungen.* Proceedings der Tagung der Gesellschaft für Archäoastronomie in Gilching 2019.
- Band 52 (2023): Wolfschmidt, Gudrun (Hg.): *Instrumente, Methoden und Entdeckungen für innovative Entwicklungen in der Astronomie.* Proceedings der Tagung des Arbeitskreises Astronomiegeschichte in der AG in Bremen 2022.
- Band 53 (2024): Wolfschmidt, Gudrun (Hg.): *Der Mensch im Kosmos: Lebenswelten und Kosmologien.* Proceedings der Tagung der Gesellschaft für Archäoastronomie in Weimar 2023.
- Band 54 (2021): Erich Meyer: *Auf den Spuren Johannes Keplers – Zu seinem 450. Geburtstag.* Bearbeitet und hg. von Gudrun Wolfschmidt.
- Band 55 (2021): Wolfschmidt, Gudrun & Susanne M. Hoffmann (ed.): *Applied and Computational Historical Astronomy.* Proceedings of the Splinter Meeting in the AG, Sept. 25, 2020.
- Band 56 (2021): Christoph Prignitz: *Zeit für Hamburg – Eine Uhr der Sternwarte und ihr historisches Umfeld.* Mit Beiträgen und herausgegeben von Gudrun Wolfschmidt.
- Band 57 (2022): Susanne M. Hoffmann & Gudrun Wolfschmidt (ed.): *Astronomy in Culture – Cultures of Astronomy.* Proceedings of the Splinter Meeting in the AG, Sept. 14–16, 2021.
- Band 58 (2023): Wolfschmidt, Gudrun (Hg.): *Wandlungen in Raum und Zeit: Himmel – Heimat – Weltverständnis.* Proceedings der Tagung der Gesellschaft für Archäoastronomie in Recklinghausen 2022.
- Band 59 (2024): Wolfschmidt, Gudrun (Hg.): *Astrophysik seit 1900 – Jubiläum von Karl Schwarzschild (1873–1916) und Ejnar Hertzsprung (1873–1967).* Proceedings des AKAG in Berlin 2023.

- Ziegler, Wolfram: *Aus dem Leben eines Franken. Dr. August Ziegler (1885–1937) – Pflanzenzüchter in Togo und Rebenzüchter in Bayern.* Hg. von Gudrun Wolfschmidt. 2017.
- *Harmony and Symmetry. Celestial regularities shaping human culture.* Ed. by SONJA DRAXLER, MAX E. LIPPITSCH & GUDRUN WOLFSCHMIDT. Hamburg: tredition (SEAC Publications, Vol. 01) 2020.

Nuncius Hamburgensis –
Beiträge zur Geschichte der Naturwissenschaften, Band 37
Gudrun Wolfschmidt (Hg.)
70 Jahre Observatorium
Hoher List
Sieben Jahrzehnte astronomische
Beobachtung in der Eifel
tredition

Nuncius Hamburgensis –
Beiträge zur Geschichte der Naturwissenschaften, Band 54
Gudrun Wolfschmidt (Hg.)
Erich Meyer:
Auf den Spuren Johannes Keplers
Zum 450. Geburtstag Johannes Keplers
tredition

Gudrun Wolfschmidt (Hg.)
Vom Abakus
zum Computer, Teil 1
tredition

Nuncius Hamburgensis –
Beiträge zur Geschichte der Naturwissenschaften, Band 48
Gudrun Wolfschmidt (Hg.)
Maß und Mythos,
Zahl und Zauber
Vermessung
von Himmel
und Erde
tredition

Nuncius Hamburgensis –
Beiträge zur Geschichte der Naturwissenschaften, Band 55
Gudrun Wolfschmidt &
Susanne M. Hoffmann (ed.)
Applied and Computational
Historical Astronomy
Angewandte und computergestützte historische Astronomie
Proceedings of the Splinter Meeting
in the Astronomische Gesellschaft, Sept. 25, 2020
tredition

Nuncius Hamburgensis –
Beiträge zur Geschichte der Naturwissenschaften, Band 49
Gudrun Wolfschmidt (Hg.)
Internationalität
in der astronomischen Forschung
des 18. bis 20. Jahrhunderts
Internationality
in the Astronomical Research
of the 18th to 20th Century
tredition

Nuncius Hamburgensis –
Beiträge zur Geschichte der Naturwissenschaften, Band 51
Gudrun Wolfschmidt (Hg.)
Himmelswelten und
Kosmovisionen
Imaginationen, Modelle,
Weltanschauungen
Proceedings der Jahrestagung der
Gesellschaft für Archäoastronomie, Gilching 2019
tredition

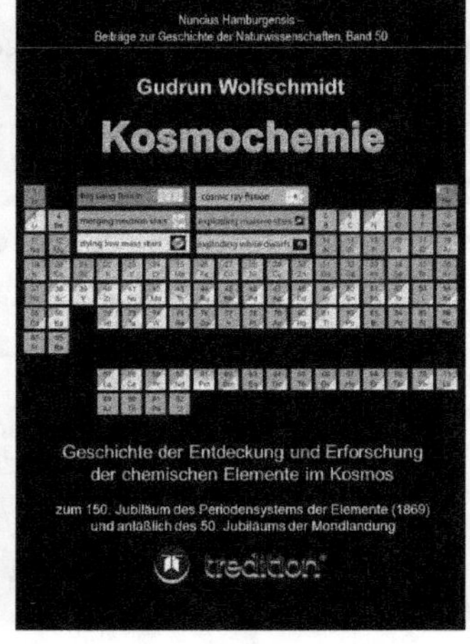
Nuncius Hamburgensis –
Beiträge zur Geschichte der Naturwissenschaften, Band 50
Gudrun Wolfschmidt
Kosmochemie
Geschichte der Entdeckung und Erforschung
der chemischen Elemente im Kosmos
zum 150. Jubiläum des Periodensystems der Elemente (1869)
und anläßlich des 50. Jubiläums der Mondlandung
tredition

Nuncius Hamburgensis –
Beiträge zur Geschichte der Naturwissenschaften, Band 57
Gudrun Wolfschmidt (Hg.)
Astronomy in Culture
Cultures of Astronomy
tredition

Personenregister

A

Agnesi, Maria Gaetana (1718–1799), 357
Anaximenes von Milet (∼610–∼546 v.Chr.), 36
Anna Amalia von Braunschweig-Wolfenbüttel (1739–1807), 12, 295, 296, 300, 303, 323, 339, 343

B

Benzenberg, Johann Friedrich (1777–1846), 355
Blauzahn → Harald I. „Blauzahn' Gormsson
Bosch, Hieronymus [Jheronimus van Aken] (∼1450–1516), 60, 62

C

Châtelet, Gabrielle-Émilie le Tonnelier du Breteuil du (1706–1749), 357
Chladni, Ernst Florens Friedrich (1756–1827), 12
Copernicus, Nikolaus (1473–1543), 349

D

Dechend, Hertha von (1915–2001), 36, 41, 50, 54, 58, 73
Diodorus Siculus (fl. erste Hälfte des 1. Jahrhunderts v.Chr.), 89, 91, 101
Dürer, Albrecht (1471–1528), 134

E

Eimmart, Georg Christoph (1638–1705), 351

F

Feininger, Lyonel (1871–1956), 17
Fell, Howard Barraclough [Barry] (1917–1994), 120

G

Gärtner, Hans (1974–1980), 66
Gabelbart → Sven I. Tveskæg „Gabelbart“
Galilei, Galileo (1564–1641 jul. / 1642 greg.), 353
Gauß, Carl Friedrich (1777–1855), 355
Ginzel, Karl Friedrich (1850–1926), 24, 278, 285
Godeffroy, Johan Cesar VI. (1813–1885), 355
Goethe, Johann Wolfgang von (1749–1832), 2, 4, 12, 13, 15, 16, 299, 300, 303, 305, 313, 316,

317, 319–323, 335, 339, 343, 353
Guericke, Otto von (1602–1686), 351

H

Haidinger, Wilhelm Karl Ritter von (1795–1871), 23
Hammurabi (∼1810–∼1750 BC), 197
Harald I. „Blauzahn“ Gormsson [König von Dänemark von ∼958 bis 987 und von Norwegen von ∼970 bis ∼975] (∼910–987), 265, 266
Hecataeus von Abdera (∼360–290 v.Chr.), 91
Herschel, Friedrich Wilhelm (1738–1822), 349
Hertz, Heinrich (1857–1894), 351, 353
Hertzsprung, Ejnar (1873–1967), 359
Hesiod (vor 700 v.Chr.), 29
Hevelius [Hewelcke], Johannes (1611–1687), 357
Homer (∼800 v.Chr.), 29, 32, 45, 61, 66, 241, 242
Hoppenhaupt, Moritz Ehrenreich (1723–1756), 147–149, 187
Hubble, Edwin Powell (1889–1953), 349
Humboldt, Alexander von (1769–1859), 39, 67

I

J

Jensch, Alfred (1912–2001), 355

K

Kepler, Johannes (1571–1630), 359
Klein, Felix Christian (1849–1925), 56
Kosmas Indikopleustes (484–577 n.Chr.), 36
Kroll, Wilhelm (1906–1939), 66
Kuffner, Moriz Edler von (1854–1939), 353

L

M

Marius, Simon (1573–1624), 353
Merz [Firma in München, 1838–1932], 357
Miller, Stanley Lloyd (1930–2007), 324, 326
Mittelhaus, Karl (1939–1946), 66
Muche, Georg (1895–1987), 16

N

Newton, Isaac (1642–1726 jul./1643–1727 greg.), 12
Numenios of Apameia (Mitte 2. Jh. v.Chr.), 46

O

P

Parmenides (∼520/515–460/455 v.Chr.), 31
Plato (428/427–348/347 v.Chr.), 40, 46, 68, 73, 75
Popper, Karl Raimund (1902–1994), 262
Porphyrios von Tyros (∼233–301/305 n.Chr.), 46
Proclus (412–485), 46
Pythagoras von Samos (570–∼495), 105, 106, 113

Q

R

Remeis, Karl (1837–1882), 355
Riekher, Rolf (1922–2020), 355

S

Santillana, Giorgio Diaz de (1902–1974), 36, 41, 50, 54, 58, 73
Schiller, Johann Christoph Friedrich [ab 1802 von Schiller] (1759–1805), 16
Schimank, Hans (1888–1979), 351
Schlosser, Wolfhard (1940–2022), 125
Schmeller, Johann Joseph (1796–1841), 339
Schöner, Johannes (1477–1547), 16, 294–297
Schwarzschild, Karl (1873–1916), 359
Scriba, Christoph J. (1929–2013), 357
Shoujing, Guo (1231–∼1215), 79
Sven I. Tveskæg „Gabelbart“ [König von Dänemark von ∼987 bis 1014] (963–1014), 265, 266

T

Tacitus, Publius Cornelius (∼58–∼120), 149, 187
Thom, Alexander (1894–1985), 110, 113, 115
Toland, John (1670–1722), 91

U

Urey, Harold Clayton (1893–1981), 324, 326

V

Van der Waerden, Bartel Leendert → Waerden, Bartel Leendert van der
Velde, Henry Clemens van de (1863–1957), 16
Vespucci, Amerigo (1451–1512), 295, 296
Virgil [Publius Vergilius Maro] (70–19 v.Chr.), 41, 42

W

Waerden, Bartel Leendert van der (1903–1996), 27, 75, 200, 202, 204, 205, 212, 243
Walther, Bernhard (1430–1504), 351
Weber, Wilhelm Eduard (1804–1891), 353
Wissowa, Georg (1890–1906), 66
Wuthenau, Carl Leberecht von [Rittmeister] (fl. 1750), 149

X

Xun, Wang (1235–1281), 79

Y

Z

Ziegler, Konrat (1946–1974), 66

Abbildung 19.1:
-Sonnenaufgang in Goethes Römischen Haus in Weimar am 23. Juni 2023 (vgl. Abb. 16.2)

(Fotos: Katrin Cura)

www.ingramcontent.com/pod-product-compliance
Lightning Source LLC
LaVergne TN
LVHW080555200726
843510LV00004B/923

* 9 7 8 3 3 8 4 1 0 3 5 0 5 *